THE MIF HANDBOOK

THE MIF HANDBOOK

edited by

Richard Bucala
Yale University School of Medicine, USA

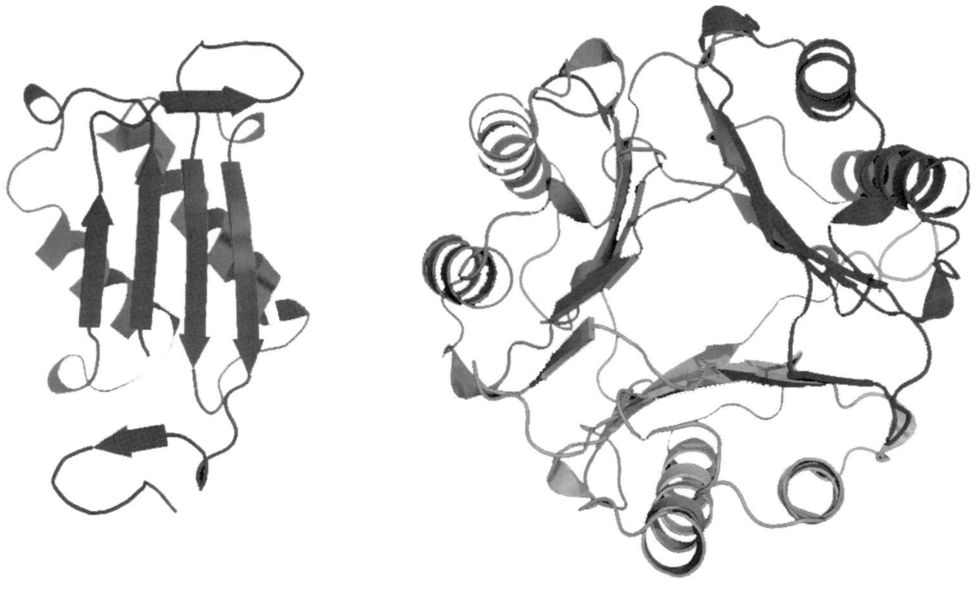

World Scientific
NEW JERSEY • LONDON • SINGAPORE • BEIJING • SHANGHAI • HONG KONG • TAIPEI • CHENNAI

Published by

World Scientific Publishing Co. Pte. Ltd.
5 Toh Tuck Link, Singapore 596224
USA office: 27 Warren Street, Suite 401-402, Hackensack, NJ 07601
UK office: 57 Shelton Street, Covent Garden, London WC2H 9HE

British Library Cataloguing-in-Publication Data
A catalogue record for this book is available from the British Library.

THE MIF HANDBOOK

ISBN-13 978-981-4335-35-5
ISBN-10 981-4335-35-5

Typeset by Stallion Press
Email: enquiries@stallionpress.com

Printed in Singapore by Mainland Press Pte Ltd.

Preface

R. Bucala, Editor

The role of the cytokine, macrophage migration inhibitory factor (MIF) in the immune response and in the immunopathogenesis of different inflammatory, autoimmune, and infectious disorders is now well-established. Recent studies continue to broaden considerably the role of MIF in both normal physiology and pathology, which range from such diverse areas as oncogenesis, cardiac physiology, and neurodevelopment. MIF's molecular mechanism of action in these contexts is becoming increasingly understood and the role of variant MIF alleles in different conditions continues to be defined. Unique structural features of the protein, such as an intrinsic catalytic activity, and the continuing elucidation of its receptor-dependent mechanism of action offer attractive opportunities for therapeutic intervention. This volume will provide a comprehensive synthesis of the current state-of-the-art of MIF science.

Contents

PART I

MIF Structure and Mechanism of Action

I-1

MIF, *MIF* Alleles, and the Regulation of the Host Response

Richard Bucala*

1. Introduction

Macrophage migration inhibitory factor (MIF) is the first cytokine activity to be discovered, although it resisted cloning and molecular characterization until relatively late in the era of cytokine discovery. Beyond its eponymous effect on macrophage mobility, MIF now is understood to be a critical upstream regulator of innate immunity that sustains activation responses by mechanisms that include counter-regulating the immunosuppressive action of glucocorticoids and inhibiting stimulus-induced apoptosis. These properties act physiologically to regulate the set-point and the magnitude of an immune response. The recent description of prevalent and functional alleles for *MIF* and their association with autoimmunity, infection, and cancer has focused attention not only on MIF's regulatory role in the host response but also on the importance of innate immune pathways in the clinical expression of disease. *MIF* alleles show significant population stratification, which may reflect the influence of selective pressure, and they likely provide an essential level of variation in innate responsiveness within the human population. Highly homologous orthologues of MIF also have been described in parasitic organisms, and early data suggest that they regulate the host-parasite interaction. First insight into the MIF receptor complex and the structural basis for MIF signal transduction has revealed unique features that hold promise for the pharmacologic modulation of MIF-dependent pathways. Recent developments in MIF biology are reviewed herein and integrated within the concept that *MIF* allele specific responses influence disease development, whether of an autoimmune, infectious or oncogenic etiology.

*Corresponding author: Department of Medicine, Pathology, and Epidemiology and Public Health, Yale University School of Medicine, The Anlyan Center for Biomedical Research, S525, 300 Cedar Street, New Haven, CT 06520-8031, USA. Email: richard.bucala@yale.edu

2. MIF Gene

The observation of an immune basis for leukocyte motility can be attributed to Arnold Rich and Margaret Lewis, who in 1932 showed that the migration of cells from the lymph nodes of an antigen-sensitized animal was impaired in the presence of antigen.[1] This *in vitro* demonstration of cellular immunity engendered significant interest among immunologists, especially after refinements in the quantification of cell movement were made in the 1950s.[2] John David and Barry Bloom attributed migration arrest to a lymphokine,[3,4] and a unique gene encoding MIF was ultimately reported in 1989.[5] Recombinant MIF protein followed the cloning of MIF from corticotrophic pituitary cells, which itself was unexpected and pointed to a systemic role for MIF in the regulation of the immunologic and neuroendocrine systems.[6] MIF circulates normally in plasma and its levels rise together with adrenocorticotrophic hormone (ACTH) in response to stress or invasive stimuli. ACTH serves to stimulate adrenal glucocorticoid production while MIF acts to counter-regulate the immunosuppressive action of glucocorticoids.[6–8]

There is a single *MIF* gene in the human genome (22q11.2) and both the exonic structure and DNA sequence of MIF are highly conserved across phylogeny. MIF transcription is constitutive in many cell types and induced transcription is regulated by proinflammatory, glucocorticoid, and hypoxic signals acting on AP-1, CREB, and HIF-1α responsive elements.[9,10]

A remarkable feature of the *MIF* gene is the presence of a microsatellite repeat $(CATT)_n$ within the 5′ promoter region.[11] This repeat unit is present in 5–8 copies and lies within a Pit-1 transcription factor binding site, which may provide for the neuroendocrine regulation of MIF expression (Fig. 1). Both gene reporter assays and human clinical studies indicate that repeat number is associated with higher MIF expression.[11,12] Repeat number or a single-nucleotide polymorphism that is in strong linkage disequilibrium with $CATT_7$ has been found in genetic epidemiology studies to be associated with increased innate responses in numerous human diseases.[13–23] The *MIF* allelic structure also shows significant population stratification, with increasing repeat number following human migration patterns and genomic diversification.[24]

3. MIF Production and Signal Transduction

Diverse activating stimuli induce the rapid release of MIF from pre-formed, cytoplasmic pools; this is followed by an upregulation in MIF mRNA expression and the replenishment of intracellular protein content[25,26] within the

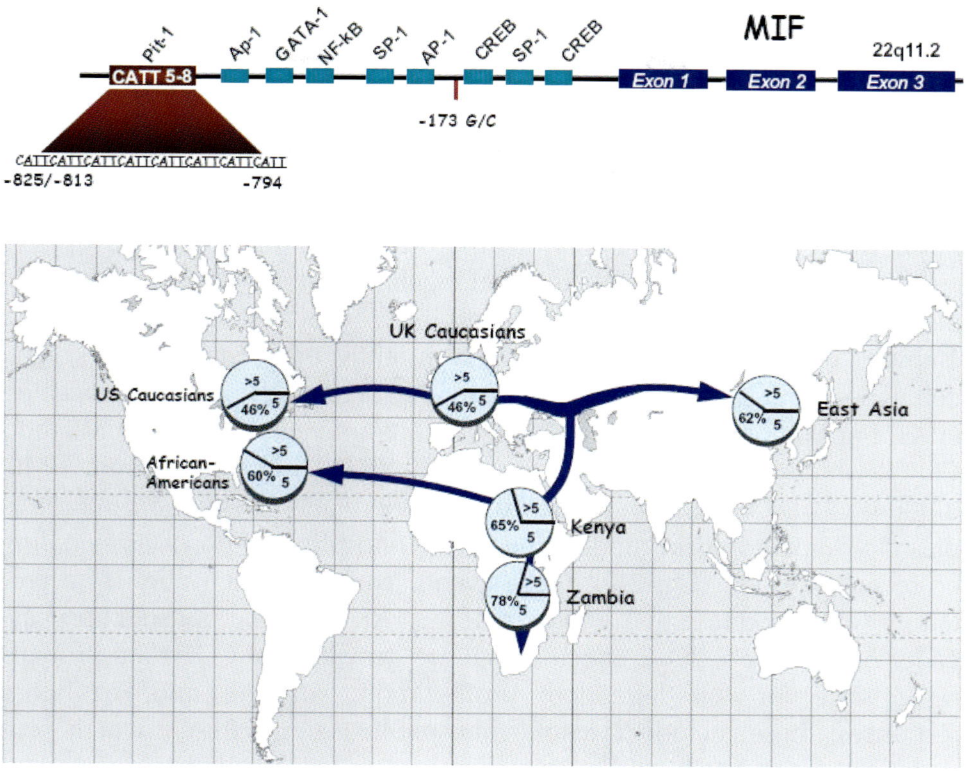

Fig. 1. Upper Panel: Diagram of the human MIF gene showing its exonic structure, putative transcription factor binding sites, and the functional –794 $CATT_{5-8}$ promoter polymorphism. A nearby promoter SNP (–173 G/C) also may contribute to functionality. Lower Panel: Prevalence of MIF CATT alleles in different populations superimposed on human migration patterns (5 = $CATT_5$, >5 = $CATT_6$, $CATT_7$, and $CATT_8$). From Ref. 93.

immune system. Such stimuli include Toll-like receptor (TLR) agonists, mitogens, and selected proinflammatory mediators.[27–29] Ischemia,[30] glucocortiocoids in low concentrations,[7] and corticotrophin-releasing hormone also induce MIF release from different cell types.[31] MIF lacks a classical signal sequence and it is secreted by an unconventional route that requires the Golgi-associated protein, p115. The inflammatory activation of macrophages by endotoxin or by intracellular infection leads to the cytoplasmic redistribution and co-export of p115 and MIF. The genetic targeting of p115 reduces MIF secretion without affecting the stimulus-induced secretion of other innate cytokines.[32]

The solution of MIF's crystal structure revealed a new structural superfamily but introduced a conundrum for the potential mechanism of MIF action.[33]

The three-dimensional structure of MIF showed significant homology with two prokaryotic tautomerases; this fueled speculation that MIF exerts immunoregulatory function(s) by an enzymatic reaction. Rorsman and colleagues established that MIF indeed tautomerizes model, and possibly physiologic, substrates[34,35]; however, most studies as well as a recent genetic knock-in mouse model suggest that MIF's enzymatic active site is most likely vestigial but nevertheless has a structural role in protein-protein interactions.[36]

MIF activates the extracellular signal regulated kinase (ERK) 1/2 subfamily of MAP kinases,[37] but in contrast to most inducers of this signal transduction pathway, which phosphorylate ERK1/2 transiently, the effect of MIF in many cell types is sustained. Expression cloning of an MIF receptor revealed ERK1/2 activation to result from a high-affinity interaction between MIF and CD74, which is the cell surface form of the Class II invariant chain.[38] MIF binding to CD74 results in the serine phosphorylation of its short intracytoplasmic domain and in modulation of the phosphorylation of its signaling co-receptor, CD44 (Fig. 2).[39] These events lead to the activation of a Src-family tyrosine kinase and initiation of the ERK1/2 signal transduction cascade. Among the ERK1/2 effectors that are activated by CD74 are proteins responsible for MIF-dependent proliferation, survival, and regulatory responses; these include cytosolic phospholipase A_2 ($cPLA_2$), which produces the arachidonic acid precursor necessary for the synthesis of prostaglandins and leukotrienes. Intracellular arachidonic acid also activates the c-Jun-N-terminal kinase (JNK) to promote the efficient translation of mRNAs for TNF and other cytokines.[40] MIF stimulates $cPLA_2$ in the face of immunosuppressive concentrations of glucocortiocoids, which is one mechanism whereby MIF overrides glucocorticoid-mediated anti-inflammatory action.[37] Follow-up studies have demonstrated that MIF inhibits the glucocorticoid-induced expression of MAP kinase phosphatase (MKP-1), which is an important mechanism by which glucocorticoids reduce the inflammatory responses initiated by the ERK1/2, JNK, and p38 MAPK pathways.[41,42] MKP-1 also promotes the translation of short-lived proinflammatory cytokine mRNAs by stabilization of 3-AU-rich elements.

The identification of the transcriptional regulator c-Jun activation domain binding protein 1 (JAB1) as a binding partner for MIF provided an explanation for MIF's noted ability to induce sustained phase ERK1/2 activation.[43] Intracellular JAB1 levels temporally regulate MIF-dependent ERK1/2 activation: high JAB1 expression inhibits sustained but not transient ERK1/2 phosphorylation, while low JAB1 levels are sufficient for transient activation.[44] This effect may be due to the known role of JAB1 in the COP9 signalosome, where it regulates the degradation of signaling components.

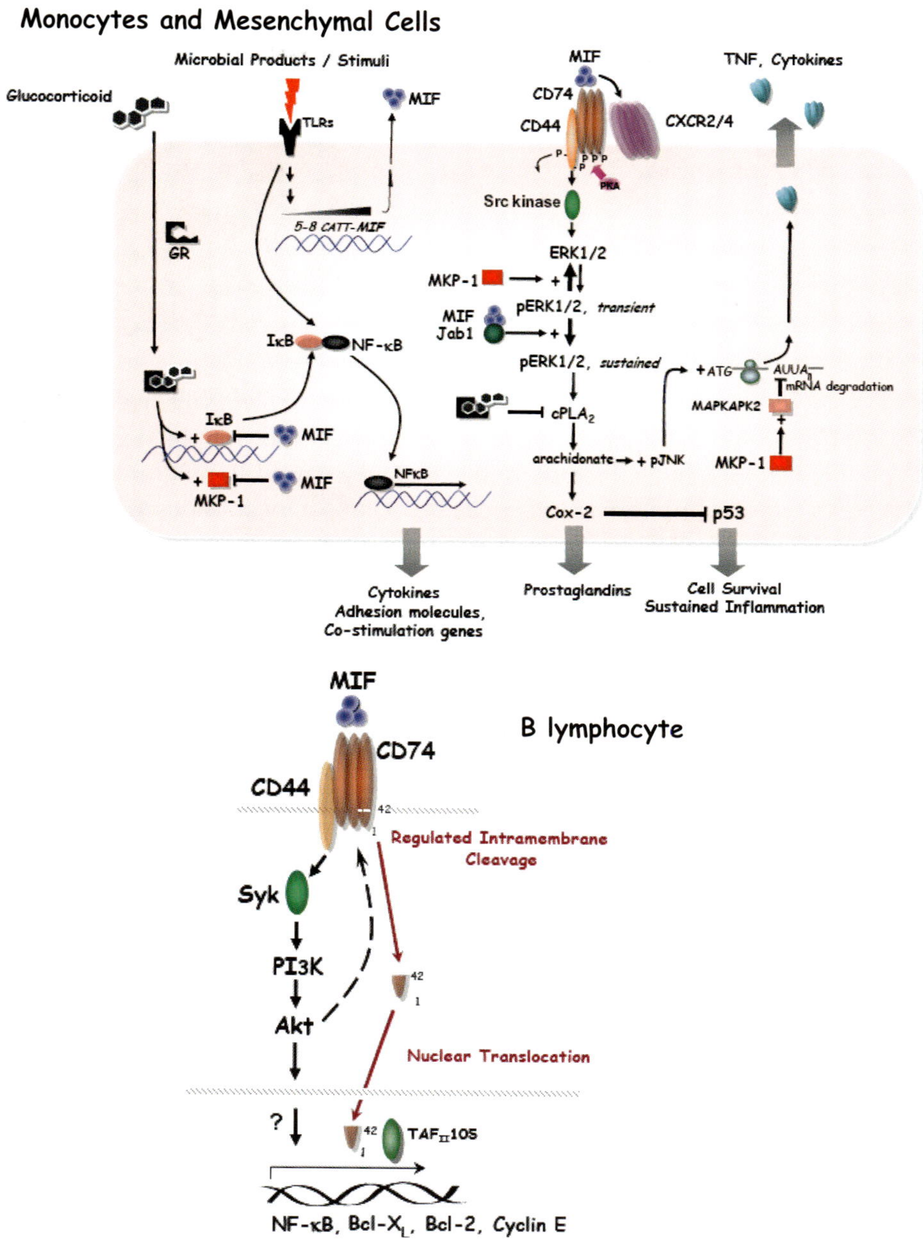

Fig. 2. Upper Panel: Integrated scheme for activating effects of MIF signal transduction and the regulation of the glucocorticoid immunosuppression based on data obtained in monocytes/macrophages and mesenchymal cells. Lower Panel: MIF signal transduction in B lymphocytes, which additionally involves the regulated intramembrane cleavage of CD74, nuclear translocation of $CD74_{1-42}$ and the activation of a transcriptional response mediated by NF-κB p65/RelA and the transcriptional coactivator $TAF_{II}105$.

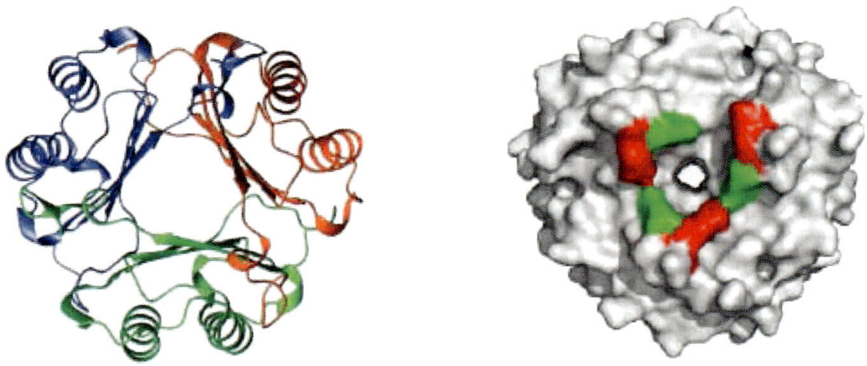

Fig. 3. Left panel: X-ray crystal structure of human MIF showing the axial symmetry of its trimeric form. Right panel: Surface model of MIF showing the position of the conformational *pseudo*-(E)LR motifs formed by the two nonadjacent residues Arg-11 (R11) and Asp-44 (D44), which reside in an ELR-like spacing in exposed neighboring loops. Members of the ELR family of chemokines, such as CXCL8, have three amino acids Glu-3 (E3), Leu-4 (L4), and Arg-5 (R5) that form an ELR motif that is essential for signaling through CXCR2.

MIF's unique structure recently led to the proposition that a topologic similarity with IL-8 may explain MIF's longstanding function as an "arrest" chemokine.[45] Activation of the chemokine receptor CXCR2 by its cognate ligand, IL-8 (CXCL8), is mediated by an N-terminal Glu-Leu-Arg (ELR) motif. MIF has a *pseudo*-ELR motif comprising two nonadjacent but appropriately spaced residues (Asp and Arg) in exposed neighboring loops that mimic the structure present in ELR chemokines[46] (Fig. 3). The biologic significance of this pathway has been affirmed by the finding that blockade of MIF but not of canonical ligands for CXCR2 or CXCR4 reduces monocyte and T-cell content in murine atherosclerotic plaques. Functional cell surface complexes form between CD74 and these two chemokine receptors, and while anti-CD74 blocks MIF induction of a CXCR4-dependent AKT survival pathway, it does not inhibit AKT phosphorylation induced by a pure CXCR4 agonist such as CXCL12.[47]

4. Innate Immunity

Studies with MIF-KO mice have confirmed an upstream activating role for MIF in diverse inflammatory responses.[48–50] Mice genetically deficient in MIF appear phenotypically normal; however, a subtle defect in lung maturation due to a delay in developmental regulation by VEGF and glucocorticoids is revealed by premature delivery. This murine phenotype recapitulates the key pathologic findings observed in the neonatal respiratory distress syndrome of prematurity.[51] MIF-KO mice manifest significant immunoregulatory defects when confronted by immunologic or invasive challenge; for instance, a reduction in the expression of innate mediators such as TNF, IL-1, IL-6, IL-12, IL-18, IFN-α, PGE_2 and NO.[48,49,52–55] These effects are in accord with MIF's role in inhibiting p53-dependent apoptosis and in stimulating AKT-dependent pathways of cell survival.[49,56] The activation of monocytes/macrophages is sustained in the presence of MIF and a robust proinflammatory response ensues. MIF is also required for the optimal expression of TLR4, which is the cell surface receptor for the MD-2/LPS complex, indicating that the earliest events for pathogen recognition are reliant on adequate MIF expression. The innate responses of barrier epithelial and endothelial cell types also are augmented by MIF, which may itself be released from these cells by activating or cytotoxic stimuli.[57–60] In the case of the gastrointestinal tract, for instance, MIF alone upregulates microfold (M cell)-mediated transport of antigen across the follicle-associated epithelium of intestinal Peyer's patches.[61] Additional innate immune cell types such as the neutrophil, the eosinophil, and the mast cell produce MIF and contribute to MIF-dependent inflammatory responses.[62–64]

Whether MIF exerts a deleterious or beneficial effect on the host varies with the nature of the invasive agent. In murine models of *E. coli* and

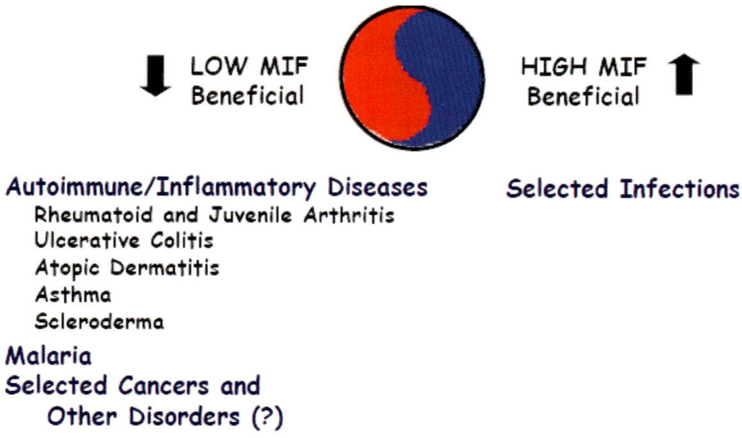

Fig. 4. Examples of genetic association studies that have revealed the influence of *MIF* alleles on different inflammatory diseases. For certain infections, such as those responsible for community-acquired pneumonia, high-expression alleles appear to confer a survival benefit.[22]

polymicrobial sepsis,[65] malaria,[18] or West Nile virus infection,[54] where MIF disrupts the blood-brain barrier, a reduction in MIF-dependent responses reduces immunopathology and promotes survival. In infections caused by intracellular *Salmonella*,[52] the protozoans *Leishmania*,[66] *Toxoplasma*,[55] *Trypanosoma*,[53] or the helminthic pathogen *Taenia crassiceps*,[67] MIF is critical for host defense. Experimental analyses of these infections have revealed important differences in the temporal or tissue-specific expression of downstream effector pathways, and these findings emphasize the important influence of MIF on the regulation of the innate response and ultimately host survival.[29]

Antigen presentation is initiated by the innate response and provides a segue into adaptive immunity. Deficient antigen presentation in the absence of MIF has been reported in murine allergic asthma.[16] The precise basis for this impairment remains unknown although antigen presentation function by monocytes/macrophages and mast cells appears more profoundly affected than that of dendritic cells.[64]

5. Adaptive Immunity

Given the central role of innate immunity in the differentiation of the adaptive response, it is not surprising that MIF influences T and B cell responses. Activated T cells produce MIF, which then acts by an autocrine pathway to enhance IL-2, IL-2R, and IFN-γ production.[27] Immunoneutralization or genetic deficiency of MIF in mice reduces T cell priming and memory responses,[27] T cell cytokine production (e.g., IL-2, IL-4, IL-5, IL-13, eotaxin, IL-13, IFN-γ),[16,64] and T cell-dependent antibody responses.[16] T cell responses appear reduced irrespective of whether a Th1 or Th2 T cell response is elicited, and a role for MIF in the differentiation of inflammatory Th17 cells has also been reported recently.[68] Immunoneutralization or genetic deletion of MIF ameliorates inflammatory tissue damage in T cell-mediated disease models, such as collagen-induced arthritis,[69,70] inflammatory bowel disease,[71] and autoimmune encephalomyelitis.[72,73]

As in the case of monocytes, MIF likely sustains the activation responses of cells of the adaptive system. The pathway that has been examined in greatest detail is in B cells (Fig. 2). Indeed, prior to the discovery of CD74 as the MIF receptor, CD74 had already been identified as a B cell survival factor. In response to forced overexpression or an activating antibody, CD74 undergoes regulated intramembrane proteolysis to create a cytosolic protein that translocates to the nucleus to activate a transcriptional response mediated by NF-κB p65/RelA and $TAF_{II}105$.[74] This sequence of events was shown recently to be recapitulated by MIF engagement, and it appears to occur downstream of CD44, which activates the Src kinase, Syk, leading to Akt phosphorylation, augmentation of Bcl-X_L and Bcl-2 transcription, and enhanced B cell survival. Within the bone marrow, the survival of recirculating B cell populations has been shown to be reliant on MIF produced by resident dendritic cells.[75]

6. MIF Integrates Immune, Metabolic, and Oncogenic Responses

Infection or tissue invasion is frequently accompanied by a derangement in host metabolism. Protein wasting, or cachexia, has long been associated with a systemic proinflammatory response, although the role of specific cytokines such as TNF *in vivo* has been controversial. MIF-deficient mice show normalized blood glucose and lactate responses during endotoxic challenge,[76] and high levels of MIF have a direct catabolic effect of muscle, independently of TNF.[77] Infiltration of macrophages in white adipose tissue is considered an important mechanism for the development of insulin resistance, Type 2 diabetes, and atherosclerosis, and all of these features are reduced in the setting of MIF deficiency.[78] Plasma lipids and adiposity are not affected, however, suggesting that the primary action of MIF is in regulating the magnitude of the systemic inflammatory response. There are emerging human data, both with respect to circulating MIF levels and *MIF* genotype, to support MIF's role in the development of insulin resistance and Type 2 diabetes.[79]

The ability of MIF to induce glucose transport in muscle recently prompted investigations into the potential importance of this effect in ischemic myocardium. Miller *et al.* found that MIF is released by ischemic cardiomyocytes, where it stimulates activation of AMP-activated protein kinase (AMPK) by engaging the CD74/CD44 receptor complex. This cardioprotective response is critically dependent on MIF, as MIF-KO mice suffer from impaired ischemic signaling and larger cardiac infarctions. This protective pathway also shows reduced activity in human cells with a low-expression *MIF* allele, suggesting that *MIF* genotype may well predict risk in patients with coronary artery disease.[30]

Multiple lines of evidence link inflammation to the development and progression of cancer. MIF's ability to regulate molecular pathways that are necessary for migration and invasion, proliferation, and evasion of apoptosis have made it a molecule of high interest for investigations of tumorigenesis. The known pathways for MIF signal transduction that include the ability to induce sustained ERK1/2 activation, which is reminiscent of oncogenic RAS, stimulation of the AKT pathway, and the regulation of JAB1 and p53 further support this notion. MIF is strongly upregulated by hypoxia inducible factor-1α,[9] and MIF promotes maximal HIF-1α expression,[80] suggesting mutually reinforcing pathways for tumor progression and adaptation. The signaling component of the MIF receptor complex, CD44, has also been shown to be strongly associated with tumor cell adhesion and invasion and to promote metastasis, and it is a feature of the tumor stem cell phenotype.[81]

7. *MIF* Alleles in Human Disease

Evidence for a pathogenetic role for MIF in different inflammatory and infectious diseases emerged quickly after the observation that immunoneutralization of MIF fully protects mice from endotoxic shock. This result placed MIF in a central regulatory node with respect not only to the expression of innate immunity but to the progression of tissue injurious pathways of inflammation.[6,29] The potential clinical importance of MIF has now been underscored by epidemiologic studies that have shown associations between functional promoter alleles and diseases with an immunologic basis (Fig. 4). In the examples of autoimmune or inflammatory disorders that have been studied, such as in adult or juvenile forms of rheumatoid arthritis,[12,13] ulcerative colitis,[14] asthma,[16] and systemic sclerosis,[20] the predominant impact of high-expression *MIF* alleles is on the severity or the clinical phenotype of disease rather than on susceptibility. In asthma, for example, the presence of a longer (and higher expression) CATT repeat was related to more severe asthma as defined by GINA (Global Initiative for Asthma) criteria,[16] and in systemic sclerosis, with a more severe clinical phenotype known as diffuse cutaneous disease.[20] Whether the actual impact of MIF is on clinical severity or if these conclusions instead reflect a statistical limitation of these studies remains to be determined. One exception to these observations is atopic dermatitis, in which high-expression *MIF* alleles confer a 3.5-fold increased susceptibility to disease development and where the role of the *MIF* locus (22q11) was identified by linkage analysis.[15] Overall, these human genetic data are consonant with experimental studies that have emphasized the amplifying effect of MIF on innate pathways of tissue and end-organ damage.

The observation that the highest prevalence of the $CATT_5$ allele occurs in Sub-Saharan Africa (Fig. 1) led to the hypothesis that low-expression allelic variants may provide a measure of protection from malaria, as this region historically has suffered the greatest mortality from this infection. Lethal malaria occurs most commonly in immunologically naïve children, with death ensuing from the complications of an excessive innate immune response that produces the clinical sequela of cerebral disease, severe anemia, and a sepsis-like syndrome. Malaria is also believed responsible for the selection and evolutionary persistence of minor hemoglobin genes such as HbS (sickle hemoglobin), which confer protection against lethal malaria. Notably, the prevalence of the HbS decreases in the southern latitudes of the African continent, which suggests that additional genes such as MIF may have a role in resistance to disease. Experimental studies in MIF-deficient mice have confirmed a role for MIF in the inflammatory complications producing severe malarial anemia.[18] Genetic epidemiology studies are now showing a relationship between longer repeats of the CATT polymorphism and the development of severe malarial anemia, which is a leading cause of death in children with malaria.[23] Whether MIF influences the development of other parasitic or chronic infections endemic to areas with a particular prevalence of *MIF* alleles is also coming under investigation.

The studies that have examined the influence of *MIF* on autoimmune inflammatory diseases, or a disease such as malaria where inflammatory complications play a lethal role, have uniformly reported a deleterious role for high-expression allelic variants. The results of a recent study of a large cohort of patients with pneumonia at risk for septic shock is therefore noteworthy because a high-expression allele — in this study, a promoter SNP (–173 C) that is closely associated with $CATT_7$ — was found to be associated with a 50% improvement in survival.[22] This result was somewhat unexpected given the prevailing hypothesis that an excessive innate response underlies the immunopathogenesis of septic shock. These human genetic data, which were obtained in the largest cohort of sepsis patients yet studied, emphasize the protective and microbicidal role of innate immunity. They also lend support to the notion that *MIF* alleles exist in a balanced polymorphism that has been maintained by the influence of different selective pressures, presumably from infections.

There is emerging interest in examining the *MIF* allelic system in the development of diseases that may not be considered nosologically to be inflammatory, but in which inflammation nevertheless makes a pathogenic contribution. In prostate cancer, for instance, the presence of inflammatory cells in tissue biopsies portends a worse prognosis, and cytokine signals have been hypothesized to contribute to tumor progression. In an initial study, patients with the high-expression $CATT_7$ allele were found to have an almost five-fold increased risk of prostate cancer recurrence.[19] In autism, which is a neurodevelopmental disorder of unknown etiology but associated with immune abnormalities, an association between functional polymorphisms in the promoter for *MIF* and autism spectrum disorder behaviors was found. Affected probands also exhibited higher circulating MIF levels than their unaffected siblings.[21]

Finally, the recent description of close orthologues of mammalian MIF in parasitic nematodes and in the protozoan pathogens responsible for malaria and leishmaniasis has raised the question of whether these proteins play a role in immune evasion.[82] There is early evidence that parasite-encoded MIFs influence cell migration and host immunity, in part by a functional interaction with the CD74 MIF receptor.[83] A *Leishmania* MIF orthologue, for instance, exhibits an antiapoptotic activity that may facilitate the intracellular persistence of the parasite within the macrophage.[84]

8. Therapeutic Implications and Future Directions

Unique structural features together with MIF's apex position in the inflammatory and cell survival pathways has made it an attractive pharmacologic target. Biologically based therapies are under development and a humanized monoclonal antibody (Milatuzumab) directed against the MIF receptor is in clinical trials for the treatment of B cell chronic lymphocytic leukemia.[85]

MIF's intrinsic tautomerase activity and substrate binding site have also attracted considerable interest, especially after the finding that selected inhibitors of MIF tautomerase activity inhibit biologic activity, presumably by imparting a conformational change on the protein that interferes with receptor interaction.[86] There has been significant progress made in the design of specific, small molecule-based receptor antagonists. Early generation compounds, including those with oral bioavailability, have shown promise in preclinical models of disease.[87,88] A recent, computationally based discovery approach examined 2.1 million compounds and uncovered several high-potency receptor antagonists.[89] Interestingly, the phosphodiesterase inhibitor, ibudilast, which is used in the treatment of asthma, exhibits additional anti-inflammatory activities that may be attributed to its ability to inhibit MIF's tautomerase, receptor binding, and biologic activities. Ibudilast was recently co-crystallized with MIF and found to occupy the MIF tautomerase site.[90] Another molecule in clinical development, 4-iodo-6-phenylpyrimidine (4-IPP), was discovered by a computational modeling strategy and covalently modifies the MIF tautomerase site.[91] 4-IPP has the interesting property of preventing cellular MIF secretion, and it appears to target a different MIF interaction[32]; MIF complexed with 4-IPP exhibits altered binding to the

Golgi-associated protein p115, which may disrupt the normal pathway of MIF secretion.

Finally, the recent crystallization of parasite-encoded MIFs has opened opportunities for the design of parasite-specific MIF inhibitors. It is noteworthy that while all MIF's identified to date have a potentially catalytically active N-terminal proline, significant differences in the dimensions and in the charge distribution within the tautomerase site exist in the MIF proteins from hookworm (*Ancyclostoma ceylonicum*)[92] and leishmania (*Leishmania major*).[84] Both proteins also show different profiles of catalytic inhibition when tested with prototypic small molecules inhibitors of human MIF.

Polymorphic genes constitute an important basis for variation in the host immune response, and MIF clearly occupies an apex position with respect to the regulation of the innate response. Experimental studies that have defined MIF's regulatory role in the clinical expression of autoimmunity, different infections, and in oncogenic progression are now being verified by human genetic studies. These broad findings are within the known mechanisms of MIF action and signal transduction, which reflect features more central to growth regulation, apoptosis, and cell cycle control than to simple inflammatory activation. The precise interplay between MIF and other genes with respect to pathogenesis and the expression of different diseases remains to be better elucidated. The genetically defined variations in human MIF expression that have been defined also offer the prospect of a natural therapeutic window that may guide pharmacologic interventions aimed at regulating MIF-directed pathways.

Acknowledgments

Studies in the author's laboratory have been supported by the NIH, the Alliance for Lupus Research, the Brookdale Foundation, and the Kanowitz Foundation. The author is grateful to the greater MIF community for their contributions and insights, to Idit Shachar (Weizmann Institute) for reviewing Fig. 2, and to Jürgen Bernhagen (RWTH Aachen University) for Fig. 3.

References

1. Rich AR, Lewis MR. (1932) The nature of allergy in tuberculosis as revealed by tissue culture studies. *Bull Johns Hopkins Hosp* **50**: 115–131.
2. George M, Vaughn JH. (1962) *In vitro* cell migration as a model for delayed hypersensitivity. *Proc Soc Exptl Biol Med* **111**: 514–521.

3. David J. (1966) Delayed hypersensitivity *in vitro*: Its mediation by cell-free substances formed by lymphoid cell-antigen interaction. *Proc Natl Acad Sci USA* **56**: 72–77.
4. Bloom BR, Bennett B. (1966) Mechanism of a reaction *in vitro* associated with delayed-type hypersensitivity. *Science* **153**: 80–82.
5. Weiser WY *et al.* (1989) Molecular cloning of a cDNA encoding a human macrophage migration inhibitory factor. *Proc Natl Acad Sci USA* **86**: 7522–7526.
6. Bernhagen J *et al.* (1993) MIF is a pituitary-derived cytokine that potentiates lethal endotoxaemia. *Nature* **365**: 756–759.
7. Calandra T *et al.* (1995) MIF as a glucocorticoid-induced modulator of cytokine production. *Nature* **377**: 68–71.
8. Flaster H, Bernhagen J, Calandra T, Bucala R. (2007) The macrophage migration inhibitory factor-glucocorticoid dyad: Regulation of inflammation and immunity. *Mol Endocrinol* **21**: 1267–1280.
9. Elsby LA *et al.* (2009) Hypoxia and glucocorticoid signaling converge to regulate macrophage migration inhibitory factor gene expression. *Arth Rheum* **60**: 2220–2231.
10. Leng L *et al.* (2009) Glucocorticoid-induced MIF expression by human CEM T cells. *Cytokine* **48**: 177–185.
11. Baugh JA *et al.* (2002) A functional promoter polymorphism in the macrophage migration inhibitory factor (MIF) gene associated with disease severity in rheumatoid arthritis. *Genes Immun* **3**: 170–176.
12. Radstake TRDJ *et al.* (2005) Correlation of rheumatoid arthritis severity with the genetic functional variants and circulating levels of macrophage migration inhibitory factor. *Arthritis Rheum* **52**: 3020–3029.
13. De Benedetti F *et al.* (2003) Functional and prognostic relevance of the -173 polymorphism of the MIF gene in systemic juvenile idiopathic arthritis. *Arthritis Rheum* **48**: 1398–1407.
14. Nohara H *et al.* (2004) Association of the -173G/C polymorphism of the macrophage migration inhibitory factor gene with ulcerative colitis. *J Gastroenterol* **39**: 242–246.
15. Hizawa N, Yamaguchi E, Takahashi D *et al.* (2004) Functional polymorphisms in the promoter region of macrophage migration inhibitory factor and atopy. *Am J Respir Crit Care Med* **169**: 1014–1018.
16. Mizue Y *et al.* (2005) Role for macrophage migration inhibitory factor (MIF) in asthma. *Proc Natl Acad Sci USA* **102**: 14410–14415.
17. Plant BJ *et al.* (2005) Cystic fibrosis, disease severity and a macrophage migration inhibitory factor polymorphism. *Am J Respir Crit Care Med* **172**: 1412–1415.

18. McDevitt MA *et al.* (2006) A critical role for the host mediator macrophage migration inhibitory factor in the pathogenesis of malarial anemia. *J Exp Med* **203**: 1185–1196.
19. Meyer-Siegler KL *et al.* (2007) Macrophage migration inhibitory factor (MIF) gene polymorphisms are associated with increased prostate cancer incidence. *Genes Immun* **8**: 646–652.
20. Wu SP *et al.* (2006) *MIF* promoter polymorphisms influence the clinical expression of scleroderma. *Arthritis Rheum* **54**: 3661–3669.
21. Grigorenko EL *et al.* (2008) Macrophage migration inhibitory factor and autism spectrum disorders. *Pediatrics* **122**: E438–E445.
22. Yende S *et al.* (2009) The influence of macrophage migration inhibitory factor (MIF) polymorphisms on outcome from community-acquired pneumonia. *FASEB J* **23**: 2403–2411.
23. Awandare GA *et al.* (2009) *MIF* promoter polymorphisms and susceptibility to severe malarial anemia. *J Infect Dis* **15**: 629–637.
24. Zhong X *et al.* (2003) Single-nucleotide polymorphism genotyping on optical thin-film biosensor chips. *Proc Natl Acad Sci USA* **100**: 11559–11564.
25. Calandra T, Bernhagen J, Mitchell RA, Bucala R. (1994) The macrophage is an important and previously unrecognized source of macrophage migration inhibitory factor. *J Exp Med* **179**: 1895–1902.
26. Bacher M *et al.* (1997) Migration inhibitory factor expression in experimentally induced endotoxemia. *Am J Pathol* **150**: 235–246.
27. Bacher M *et al.* (1996) An essential regulatory role for macrophage migration inhibitory factor in T-cell activation. *Proc Natl Acad Sci USA* **93**: 7849–7854.
28. Bernhagen J *et al.* (1996) An essential role for macrophage migration inhibitory factor in the tuberculin delayed-type hypersensitivity reaction. *J Exp Med* **183**: 277–282.
29. Calandra T, Roger T. (2003) Macrophage migration inhibitory factor: A regulator of innate immunity. *Nat Rev Immunol* **3**: 791–800.
30. Miller EJ *et al.* (2008) Macrophage migration inhibitory factor stimulates AMP-activated protein kinase in the ischaemic heart. *Nature* **451**: 578–582.
31. Nishino T *et al.* (1995) Localization of macrophage migration inhibitory factor (MIF) to secretory granules within the corticotrophic and thyrotrophic cells of the pituitary gland. *Mol Med* **1**: 781–788.
32. Merk M *et al.* (2009) The Golgi-associated protein p115 mediates the secretion of macrophage migration inhibitory factor. *J Immunol* **182**: 6896–6906.
33. Suzuki M *et al.* (1996) Crystal structure of the macrophage migration inhibitory factor from rat liver. *Nat Struct Biol* **3**: 259–266.
34. Rosengren E *et al.* (1996) The immunoregulatory mediator macrophage migration inhibitory factor (MIF) catalyzes a tautomerization reaction. *Mol Med* **2**: 143–149.

35. Rosengren E *et al.* (1997) The macrophage migration inhibitory factor MIF is a phenylpyruvate tautomerase. *FEBS Lett* **417**: 85–88.
36. Fingerle-Rowson G *et al.* (2009) A tautomerase-null macrophage migration-inhibitory factor (MIF) gene knock-in mouse model reveals that protein interactions and not enzymatic activity mediate MIF-dependent growth regulation. *Mol Cell Biol* **29**: 1922–1932.
37. Mitchell RA, Metz CN, Peng T, Bucala R. (1999) Sustained mitogen-activated protein kinase (MAPK) and cytoplasmic phospholipase A2 activation by macrophage migration inhibitory factor (MIF). Regulatory role in cell proliferation and glucocorticoid action. *J Biol Chem* **274**: 18100–18106.
38. Leng L *et al.* (2003) MIF signal transduction initiated by binding to CD74. *J Exp Med* **197**: 1467–1476.
39. Shi X *et al.* (2006) CD44 is the signaling component of the macrophage migration inhibitory factor-CD74 receptor complex. *Immunity* **25**: 595–606.
40. Swantek JL, Cobb MH, Geppert TD. (1997) Jun N-terminal kinase/stress activated protein kinase (JNK/SAPK) is required for lipopolysaccharide stimulation of tumor necrosis factor alpha (TNF-a) translation: Glucocorticoids inhibit TNF-a translation by blocking JNK/SAPK. *Mol Cell Biol* **17**: 6274–6282.
41. Roger T, Chanson AL, Knaup-Reymond M, Calandra T. (2005) Macrophage migration inhibitory factor promotes innate immune responses by suppressing glucocorticoid-induced expression of mitogen-activated protein kinase phosphatase-1. *Eur J Immunol* **35**: 3405–3413.
42. Aeberli D *et al.* (2006) Endogenous macrophage migration inhibitory factor modulates glucocorticoid sensitivity in macrophages via effects on MAP kinase phosphatase-1 and p38 MAP kinase. *FEBS Lett* **580**: 974–981.
43. Kleemann R *et al.* (2000) Intracellular action of the cytokine MIF to modulate AP-1 activity and the cell cycle through Jab1. *Nature* **408**: 211–216.
44. Lue H, Kapurniotu A, Fingerle-Rowson G *et al.* (2006) Rapid and transient activation of the ERK MAPK signalling pathway by macrophage migration inhibitory factor (MIF) and dependence on JAB1/CSN5 and Src activity. *Cell Signal* **18**: 688–703.
45. Bernhagen J *et al.* (2007) MIF is a noncognate ligand of CXC chemokine receptors in inflammatory and atherogenic cell recruitment. *Nat Med* **13**: 587–596.
46. Weber C *et al.* (2008) Structural determinants of MIF functions in CXCR2-mediated inflammatory and atherogenic leukocyte recruitment. *Proc Natl Acad Sci USA* **105**: 16278–16283.
47. Schwartz V *et al.* (2009) A functional heteromeric MIF receptor formed by CD74 and CXCR4. *FEBS Lett* **583**: 2749–2757.

48. Bozza M *et al.* (1999) Targeted disruption of migration inhibitory factor gene reveals its critical role in sepsis. *J Exp Med* **189**: 341–346.
49. Mitchell RA *et al.* (2002) Macrophage migration inhibitory factor (MIF) sustains macrophage proinflammatory function by inhibiting p53: Regulatory role in the innate immune response. *Proc Natl Acad Sci USA* **99**: 345–350.
50. Fingerle-Rowson G *et al.* (2003) The p53-dependent effects of macrophage migration inhibitory factor revealed by gene targeting. *Proc Natl Acad Sci USA* **100**: 9354–9359.
51. Kevill KA *et al.* (2008) A role for macrophage migration inhibitory factor in the neonatal respiratory distress syndrome. *J Immunol* **180**: 601–608.
52. Koebernick H *et al.* (2002) Macrophage migration inhibitory factor (MIF) plays a pivotal role in immunity against *Salmonella* Typhimurium. *Proc Natl Acad Sci USA* **99**: 13681–13686.
53. Reyes JL *et al.* (2006) Macrophage migration inhibitory factor contributes to host defense against acute *Trypanosoma cruzi* infection. *Infect Immun* **74**: 3170–3179.
54. Arjona A *et al.* (2007) Abrogation of macrophage migration inhibitory factor decreases West Nile virus lethality by limiting riral neuroinvasion. *J Clin Invest* **117**: 3059–3066.
55. Flores M. *et al.* (2008) Macrophage migration inhibitory factor (MIF) is critical for the host resistance against *Toxoplasma gondii. FASEB J* **22**: 3661–3671.
56. Lue H *et al.* (2007) Macrophage migration inhibitory factor (MIF) promotes cell survival by activation of the Akt pathway and role for CSN5/JAB1 in the control of autocrine MIF activity. *Oncogene* **26**: 5046–5059.
57. Matsuda N, Nishihira J, Takahashi Y *et al.* (2006) Role of macrophage migration inhibitory factor in acute lung injury in mice with acute pancreatitis complicated by endotoxemia. *Am J Respir Cell Mol Biol* **35**: 198–205.
58. Amin MA *et al.* (2006) Migration inhibitory factor up-regulates vascular cell adhesion molecule-1 and intercellular adhesion molecule-1 via Src, PI3 kinase, and NFkappa B. *Blood* **107**: 2252–2261.
59. Marsh LM *et al.* (2009) Surface expression of CD74 by type II alveolar epithelial cells: A potential mechanism for macrophage migration inhibitory factor-induced epithelial repair. *Am J Physiol Lung Cell Mol Physiol* **296**: L442–L452.
60. Vera PL, Wang X, Bucala RJ, Meyer-Siegler KL. (2009) Intraluminal blockade of cell-surface CD74 and glucose regulated protein 78 prevents substance P-induced bladder inflammatory changes in the rat. *PLoS One* **4**: e5835.
61. Man AL *et al.* (2008) Macrophage migration inhibitory factor plays a role in the regulation of microfold (M) cell-mediated transport in the gut. *J Immunol* **181**: 5673–5680.

62. Riedemann NC *et al.* (2004) Regulatory role of C5a on macrophage migration inhibitory factor release from neutrophils. *J Immunol* **173**: 1355–1359.
63. Rossi AG *et al.* (1998) Human circulating eosinophils secrete macrophage migration inhibitory factor (MIF). Potential role in asthma. *J Clin Invest* **101**: 2869–2874.
64. Wang B *et al.* (2006) Cutting edge: Deficiency of macrophage migration inhibitory factor impairs murine airway allergic responses. *J Immunol* **177**: 5779–5784.
65. Calandra T *et al.* (2000) Protection from septic shock by neutralization of macrophage migration inhibitory factor. *Nat Med* **6**: 164–169.
66. Satoskar A, Bozza M, Rodriguez-Sosa M *et al.* (2001) Migration inhibitory factor gene-deficient mice are susceptible to cutaneous *Leishmania major* infection. *Infect Immun* **69**: 906–911.
67. Rodriguez-Sosa M *et al.* (2003) Macrophage migration inhibitory factor plays a critical role in mediating protection against the helminth parasite *Taenia crassiceps*. *Infect Immun* **71**: 1247–1254.
68. Stojanovic I, Cvjeticanin T, Lazaroski D *et al.* (2009) Macrophage migration inhibitory factor stimulates interleukin-17 expression and production in lymph node cells. *Immunology* **126**: 74–83.
69. Mikulowska A, Metz CN, Bucala R, Holmdahl R. (1997) Macrophage migration inhibitory factor is involved in the pathogenesis of collagen type II-induced arthritis in mice. *J Immunol* **158**: 5514–5517.
70. Santos LL, Dacumos A, Yamana J *et al.* (2008) Reduced arthritis in MIF deficient mice is associated with reduced T cell activation: Down-regulation of ERK MAP kinase phosphorylation. *Clin Exp Immunol* **152**: 372–380.
71. de Jong YP *et al.* (2001) Development of chronic colitis is dependent on the cytokine MIF. *Nat Immunol* **2**: 1061–1066.
72. Denkinger CM, Denkinger M, Kort JJ *et al.* (2003) *In vivo* blockade of macrophage migration inhibitory factor ameliorates acute experimental autoimmune encephalomyelitis by impairing the homing of encephalitogenic T cells to the central nervous system. *J Immunol* **170**: 1274–1282.
73. Powell ND *et al.* (2005) Macrophage migration inhibitory factor is necessary for progression of experimental autoimmune encephalomyelitis. *J Immunol* **175**: 5611–5614.
74. Matza D, Kerem A, Medvedovsky H *et al.* (2002) Invariant chain-induced B cell differentiation requires intramembrane proteolytic release of the cytosolic domain. *Immunity* **17**: 549–560.
75. Sapoznikov A *et al.* (2008) Perivascular clusters of dendritic cells provide critical survival signals to B cells in bone marrow niches. *Nat Immunol* **9**: 388–395.

76. Atsumi T *et al.* (2007) The proinflammatory cytokine macrophage migration inhibitory factor (MIF) regulates glucose metabolism during systemic inflammation. *J Immunol* **179**: 5399–5406.
77. Benigni F *et al.* (2000) The proinflammatory mediator macrophage migration inhibitory factor (MIF) induces glucose catabolism in muscle. *J Clin Invest* **106**: 1291–1300.
78. Verschuren L *et al.* (2009) MIF deficiency reduces chronic inflammation in white adipose tissue and impairs the development of insulin resistance, glucose intolerance, and associated atherosclerotic disease. *Circ Res* **105**: 99–107.
79. Herder C *et al.* (2008) Effect of macrophage migration inhibitory factor (MIF) gene variants and MIF serum concentrations on the risk of type 2 diabetes: Results from the MONICA/KORA Augsburg Case-Cohort Study, 1984–2002. *Diabetologia* **51**: 276–284.
80. Winner M, Koong AC, Rendon BE *et al.* (2007) Amplification of tumor hypoxic responses by macrophage migration inhibitory factor-dependent hypoxia-inducible factor stabilization. *Cancer Res* **67**: 186–193.
81. Hurt EM, Kawasaki BT, Klarmann GJ *et al.* (2008) CD44(+)CD24(–) prostate cells are early cancer progenitor/stem cells that provide a model for patients with poor prognosis. *Br J Cancer* **98**: 756–765.
82. Vermeire JJ, Cho Y, Lolis E *et al.* (2008) Orthologs of macrophage migration inhibitory factor from parasitic nematodes. *Trends Parasitol* **24**: 355–363.
83. Augustijn KD *et al.* (2007) Functional characterization of the *Plasmodium falciparum* and P-berghei homologues of macrophage migration inhibitory factor. *Infect Immun* **75**: 1116–1128.
84. Kamir D *et al.* (2008) A *Leishmania* ortholog of macrophage migration inhibitory factor modulates host macrophage responses. *J Immunol* **180**: 8250–8261.
85. Stein R *et al.* (2007) CD74: A new candidate target for the immunotherapy of B-cell neoplasms. *Clin Cancer Res* **13**: 5556S–5563S.
86. Senter PD *et al.* (2002) Inhibition of macrophage migration inhibitory factor (MIF) tautomerase and biological activities by acetaminophen metabolites. *Proc Natl Acad Sci USA* **99**: 144–149.
87. Cvetkovic I *et al.* (2005) Critical role of macrophage migration inhibitory factor activity in experimental autoimmune diabetes. *Endocrinology* **146**: 2942–2951.
88. Dagia NM *et al.* (2009) A fluorinated analog of ISO-1 blocks the recognition and biological function of MIF and is orally efficacious in a murine model of colitis. *Eur J Pharmacol* **607**: 201–212.
89. Cournia Z *et al.* (2009) Discovery of human macrophage migration inhibitory factor (MIF)-CD74 antagonists via virtual screening. *J Med Chem* **52**: 416–424.

90. Cho Y *et al.* (2010) Allosteric inhibition of macrophage migration inhibitory factor revealed by ibudilast. *Proc Natl Acad Sci USA* **107**: 11313–11318.
91. Winner M *et al.* (2008) A novel, macrophage migration inhibitory factor suicide substrate inhibits motility and growth of lung cancer cells. *Cancer Res* **68**: 7253–7257.
92. Cho YS *et al.* (2007) Structural and functional characterization of a secreted hookworm macrophage migration inhibitory factor (MIF) that interacts with the human MIF receptor CD74. *J Biol Chem* **282**: 23447–23456.
93. Zhong XB *et al.* (2005) Simultaneous detection of microsatellite repeats and SNPs in the macrophage migration inhibitory factor (MIF) gene by thin-film biosensor chips and application to rural field studies. *Nucleic Acids Res* **33**: 121–129.

I-2

MIF and the Chemokine Axis

Sandra Kraemer*, Christian Weber† and Jürgen Bernhagen*,‡

1. Introduction

Macrophage migration inhibitory factor (MIF) is a pleiotropic inflammatory cytokine with unique structural and functional properties that has important roles in physiology and pathophysiology. MIF is an upstream regulator of innate immunity and controls cell homeostasis, but owing to its overall inflammatory activity profile, it is a pivotal mediator of acute and chronic inflammatory diseases, cardiovascular disease, and cancer. Recent work also has shown that MIF has critical chemokine-like functions (CLF) that underlie its leukocyte recruitment activities in atherosclerotic pathologies. At the molecular level, CLF functions of MIF are based on a (remote) 3D architectural homology between the MIF monomer and dimers of CXC family chemokines and on high affinity, noncognate interactions between MIF and the chemokine receptors CXCR2 and CXCR4. In this chapter, we summarize the biochemical and structural evidence that has led to the classification of MIF as a CLF chemokine, discuss other links between MIF and chemokine functions, and evaluate the role of the MIF/CXC chemokine receptor axis in inflammation and cardiovascular pathology. Approaches on how to exploit the knowledge about the MIF/chemokine receptor binding interface for potential novel therapeutic avenues will be briefly discussed as well.

MIF was originally discovered as a T lymphocyte-derived protein mediator that inhibited the random migration of macrophages out of capillary tubes.[1] Some 25 years later, it was discovered that MIF is produced by pituitary cells and macrophages[2,3] and it was redefined as a pleiotropic inflammatory

*Institute of Biochemistry and Molecular Cell Biology, RWTH Aachen University, Aachen, Germany.

†Institute for Cardiovascular Prevention, Ludwig-Maximilians University, Munich, Germany.

‡Corresponding author. Email: jbernhagen@ukaachen.de

cytokine that is involved in the pathogenesis of a number of acute and chronic inflammatory and immune diseases[4] including Gram-negative septic shock,[2,5] colitis,[6] delayed-type hypersensitivity,[7] glomerulonephritis,[8] inflammatory lung disease,[9] and rheumatoid arthritis.[10] In such conditions, MIF is an upstream regulator of excessive innate immune and inflammation reactions and triggers inflammatory signal transduction pathways to promote macrophage survival and inflammatory cytokine expression.[4] Its proinflammatory activity is also based on its unique property of counter-regulating the immunosuppressive and anti-inflammatory activity of endogenous glucocorticoids.[4,11,12] Moreover, MIF's pivotal role in the pathogenesis of cardiovascular disease involves the proinflammatory recruitment of leukocytes, inflammatory activation of intimal macrophages, foam cell formation, and plaque destabilization.[13–17] As introduced above, MIF shares structural and functional features with chemokines and has been classified as a CLF chemokine.[16,18,19] In fact, CLF properties of MIF have emerged to be pivotal for its contribution to the pathogenesis of atherosclerosis.

Chemokines play a crucial role in driving atherosclerotic inflammatory processes as well as numerous other inflammatory conditions. At first sight, the chemokine system appears highly redundant, given that around 50 human chemokines can interact with about 20 chemokine receptors. In addition, many chemokines are produced from several different cell types and are able to bind several receptors. However, specificities in cell type-, site- and disease stage-specific capacities have been identified, in addition to a preferential involvement of chemokines in specific steps of the inflammatory leukocyte recruitment cascade.[20] Thus, the apparent redundancy of the chemokine system could also be viewed as a feature of elaborate robustness that ensures a highly orchestrated recruitment of inflammatory cells into inflamed vessels and sub-vessel tissues. Likewise, this redundancy guarantees the various homeostatic functions that chemokines fulfill in cell homing, angiogenesis, and tissue organization.[21,22]

2. Chemokines: Classes, Structure, Function, and Receptors

Chemokines are small chemotactic cytokines of a molecular weight of 8 kDa. Since their discovery over 20 years ago, they have emerged as the most important regulators of leukocyte trafficking. Chemokines orchestrate the activation and recruitment of leukocytes during immune surveillance and inflammation, but also are pivotal regulators of various other cell types. Due to their plethora of activities towards fibroblasts, endothelial cells, other parenchymal cells, and tumor cells, chemokines have numerous functions in normal physiology, tissue remodeling, atherosclerosis, and tumor progression.[21,23,24]

Chemokines bind to a subclass of the large family of seven-transmembrane (TM)-helix receptors, the chemokine receptors. Most chemokine receptors are coupled to heterotrimeric proteins of the G_i type. The predominant cellular response to chemokine exposure is directed migration or chemotaxis. If the responder cells are leukocytes, chemokine-driven chemotaxis is often intimately coupled to haptotaxis, a process through which leukocytes marginate from the bloodstream by following an endothelial cell surface-bound chemokine gradient. This enables leukocytes to reach their destinations at sites of inflammation in a coordinated manner. In addition, leukocyte activation by chemokines triggers a variety of signaling responses. The specificity and selectivity of the response depends on the specific chemokine-chemokine receptor pair involved as well as the engagement of various cross-talk mechanisms.[21,22,24–33]

The structural classification of chemokines and their G protein-coupled receptors (GPCRs) has been well defined. Some 50 chemokines bind to some 20 receptors, which constitutes a redundancy on the side of the ligands. Chemokines may be classified into four structural classes: CC chemokines, CXC chemokines, C chemokines, and CX_3C chemokines. This classification is based on the localization of one or more cysteine residues in the N-terminal sequence region that are spaced by zero, one or two other amino acids. The CXC/CC motif (the CXC and CC chemokines are the two most prominent chemokine classes) connects the functionally important N-loop and N-terminal residues, and also plays a structural role by forming disulfide bonds that connect the N-terminal and N-loop residues to the protein core. The chemokines of these various classes bind to corresponding GPCR-type receptors, which are also grouped according to the CXC/CC/C/CX_3C terminology. Classical chemokine structures also feature a specific structural fold, the so-called chemokine fold, which in a chemokine monomer is made of a three-stranded antiparallel β-sheet and an α-helix. Binding of chemokines to their receptors and receptor activation involves two interactions: (i) between the ligand N-loop and receptor N-domain residues (site I), and (ii) between the ligand N-terminal sequence stretch and receptor extracellular/transmembrane residues (site II), i.e., extracellular loop regions (ELs). In the case of the subclass of ELR+ CXC chemokines (featuring the residue motif Glu-Leu-Arg at position 4–6), the site II interaction is constituted by the ELR motif and the CXC chemokine receptor ELs (for details see Sec. 7 of this chapter). The wealth of information available today on the structural properties of chemokines and the mechanisms by which they activate their receptors has been comprehensively gathered in a number of excellent and fairly recent review articles.[21,22,34–40]

In addition, mediators with chemokine-like functions but remotely related structural characteristics (CLF chemokines) have been defined. CLF chemokines have been demonstrated to bind to a chemokine receptor of one of the four classes[18,41–45] (for details see Sec. 5 of this chapter).

Chemokines adopt a dimeric structure in X-ray crystallographic analysis, presumably due to the high millimolar concentrations used for this analysis method. In contrast, at nanomolar concentrations it is thought that chemokines predominantly exist as monomers. It is currently unclear whether a chemokine monomer or dimer is the physiologically relevant species interacting with a corresponding chemokine receptor. While for some chemokines and in some reports the monomer has been demonstrated to be the predominant receptor-interacting species, inducing cell migration responses, it has been argued that the dimer is important for haptotactic gradient formation, protection from proteolysis, and signaling related to processes distinct from migration.[46–48] Of note, even higher-order and heteromeric chemokine oligomerization has been regarded as an important fine-tuning mechanism for chemokine activities in certain tissue microenvironments.[49] The complexity of chemokine oligomerization and its functional implications are probably best documented by the available literature on the activities of the monomeric versus dimeric forms of the protagonistic CXC chemokines CXCL8 (also known as interleukin-8, IL-8) and CXCL12 (also known as stromal cell-derived factor-1α, SDF-1α). At physiologically relevant nanomolar concentrations, CXCL12 is a monomer, but binding partners of CXCL12 promote its self-association at higher concentrations to form a typical CXC chemokine homodimer. Yet the monomeric structure was recently shown to mediate the cardioprotective function of this chemokine,[50,51] while others have reported that the dimer is necessary for CXCL12-mediated cell migration.[52] Under physiological conditions, CXCL8 can exist as monomers, dimers or a mixture of monomers and dimers, but the monomer seems to be the high-affinity receptor-interacting species. Nevertheless, both forms of CXCL8 can interact with the CXCL8 receptors CXCR1 and CXCR2, although the dimer with lower affinities. Moreover, functional studies indicate that both the CXCL8 monomer and dimer are able to trigger CXCR1- or CXCR2-driven cellular responses but act differentially in regulating CXCR1 and CXCR2 receptors and potencies with respect to the downstream signaling pathways.[47,53] The structural information about the CXCL8 and CXCL12 interactions with their receptors is important in the context of this chapter, because we will deal with the structural requirements of the identified interaction between MIF and CXCR2 and CXCR4 later on in this chapter. It should also be noted that chemokine dimerization is also critical for CC-only chemokine oligomers. For example, homo-oligomerization is required for the recruitment functions of CCL5 (also known as RANTES).[54] CCL2/MCP-1 also forms dimers.[55]

Chemokines bind to specific 7-TM receptors linked to heterotrimeric G_i proteins. Thus, chemokine receptors belong to the rhodopsin-like GPCR receptor class. GPCRs constitute the largest transmembrane receptor family; most GPCRs are activated by small molecule ligands such as amino acid derivatives or small peptides, but some GPCR subclasses, such as the chemokine receptors, interact with polypeptides or even larger proteins. In addition to their 7-TM nature and cytosolic G protein binding domain, chemokine receptors show all the typical properties of GPCRs, including a relatively short extracellular N-terminus critically involved in ligand binding and N-glycosylation. However, chemokine receptors have a considerably smaller C-terminal tail than most other GPCRs and thus are smaller in size, with molecular weights in the range of 30–40 kDa. Also, chemokine receptors generally activate "Gα and $\beta\gamma$ responses" rather than Gα s signaling. Only a few GPCR structures have been elucidated by X-ray crystallography, but very recently the X-ray structure of CXCR4 has been resolved.[56] This study has confirmed prior predictions according to which chemokine receptors would form typical 7-TM structures based on their analogy to the rhodopsin and α-adrenergic receptors. Similar to the adrenergic or opioid receptors, it had been assumed that chemokine receptors are activated by a two-site binding model, with the site I interaction making the initial ligand-receptor contact and inducing a conformational change in the receptor protein that enables site II interaction and receptor activation (see also above).[40] The recent elucidation of the structure of CXCR4, which was obtained in a complex with an antagonist small molecule and a cyclic peptide at 2.5–3.2-Å resolution, fully confirmed these predictions and additionally unraveled valuable structural details which help to understand the nature of distinct CXCL12/CXCR4 interactions but which, based on the pinpointed binding interface between CXCR4 and HIV-gp120, also hinted at structural sites involved in promiscuous interactions with nonclassical CLF chemokines such as MIF.[56]

It is well accepted that chemokine receptors homodimerize, and obviously, receptor homo-oligomerization would mirror the complexity on the ligand side. For example, the X-ray structure elucidation of CXCR4 unanimously confirmed various prior evidence that this chemokine receptor as well as likely several other chemokine receptors occur as dimers.[56] As discussed extensively by Thelen and others,[21] GPCR homodimerization or homo-oligomerization likely serves to fine-tune and diversify signaling and/or to terminate such responses by inducing refractory behavior through desensitization and internalization. In principal, 7-TM-domain receptors such as chemokine receptors operate as conformational switches that possess an "off" and an "on" state to allow, for example, for rapid Ca^2+ mobilization. Similarly, phosphoinositide-3-kinase/phosphatidylinositol-3-phosphate (PI3K/PIP3)-mediated chemokine signaling responses have been suggested to be mediated by monomeric and thus "fast-track" receptor states. On the other hand, GPCRs can operate as oligomers. Receptor dimerization is probably too slow to be a prerequisite for rapid calcium mobilization but can support protein kinase B (AKT/PKB), extracellular signal-regulated kinase (ERK), Src kinase (SRC), and Janus kinase signaling responses. Higher-order receptor clustering and oligomerization has been suggested to promote arrestin binding and internalization of receptors. Whether all of these properties also apply to several or all of the chemokine receptors is likely, but there has only been sparse experimental evidence. It is debated whether chemokine receptor dimerization is ligand-inducible or occurs inherently during trafficking of the receptor proteins through the ER/Golgi system. In fact, ligand-independent CXCR2 homodimerization was shown when differentially tagged CXCR2 receptor species were co-expressed in HEK293 cells.[57] Dimer formation most likely already occurred during biosynthesis in the ER. Moreover, homodimer formation of CXCR2 proved to be important for ERK1/2 and AKT phosphorylation, as well as for CXCL8-mediated chemotaxis. In contrast, for CXCR1, there is doubt as to whether dimerization is relevant.[58]

3. Structural Classification of MIF

MIF is an evolutionarily conserved protein that is abundantly expressed in humans and nonprimate mammals. MIF orthologues are also found in unicellular parasites. The cloning of human and mouse MIF and the elucidation of the three-dimensional (3D) structure of human MIF by X-ray crystallography and NMR revealed that the sequence and structure of MIF are unique within the family of cytokines.[2,59–62] The MIF cDNA predicts a protein size of 115 residues, but due to processing of the N-terminal methionine, which occurs in essentially every cell type including during overexpression in *E. coli*, human MIF consists of 114 amino acids and has a molecular weight of 12,345 Da. The 3D structure of human MIF shows that MIF crystallizes as a trimer of three identical subunits. Each monomer consists of a four-stranded β-sheet placed above two antiparallel α-helices. This structural module thus grossly resembles the sheet/helix fold of a chemokine dimer, although the topology is different. Two or three remaining β-strands in MIF are part of intertwining loops and contribute to the stabilization of the MIF trimer by forming interactions with β-sheets of adjacent subunits.[61,63] In contrast to the trimer determined by crystallography, biochemical studies relying on soluble MIF have suggested that monomers, dimers, and trimers exist in an equilibrium, with monomers and dimers representing the major species at physiological concentrations in the ng/ml range, whereas the trimer is the predominant species at higher MIF concentrations >10 μg/m.[64–67] The amino acid sequence and 3D structure of MIF is also similar to that of glycosylation inhibition factor (GIF).[68] The barrel structure of MIF/GIF in part resembles that of "trefoil" cytokines such as interleukin-1β (IL-1β) and fibroblast growth factor (FGF). As indicated above, the sheet/helix structural fold of the MIF monomer also shows remote similarities to that of the chemokine interleukin-8 (IL-8, CXCL8). However, these structural similarities do not allow for a classification of MIF into the known structural cytokine families or the CXC chemokine family (but see the paragraph below on CLF chemokines). The structure of MIF/GIF also has an interesting similarity to the peptide binding domain of major histocompatibility complex class I (MHC I) protein, although the topology of the polypeptide chain again is quite different. Nevertheless, this similarity is notable because CD74, the membrane form of invariant chain (Ii), which regulates peptide antigen loading into MHC class II complexes, has been identified as a receptor for MIF (see below).

The 3D architecture of MIF is very similar to that of a family of bacterial isomerases/tautomerases as well as to human D-dopachrome tautomerase, indicating that these proteins may have evolved from a joint ancestral gene.[64] In fact, uniquely among cytokines, the MIF trimer shares with these enzymes a catalytic pocket which is formed by the N-terminal proline residue of one subunit and several residues of the same and a second subunit and which enables MIF to catalyze certain isomerization/tautomerization reactions. It has remained controversial whether the catalytic isomerase activity of MIF is functionally related to its numerous inflammatory properties.[64] Notwithstanding, the identification of a catalytic site in MIF has triggered several studies to develop small molecular weight (SMW)-based inhibitors that could target this site, and it has been hoped that for the first time this would make a soluble cytokine amenable to targeting by SMW drugs.[10,69] In addition, the MIF protein features a second conserved catalytic site. This is a Cys-Xaa-Xaa-Cys motif and MIF shares this motif with thiol-protein oxidoreductases (TPORs). Accordingly, MIF has been demonstrated to have redox activity *in vitro* and to be involved in the regulation of cellular redox homeostasis.[70,71]

4. Nonclassical MIF Secretion and Link to Secretion of CLF Chemokines

Upon protein biosynthesis, MIF is stored in the cytosol, where on the one hand it engages in interactions with several intracellular proteins[70,72] (for details see other chapters) and on the other hand constitutes preformed stores serving as a pool for rapid secretion.

MIF was originally discovered as T cell-derived cytokine. Today, it is clear that mononuclear phagocytes represent the main MIF-producing cell type in inflammation. However, several other inflammatory cells such as eosinophils as well as endothelial cells, numerous epithelial, parenchymal, and tumor cells can produce MIF in substantive amounts.[4] Thus, the expression of MIF is fairly ubiquitous. In contrast, its secretion is tightly regulated and only occurs upon distinct inflammatory, immune or stress stimulation of a MIF-producing cell. MIF secretion from monocytes/macrophages is triggered by endotoxin and inflammatory cytokines such as tumor necrosis factor-a (TNF-α).[4] Endothelial cells secrete MIF upon stimulation with oxidized low-density lipoprotein (LDL) or during hypoxia. Cardiomyocytes secrete MIF upon ischemia/reperfusion (I/R) challenge, i.e., following combined hypoxic and hyperoxic stress.[73–75] Tumor cells secrete MIF upon stress and growth stimulation.[76,77] The precise mechanism of MIF secretion has only been partially unraveled. MIF secretion follows a so-called nonclassical ER/Golgi-independent pathway, but we have only begun to uncover the molecular components of this pathway, which is also likely to differ between MIF-secreting cell types. The first evidence for a nonconventional secretion pathway of MIF arose when the molecular cloning of MIF revealed that the N-terminus was lacking a signal sequence, suggesting that during secretion MIF does not enter the ER/Golgi compartment.[2,59] This notion was supported by pharmacological inhibitor-based evidence that the endotoxin-triggered secretion of MIF from monocytes/macrophages depends on ATP-binding cassette (ABC) transporter-mediated membrane translocation at some stage during secretion.[78] Hypoxia-induced secretion of MIF from endothelial cells also involves ABC transporter activity.[79] In addition, at least two of the known cytosolic interaction partners of MIF are functionally involved in MIF secretion. Silencer RNA-mediated knock-down of the Golgi-associated protein p115 leads to a drastic reduction of MIF secretion in endotoxin-stimulated monocytes/macrophages, suggesting that p115, which is situated on the cytosolic side of the cis-Golgi and which is involved in Golgi vesicle trafficking, is a component of the MIF secretion pathway.[80] In contrast, protein-protein interaction between MIF and JAB1/CSN5 appears to convey inhibitory signals within the MIF secretion pathway.[81]

The cytosolic "intracellular localization" of MIF and its nonclassical mode of secretion are reminiscent of those of other protein mediators to which both intra- and extracellular functions have been ascribed. Depending on their precise origin and identity, mode of processing, and release mechanism, these mediators have been termed alarmins, microchemokines, redoxkines or chemokine-like function (CLF) chemokines. Classification of such proteins also has been leveled by whether or not these mediators exert classical chemokine activities and whether or not their target cell activity is mediated by interaction with a *bona fide* chemokine receptor. High mobility group binding protein-1 (HMGB-1),[82] Y-box protein-1 (YB-1)[83] or thioredoxin (TRX)[84] are secreted by nonclassical export and/or following cell apoptosis, exert chemokine-like pro-migratory activities, but do not interact with chemokine receptors. On the other hand, the β-defensins (HBDs), which have been categorized as typical alarmins, exhibit chemokine-like properties, interact with chemokine receptors, but are usually released upon cell death.[41–43,45] Herein, the classification "chemokine-like function" (CLF) chemokine is used for protein mediators, which (i) are released during inflammation, infection, or cell stress by nonclassical export or due to apoptosis or cell death; which (ii) do not strictly share with the classical chemokines the typical chemokine fold and the N-terminal cysteines; but which (iii) exhibit chemokine-like activities in (iv) a chemokine receptor-dependent manner.

5. Chemokine-like Function (CLF) Chemokines and MIF as a CLF Chemokine

The N-terminal cleavage fragment of tyrosyl-tRNA synthetase (TyrRS) has been shown to act as a noncanonical ligand for CXCR1, and this interaction is based to a good extent on the presence of an ELR motif in TyrRS.[85] Although aminoacyl-tRNA synthetases (AaaRS) normally exhibit intracellular functions, their cleavage, i.e., by the extracellular protease leukocyte elastin, which comes into action upon its secretion under apoptotic conditions, generates AaaRS fragments with chemokine-like activities. Whereas the endothelial monocyte-activating polypeptide II-like C-terminal domain functions as a potent monocyte and neutrophil chemoattractant and induces the production of myeloperoxidase, TNF-α, and tissue factor, the N-terminal "mini-TyrRS" containing both the catalytic and anticodon-recognition domain is proangiogenic and induces neutrophil chemotaxis.[85] These functions of mini-TyrRS are manifested through its interaction with CXCR1, but not CXCR2, and structurally depend on an ELR motif in TyrRS.[86] Similarly, specific autoantigenic AaaRS, released under apoptotic conditions, exert leukocyte recruitment by triggering certain CC chemokine receptors. Both histidyl-RS (HisRS) and its N-terminal fragment are chemoattractants for T cells, activated monocytes, and immature dendritic cells (iDCs) via stimulation of CCR5, the *bona fide* receptor for RANTES/CCL5, eotaxin-2/CCL24, and MCP-2/CCL8 receptor. The same cell types are also recruited by asparaginyl-RS (AsnRS) via its interaction with CCR3, another promiscuous CC chemokine receptor which engages in cognate interactions with CCL5, CCL7, CCL8, CCL13, and CCL15.[36,87]

In addition to the AaaRS and their fragments generated during cell stress and death in inflammation, the human antimicrobial peptides b-defensin-1 and -2 (HBD-1 and -2) can be classified as CLF chemokines.[87,88] HBD-1 and -2 were identified as noncognate ligands for CCR6, mediating CCR6-dependent chemotaxis of immature dendritic cells and memory T cells.[89] By recruiting these cell types to sites of microbial infection, HBD-1 and -2 may thus promote adaptive immune responses in addition to their prominent role in innate immunity.

Thus, despite the absence of typical structural chemokine motifs, an increasing number of host proteins involved in inflammatory and immune processes seem to act via direct — *noncognate* — interaction with chemokine receptors that leads to chemokine receptor triggering via the mechanism of molecular mimicry. It should be mentioned at this point that viruses and other parasites also prominently capitalize on this principle of "molecular hijacking," e.g., to facilitate their dissemination and evasion from the host system. Examples are *Toxoplasma gondii*-derived cyclophilin-18, which potently stimulates IL-12 production by DCs through triggering CCR5,[90] and HIV-1 gp120, which interacts with host CXCR4 (and CCR5) to infect leukocytes.[91] As mentioned above, MIF cannot be grouped into any of the four classical chemokine subfamilies.[35] Instead, MIF has been classified as a CLF chemokine.[18,92]

Contrary to its eponymous name "macrophage migration inhibitory factor," MIF has pro-rather than antimigratory activities towards leukocytes. In conjunction, the following properties have led to the characterization of MIF as CLF chemokine:

(i) MIF promotes the chemotactic migration of monocytes, T cells, neutrophils, endothelial progenitor cells (EPCs) as well as certain tumor cells: various studies have shown that MIF promotes true chemotaxis rather than chemokinesis or enhancement of random migration of monocytes, T cells, neutrophils, and EPCs. This evidence has come from experiments employing purified, endotoxin-free recombinant MIF (rMIF) and neutralizing monoclonal antibodies in combination with chemotaxis migration chambers such as Transwell or modified Boyden chambers. In addition, promotion of tumor cell migration by MIF may involve the intermediate release of CXCL8.[13,32,74,93–97]

(ii) MIF desensitizes monocytes against chemotactic stimulation by CCL2/MCP-1: heterologous desensitization by rMIF of CCL2-mediated monocyte chemotaxis was achieved in modified Boyden chambers. In addition, rMIF desensitizes monocyte chemotaxis towards MIF gradients, constituting homologous desensitization.[13,98]

(iii) MIF exhibits both intra- and extracellular functions and is secreted from cytosolic stores by a nonclassical secretion pathway; see above for details.[70]

(iv) MIF engages in high affinity, noncognate binding to the CXC chemokine receptors CXCR2, CXCR4 and CXCR7: high-affinity interaction between MIF and the CXC receptors CXCR2 and CXCR4 was shown by several biochemical methods. These include receptor competition assay applying radiolabeled cognate ligand tracer, receptor internalization assays, co-immunoprecipitation, as well as flow cytometry measurements. Most recently, evidence for an interaction between MIF and the decoy receptor CXCR7 was obtained as well.[13,99–101]

(v) MIF induces the expression and/or secretion of the chemokines CXCL8/IL-8 and CCL2 from B cells, T cells, tumor cells, and endothelial cells, respectively: long-term (overnight) stimulation of B cells, T cells, or tumor cells with rMIF triggers the expression and secretion of CXCL8, while mid- to long-term (four hours to overnight) stimulation of endothelial cells triggers the secretion of CCL2.[15,102–105]

(vi) MIF triggers atherogenic and inflammatory leukocyte recruitment responses *in vivo*: MIF promotes monocyte and T cell arrest to atherogenic endothelium and enhances transmigration of leukocytes into the subendothelial space in both chronic and acute atherogenic lesions. Evidence for these effects has come from proatherogenic mouse models (*ApoE*$^{-/-}$ and *Ldlr*$^{-/-}$) *in vivo* that were crossed with *Mif*$^{-/-}$ mice as well as from studies using neutralizing anti-MIF antibodies. MIF-dependent luminal and intimal macrophage accumulation was evidenced by intravital microscopy (IVM). Inflammatory leukocyte recruitment triggered by MIF was also observed in models of rheumatoid arthritis, systemic lupus erythematodes (SLE) and in the inflamed cremasteric microvasculature.[10,13–15,17,106–108]

(vii) MIF shares structural similarities with *bona fide* chemokines that account for its affinity to chemokine receptors. The structural aspects that have led to the classification of MIF as a CLF chemokine are discussed in detail in Sec. 2 of this chapter.[13,62,68]

6. MIF/Chemokine Receptor Interactions and Functional Role in Inflammation and Atherogenesis

Upon secretion, MIF exerts specific cytokine/chemokine functions in the extracellular space. Because MIF does not belong to any of the known structural cytokine or chemokine families, homology-based cloning strategies could not be applied to identify potential MIF receptors. Accordingly, a receptor for MIF had been elusive for more than a decade after MIF cloning. However, today it is clear that the various potent, mostly proinflammatory, activities of extracellular MIF are mediated by MIF surface receptors. The first plasma membrane receptor for MIF to be identified was a protein known as CD74.[109] CD74 is the name for the membrane-expressed portion of invariant chain (Ii). CD74/Ii functions as a MHC class II chaperone, plays a critical role in class II trafficking between the endoplasmic reticulum (ER) and endolysosomal compartment, and regulates antigenic peptide loading into class II protein through its CLIP domain.[110] CD74 expression was initially assumed to be restricted to class II-positive cells such as B cells, monocytes/macrophages, and dendritic cells. However, it turned out that under inflammatory conditions as well as in activated tumor cells, CD74 is found to be upregulated and expressed even in the absence of measurable class II expression.[110–112]

That membrane-expressed CD74/Ii can additionally function as a cytokine receptor was first recognized on B lymphocytes. It was demonstrated that in these cells, CD74 promotes proliferation and survival by triggering a nuclear factor kappa-B (NF-κB)/TAp63/CXCL8-dependent pathway.[102,113–115] MIF binding to CD74 only became apparent when monocytes were pre-activated to upregulate MHC class II/CD74, i.e., when they expressed sufficient receptor molecules on the surface that could be trapped by flow cytometry-based panning/binding studies. Subsequent biophysical receptor binding studies then revealed that MIF binds to CD74 with high affinity in the low nanomolar range.[116] CD74 does not feature a conventional cytoplasmic signal-transducing domain. Instead, signaling by MIF-activated CD74 requires the recruitment of signaling-competent co-receptors such as CD44[117] or the chemokine receptors CXCR2 and CXCR4.[13] Alternatively, it was observed that in B cells, the short cytoplasmic domain of CD74 can be cleaved by regulated intramembrane proteolysis (RIP) and that the short cytoplasmic fragment traffics into the nucleus to promote NF-κB signaling by functioning as a transcriptional co-activator.[110,118–121] The observation that CD74 does not contain a typical cytoplasmic signaling domain, as well as the notion that CD74 is not expressed on key MIF-responsive cells such as neutrophils or experimental MIF-responsive cell lines such as HEK293, triggered efforts to identify additional MIF receptors.

The observed gross architectural similarity between the MIF monomer and the CXCL8 dimer[61,68] elicited biochemical investigations to study potential interactions between MIF and the *bona fide* CXCL8 receptor CXCR2. Such studies were also incited by the observation that MIF-mediated monocyte arrest on aortic endothelial cells occurred within two hours, did not involve CXCL8/KC production by the endothelial cells, and could be blocked by neutralizing anti-CXCR2 antibodies.[13,17] Receptor binding studies by tracer competition and internalization experiments, co-immunoprecipitation, and flow cytometry then revealed that MIF engages in a noncognate, high-affinity (K_D 1.5 nM) interaction with CXCR2.[13] CXCR2 is the main receptor mediating the inflammatory and angiogenic activities of CXCL8. However, as discussed above, CXCR2 is promiscuous because it also binds to six other ELR+ CXC chemokines, such as CXCL7 and CXCL1. Noncognate binding of MIF to CXCR2 thus expands the ligand spectrum of CXCR2 and adds a CLF chemokine to the ligand list of this important inflammatory chemokine receptor.[13,19] The signal transduction pathways triggered by MIF/CXCR2 have not been studied systematically, but initial investigations show that in monocytes and neutrophils, MIF binding to CXCR2 involves G_i coupling and elicits calcium transients. The identification of the noncognate interaction between MIF and CXCR2 provoked the question whether MIF would bind to other promiscuous chemokine receptors as well. A limited, rationally guided receptor screen, which took into account that notion that T cells, which do not express CXCR2 and only express low levels of CD74, revealed that MIF did not interact with CXCR1, CXCR3 or CCR5.[13] However, it also led to the surprising discovery that MIF also serves as a noncognate ligand of CXCR4.[13] Until recently, CXCR4 had been considered a highly specific chemokine receptor which would engage in a monogamous ligand/receptor relationship with CXCL12/SDF-1α only.[40] However, this dogma was challenged when it turned out that CXCL12 also interacts with CXCR7, the receptor for I-TAC/CXCL11, which is now considered a chemokine decoy receptor due to its failure to activate G_i coupling.[122–124] Receptor binding and internalization studies as well as co-immunoprecipitation experiments have shown that MIF specifically and strongly interacts with CXCR4, although the MIF/CXCR4 interaction is less affine (K_D 19 nM) than that between MIF and CXCR2 or

CD74, but still in the nanomolar range.[13,125] In addition, very recent evidence supports the notion that in rhabdomyosarcoma cells MIF also interacts with CXCR7,[101] although it is currently unclear whether this occurs directly or indirectly through CXCR4. Interestingly, CXCR4/CXCR7 heterodimeric receptor complexes have been reported to form.[124,126]

While this latter interaction appears to play a role in rhabdomyosarcoma formation,[101] and the MIF/CXCR4 interaction also was implicated in tumorigenesis,[127] it seems that MIF/CXCR interactions mainly play a role in inflammatory and atherogenic leukocyte chemotaxis and recruitment. Both CXCR2 and CXCR4 have been amply implicated in atherogenesis.[128,129] In addition to triggering G_i activation and calcium influx, binding of MIF to CXCR2 or CXCR4 induces rapid (within 1–5 min) integrin activation ($\alpha_L\beta_2$/LFA-1 in monocytes and $\alpha_4\beta_1$/VLA-4 in T cells) in conjunction with monocyte and T cell adhesion on endothelial monolayers, respectively.[13] Triggering of MIF-induced monocyte adhesion via rapid activation of $\alpha_L\beta_2$- or $\alpha_4\beta_1$-integrin is similar to the arrest effects of CXCL8.[13,130] G_i coupling of MIF-activated CXCR receptors was direct, as demonstrated by pertussis toxin blockade and failure of the neutralization of the CXCR2 ligands CXCL1 and CXCL8 to interfere with the MIF-stimulated effect on leukocyte recruitment. Interestingly, and similar to CXCL1, MIF was found to be immobilized and presented on the endothelial surface (Ref. 13 and S. Tillmann and J. Bernhagen, *unpublished observations*). The mechanism of MIF-induced monocyte arrest has been confirmed in atherogenic leukocyte recruitment models *ex vivo* and *in vivo*. In *ex vivo*-perfused carotid arteries of hyperlipidemic $ApoE^{-/-}$ mice, blockade of CXCR2 or MIF significantly reduced monocyte adhesion, suggesting that MIF induces atherogenic leukocyte arrest through CXCR2.[13] CD74 also seems to be involved in these effects, suggesting the existence of functional heteromeric MIF receptors encompassing CXCR and CD74 (see Sec. 7 of this chapter). An arrest chemokine function on early atherosclerotic endothelium as seen for MIF has previously been described for the cognate CXCR2 ligand KC/CXCL1.[131] Interestingly, KC-deficiency only led to a 50% reduction in arrest activity, whereas blockade of CXCR2 completely abolishes the arrest function of KC. This observation suggested the existence of an alternative CXCR2-dependent arrest factor.[128,132] It now appears that MIF could be the sought for alternative CXCR2 arrest ligand.

In carotid arteries from $Mif^{-/-}/Ldlr^{-/-}$ mice, monocyte adhesion under flow conditions was clearly reduced compared with $Mif^{+/+}/Ldlr^{-/-}$ mice. In $Mif^{-/-}/Ldlr^{-/-}$ mice, inhibition of CXCR2 did not further impair monocyte adhesion. After loading exogenous MIF, the inhibitory effect of the

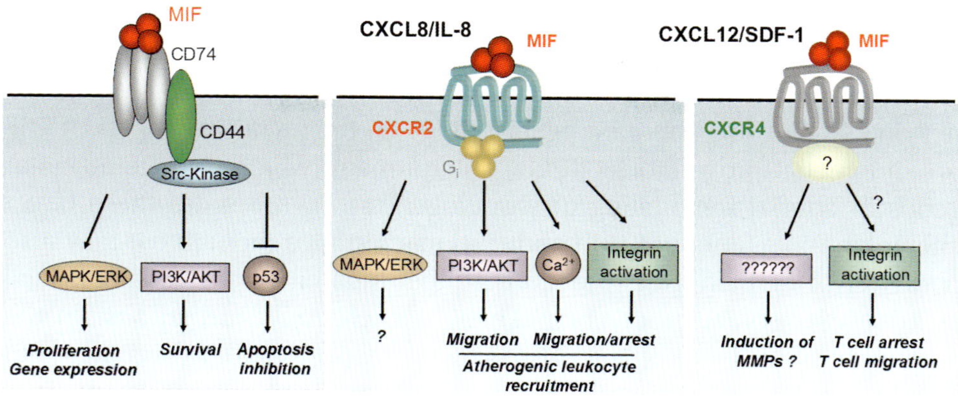

Fig. 1. Schematic summarizing the most prominent effects mediated by the three MIF receptors. CXCR2 is also the cognate receptor for CXCL8. CXCR4 is the cognate receptor for CXCL12. For CD74, MIF is the only known ligand. For simplicity, MIF is depicted as a trimer, although the receptor-activating oligomeric species of MIF have not been specifically determined.

CXCR2 antibody was fully restored in *Mif*$^{-/-}$/*Ldlr*$^{-/-}$ mice.[13] Furthermore, impaired leukocyte recruitment has also been observed in acute vascular inflammation after TNF-α treatment in irradiated *Mif*$^{-/-}$ mice supplemented with CXCR2-deficient bone marrow cells (from *Cxcr2*$^{-/-}$ mice). Together, these functional studies strongly support the important role that the MIF/CXCR2 axis plays in atherogenic leukocyte recruitment. Highly specialized functions have been ascribed to different chemokines in atherogenic monocyte and T cell recruitment.[133] In addition to monocyte recruitment, MIF induces T cell adhesion and transmigration through interaction with CXCR4.[13] This mechanism of MIF-mediated T cell recruitment may be responsible for the reduced T cell content in advanced atherosclerotic plaques after MIF blockade. Figure 1 summarizes the effects of the three MIF receptors CD74, CXCR2, and CXCR4, with a focus on the signaling pathways and primary cell responses triggered by the different MIF/receptor axes.

Together, MIF is involved in atherogenic leukocyte recruitment in a more universal sense: it has slightly reduced efficacy as compared with the cognate ligand chemokines, but through interaction with *both* CXCR4 and CXCR2, MIF promotes the recruitment of both monocytes and T cells.[13] This is in contrast to the more cell-restricted action of CXCL8 and CXCL12. Thus, the atherogenic function of MIF as a dual agonist of CXCR2 and CXCR4 could explain the striking effects of MIF inhibition on limiting atheroprogression and on mediating plaque regression.[13] On the other hand, these data would imply that MIF may retain a potent and predominant CXCR2 activity in atherogenesis, when the CXCR4/CXCL12 axis is blocked.[13] Conversely, blockade of CXCR2 would be expected to allow for maintaining MIF activities on CXCR4. This could be beneficial, given the identified effects of MIF on endothelial progenitor cell (EPC) recruitment, which may be advantageous in arterial remodeling or vasculogenesis.[74,134] Besides these direct chemokine-like functions, MIF-induced macrophage recruitment may also partly rely on enhanced expression of other inflammatory mediators, such as adhesion molecules or chemokines,[8] which could further sustain the mononuclear cell influx (see also above). In post-capillary venules and arthritic inflammation, MIF-induced monocyte recruitment critically depends on the CCL2/CCR2 axis, possibly due to enhanced endothelial release of CCL2 by MIF.[15,109] In fact, as established for other chemokines involved in atherogenic leukocyte recruitment,[133] MIF might sequentially regulate macrophage recruitment: first a rapid activation of integrin-dependent monocyte adhesion may occur through direct interaction with CXCR2 (or CXCR2/CD74 complexes; see below), followed by MIF-induced CCL2 secretion from endothelium, which would primarily promote monocyte transmigration.[15,107]

7. Structural Basis Underlying MIF/Chemokine Receptor Interactions

Structure function analyses suggest that chemokine/chemokine receptor pairs interact through two binding sites. These involve the N-domain and extracellular loops (ELs) of the receptor and the N-terminus and N-loop of the ligand.[40,135–138] The following two-site binding mechanism has been proposed: a site I interaction takes place between the chemokine N-loop and the receptor N-domain, while the chemokine N-terminus and the receptor exoloops and/or a transmembrane region form the site II binding interface.[40] The two-site binding model has been extensively studied for CXCL8, which binds to both CXCR1 and CXCR2. Site I binding involves the N-loop of CXCL8 and the receptor N-domain and is responsible for receptor selectivity and binding affinity. The ELR motif and EL2 and EL3 form the site II interaction site, which is essential for receptor activation.[39,137] Naturally, the two binding events are not independent but are coupled to each other.[40] Binding to one site, therefore, positively or negatively influences the binding affinity of the second site. It has been suggested that site I binding leads to a conformational change both on the ligand and receptor side, which affects site II interaction.[40]

Gross inspection of the MIF sequence showed that MIF does not have an ELR motif. However, bioinformatic analysis indicated that MIF could have an ELR-like motif consisting of amino acids Arg-12 and Asp-45.[1] These amino acids reside in neighboring loops in the MIF monomer and mimic an ELR motif in 3D space. The motif was thus termed pseudo-(E)LR motif.[100] Putting the "E" residue in parenthesis (E) accounts for the fact that in the MIF sequence the "E" residue of the ELR+ chemokines is substituted for aspartic acid (D) by conserved exchange. Involvement of the pseudo-(E)LR motif in MIF's chemokine-like functions was then analyzed by generating the mutants R12A-MIF, D45A-MIF, and R12A-D45A-MIF. The pseudo-(E)LR mutants could be expressed and purified like wildtype MIF, and their overall structural integrity as well as bioactivity was confirmed by biochemical methods. Importantly, in a direct receptor competition assay, the mutants only show a weak binding affinity to CXCR2, demonstrating that the pseudo-(E)LR motif is critical for MIF/CXCR2 binding. In addition, MIF's *in vitro* monocyte recruitment function is abolished following mutation of the pseudo-(E)LR residues. Especially Arg-12 is also significantly involved in promoting MIF-mediated chemotaxis of monocytes and in enhancing monocyte arrest in *ex vivo* perfused mouse carotids. Thus, the pseudo-(E)LR motif is essentially involved in MIF's atherogenic leukocyte recruitment effects.[100] Moreover, using a peptide spot array in which small peptides of the extracellular regions of CXCR2 were synthesized onto a nitrocellulose membrane, the interaction between the pseudo-(E)LR motif and EL2 and EL3 was demonstrated. In agreement with data available for CXCL8/CXCR2, this interaction could therefore represent site II binding in a putative two-site binding model for MIF/CXCR2 interactions.[99,100]

In addition, MIF features an N-like loop that is functionally similar to the N-loop of the ELR+ CXC chemokines.[99,100,139] Evidence for the existence of an N-like loop in MIF first came from studies with short peptides that represented this region and that ranged from amino acids 40–60 of MIF. These peptides were used as competitors of MIF in various assays. Peptides 40–49 and 47–56 proved to be potent MIF antagonists and were able to inhibit MIF effects. Importantly, they blocked MIF-induced monocyte arrest under flow conditions *in vitro* and inhibited MIF/CXCR2 interactions in a receptor competition assay.[99] In addition, MIF-triggered reduction of intracellular cAMP via activation of $G\alpha_i$ was also significantly blocked (Kraemer, Kim and

[1] Numbering of the MIF protein sequence is according to the cDNA sequence of MIF and thus starts with Met-1.

Bernhagen, *unpublished observations*). Intraperitoneal injection of peptides 47–56 resulted in a significant reduction of *in vivo* recruited monocytes in carotids of atherosclerotic mice.[99] Biophysical measurements and again peptide spot array analysis revealed the CXCR2 N-domain as a further MIF interaction site. Thus, the site I interaction between MIF and CXCR2 therefore consists of the MIF N-like loop and the CXCR2 N-terminal region. Interestingly, the newly identified N-like loop in MIF is structurally somewhat comparable to the N-loop of classical CXC chemokines, consisting of 10 amino acids and sharing with ELR+ chemokines conserved residues at the N-terminal and C-terminal end. The studies thus indicated that MIF, similar to the cognate ligand CXCL8, binds to CXCR2 via a two-site binding mechanism which is based on the pseudo-(E)LR and N-like loop regions of MIF. Promiscuous usage of CXCR2 by MIF therefore appears to rely on similar molecular patterns in the ligand structure, while the individual residues and charge properties at the binding interface differ significantly between MIF and CXCL8.[99,100]

Whether the binding interface between MIF and CXCR4 encompasses similar molecular parameters has not been addressed. Initial experiments indicate that the N-like loop of MIF also supports an interaction with the N-domain of CXCR4 and could thus be part of the structural basis of the MIF/CXCR4 interaction (Kraemer and Bernhagen, *unpublished observations*). However, the CXCR4 ligand CXCL12 does not belong to the subfamily of ELR+ CXC chemokines, suggesting the site II interface between MIF and CXCR4 may be governed by yet different residues and molecular patterns. Figure 2 summarizes the molecular details of the MIF/CXCR2 interface as predicted by the various biophysical and biochemical analyses and provides a hypothetical model of the MIF/CXCR4 interface as well.

Obviously, precise knowledge of the molecular determinants of the MIF/CXCR2 interface could be of significant biomedical relevance when attempting to tackle MIF's pathophysiologic functions in inflammation and atherosclerosis. Knowledge about the interface could both be used to devise peptide scaffolds for small molecule anti-MIF drugs or to map and identify potent MIF-neutralizing antibody epitopes. In fact, knowledge about chemokine/chemokine receptor binding interfaces has been successfully exploited to devise therapeutic inhibitors in disease. For example, the CXCL12/CXCR4 interface, which is mimicked by human immunodeficiency virus (HIV) protein gp120, has been extensively studied, and numerous peptide and small-molecular-mass inhibitors have been developed with potential applications in HIV infection and inflammation.[140]

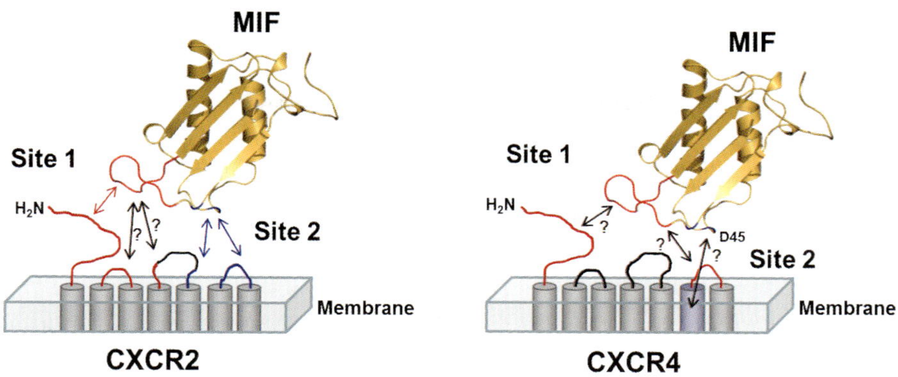

Fig. 2. Schematic depicting the binding interface between MIF and CXCR2 (left) and MIF and CXCR4 (right). Available data[99,100] for the MIF/CXCR2 interface suggest that it is composed of a site I and site II interaction. Site I binding is supported by the receptor N-domain and the MIF N-like-loop (red). Site II binding (blue) is supported by the pseudo-(E)LR motif of MIF and the extracellular loops EL2 and EL3 of CXCR2. Question marks indicate that site I interaction could additionally encompass contacts between the N-like loop and EL1 and EL2. The scheme on the right is a purely hypothetic model depicting suggesting that MIF/CXCR4 binding could be accomplished by a similar two-site mechanism. Residue D45 is proposed to contribute to site II because of the hyper-agonistic effect seen for the D45A mutant in MIF-driven peritoneal leukocyte infiltration.[100]

8. MIF Receptor Complexes

As discussed above, chemokine receptor signaling is further complicated and fine-tuned by receptor homo- and heterodimerization, which can serve to activate or inhibit, accelerate or attenuate, sensitize or desensitize the response. Chemokine receptor heterodimers for which a functional role has been demonstrated have been identified in both the CC and CXC families. In addition, cross-family heterodimers, i.e., CXCR4/CXCR7, CXCR1/CXCR2, CCR2/CCR5, CXCR4/CCR5, CXCR4/CCR2b have been found.[21,22,141–143]

Emerging evidence now also suggests that the identified MIF receptors may cluster and form heteromeric complexes. Specifically, this means that CD74 forms heteromeric complexes with either CXCR2 or CXCR4.[13,144] This is intriguing because such complexes represent receptor complexes between a single helix-spanning type II membrane protein (i.e., CD74) and a seven helix-spanning G protein-coupled receptor (GPCR) (i.e., CXCR2 or CXCR4). Yet such cross-receptor family complexes are not unique to CD74/CXCR, as a dopamine receptor/CXCR4 complex was also observed.[142]

Also, heterologous desensitization of CCL2/CCR2-mediated monocyte chemotaxis by MIF has been observed, suggesting that MIF receptor complexes functionally interact with other chemokine receptor complexes to control leukocyte migration.[98] That heteromeric CXCR/CD74 complexes are functional first became evident by blockade experiments in which anti-CXCR2 and anti-CD74 antibodies clearly impaired MIF-induced monocyte transmigration, indicating a crucial role for both receptors in MIF-induced leukocyte chemotaxis.[13]

In fact, in addition to CXCR2, CD74 is required for efficient MIF-induced monocyte adhesion on CHO cells, engineered to express adhesion molecules.[13] Since co-expression of CXCR2 and CD74 results in the formation of heteromeric CXCR2/CD74 complexes,[13] MIF signaling in leukocyte recruitment processes may be mediated by this receptor complex. Yet MIF can bind to CXCR2 in the absence of CD74 and *vice versa*. Thus, alternatively, adjunctive binding of MIF to CD74 may facilitate GPCR activation and formation of a receptor signaling complex including SRC kinases, similar to the role of CD44 as an auxiliary receptor for CCL5.[145] An auxiliary function by CD74 in MIF/CXCR-mediated leukocyte recruitment was best evidenced by chemotaxis experiments with neutrophils. Ectopic CD74 expression in CD74-deficient promyelocytic neutrophilic cells indicated that CXCR2-mediated MIF-induced neutrophil chemotaxis is reinforced by additional CD74-dependent signaling pathways.[13] CD74/CXCR2 complex formation was not only observed after experimental overexpression of the individual receptors but also under endogenous conditions.[144]

Moreover, MIF-responsive CXCR4/CD74 receptor complexes were identified. These were found in monocytes, T cells, and fibroblasts.[105,144] Given the observations that the CD74 can also complex with CD44 and c-Met, it will be important to find out whether or not hetero-trimeric CXCR2/CXCR4/CD74, heterotetrameric CXCR2/CXCR4/CD74/CD44, or even heteropentameric complexes exist, or whether only two or three of the involved receptor proteins can form complexes in a given cellular context.[146] Thus, the specific stoichiometry of such receptor complexes needs further evaluation. This concept is in line with the increasing appreciation that both chemokine and chemokine receptor oligomerization constitutes an approach of nature to fine-tune leukocyte arrest in order to ensure a highly controlled cell-, site- and disease stage-specific inflammatory cell adhesion.[20,147] Also, receptor complex formation of CXCR2 or CXCR4 with CD74 could be viewed as the linkage of more promiscuous chemokine receptors that also interact with other chemokines, with a MIF-specific receptor. This would bias the receptor complex towards an MIF-driven, as opposed to a CXCL8- or CXCL12-driven, response.

9. Conclusion

In conclusion, MIF features several mechanistic links to the chemokine and chemokine receptor system. Functionally, MIF behaves as a CLF chemokine, and structurally, this is based on interactions with CXCR2 and CXCR4, possibly CXCR7 as well, and the formation of heteromeric receptor complexes between CD74 and the MIF chemokine receptors.

10. Acknowledgments

The authors are grateful to numerous friends and collaborators for stimulating discussions and fruitful interactions on the various exciting aspects of the biology, biochemistry and biomedical implications of MIF. Authors are also indebted to the various funding agencies for supporting this work, most notably grants BE 1977/4-1, 4-2, SFB-TRR57, DFG-FOR 809, IRTG 1508 to J.B. and C.W.

References

1. David JR. (1966) Delayed hypersensitivity *in vitro*: Its mediation by cell-free substances formed by lymphoid cell-antigen interaction. *Proc Natl Acad Sci USA* **56**: 72–77.
2. Bernhagen J, Calandra T, Mitchell RA *et al.* (1993) MIF is a pituitary-derived cytokine that potentiates lethal endotoxaemia. *Nature* **365**: 756–759.
3. Calandra T, Bernhagen J, Mitchell RA, Bucala R. (1994) The macrophage is an important and previously unrecognized source of macrophage migration inhibitory factor. *J Exp Med* **179**: 1895–1902.
4. Calandra T, Roger T. (2003) Macrophage migration inhibitory factor: A regulator of innate immunity. *Nat Rev Immunol* **3**: 791–800.
5. Calandra T, Echtenacher B, Le Roy D *et al.* (2000) Protection from septic shock by neutralization of macrophage migration inhibitory factor (MIF). *Nat Med* **6**: 164–169.
6. de Jong YP, Abadia-Molina AC, Satoskar AR *et al.* (2001) Development of chronic colitis is dependent on the cytokine MIF. *Nat Immunol* **2**: 1061–1066.
7. Bernhagen J, Bacher M, Calandra T *et al.* (1996) An essential role for macrophage migration inhibitory factor in the tuberculin delayed-type hypersensitivity reaction. *J Exp Med* **183**: 277–282.
8. Lan HY, Bacher M, Yang N *et al.* (1997) The pathogenic role of macrophage migration inhibitory factor in immunologically induced kidney disease in the rat. *J Exp Med* **185**: 1455–1465.

9. Donnelly SC, Haslett C, Reid PT *et al.* (1997) Regulatory role for macrophage migration inhibitory factor in acute respiratory distress syndrome. *Nat Med* **3**: 320–323.
10. Morand EF, Leech M, Bernhagen J. (2006) MIF: A new cytokine link between rheumatoid arthritis and atherosclerosis. *Nat Rev Drug Discov* **5**: 399–410.
11. Calandra T, Bernhagen J, Metz CN *et al.* (1995) MIF as a glucocorticoid-induced modulator of cytokine production. *Nature* **377**: 68–71.
12. Flaster H, Bernhagen J, Calandra T, Bucala R. (2007) The MIF-glucocorticoid dyad: Regulation of inflammation and immunity. *Mol Endocrinol* **21**: 1267–1280.
13. Bernhagen J, Krohn R, Lue H *et al.* (2007) MIF is a noncognate ligand of CXC chemokine receptors in inflammatory and atherogenic cell recruitment. *Nat Med* **13**: 587–596.
14. Gregory JL, Leech MT, David JR *et al.* (2004) Reduced leukocyte-endothelial cell interactions in the inflamed microcirculation of macrophage migration inhibitory factor-deficient mice. *Arthritis Rheum* **50**: 3023–3034.
15. Gregory JL, Morand EF, McKeown SJ *et al.* (2006) Macrophage migration inhibitory factor induces macrophage recruitment via CC chemokine ligand 2. *J Immunol* **177**: 8072–8079.
16. Zernecke A, Bernhagen J, Weber C. (2008) Macrophage migration inhibitory factor in cardiovascular disease. *Circulation* **117**: 1594–1602.
17. Schober A, Bernhagen J, Thiele M *et al.* (2004) Stabilization of atherosclerotic plaques by blockade of macrophage migration inhibitory factor after vascular injury in apolipoprotein E-deficient mice. *Circulation* **109**: 380–385.
18. Noels H, Bernhagen J, Weber C. (2009) Macrophage migration inhibitory factor: A noncanonical chemokine important in atherosclerosis. *Trends Cardiovasc Med* **19**: 76–86.
19. Schober A, Bernhagen J, Weber C. (2008) Chemokine-like functions of MIF in atherosclerosis. *J Mol Med* **86**: 761–770.
20. Zernecke A, Weber C. (2010) Chemokines in the vascular inflammatory response of atherosclerosis. *Cardiovasc Res* **86**: 192–201.
21. Thelen M. (2001) Dancing to the tune of chemokines. *Nat Immunol* **2**: 129–134.
22. Thelen M, Stein JV. (2008) How chemokines invite leukocytes to dance. *Nat Immunol* **9**: 953–959.
23. Baggiolini M. (2001) Chemokines in pathology and medicine. *J Intern Med* **250**: 91–104.
24. Mackay CR. (2001) Chemokines: Immunology's high impact factors. *Nat Immunol* **2**: 95–101.
25. Baggiolini M, Dewald B, Moser B. (1994) Interleukin-8 and related chemotactic cytokines — CXC and CC chemokines. *Adv Immunol* **55**: 97–179.

26. Charo IF, Ransohoff RM. (2006) The many roles of chemokines and chemokine receptors in inflammation. *N Engl J Med* **354**: 610–621.
27. Charo IF, Taubman MB. (2004) Chemokines in the pathogenesis of vascular disease. *Circ Res* **95**: 858–866.
28. Hansson GK. (1999) Inflammation and immune response in atherosclerosis. *Curr Atheroscler Rep* **1**: 150–155.
29. Moser B, Loetscher P. (2001) Lymphocyte traffic control by chemokines. *Nat Immunol* **2**: 123–128.
30. Rot A, von Andrian UH. (2004) Chemokines in innate and adaptive host defense: Basic chemokinese grammar for immune cells. *Annu Rev Immunol* **22**: 891–928.
31. Schober A, Weber C. (2005) Mechanisms of monocyte recruitment in vascular repair after injury. *Antioxid Redox Signal* **7**: 1249–1257.
32. Zernecke A, Shagdarsuren E, Weber C. (2008) Chemokines in atherosclerosis: An update. *Arterioscler Thromb Vasc Biol* **28**: 1897–1908.
33. Sallusto F, Baggiolini M. (2008) Chemokines and leukocyte traffic. *Nat Immunol* **9**: 949–952.
34. Ley K. (2003) Arrest chemokines. *Microcirculation* **10**: 289–295.
35. Murphy PM, Baggiolini M, Charo IF *et al.* (2000) International union of pharmacology. XXII. Nomenclature for chemokine receptors. *Pharmacol Rev* **52**: 145–176.
36. Luster AD. (1998) Chemokines — Chemotactic cytokines that mediate inflammation. *N Engl J Med* **338**: 436–445.
37. Fernandez EJ, Lolis E. (2002) Structure, function, and inhibition of chemokines. *Annu Rev Pharmacol Toxicol* **42**: 469–499.
38. Joseph PR, Sarmiento JM, Mishra AK *et al.* (2010) Probing the role of CXC motif in chemokine CXCL8 for high affinity binding and activation of CXCR1 and CXCR2 receptors. *J Biol Chem* **285**: 29262–29269.
39. Clark-Lewis I, Kim KS, Rajarathnam K *et al.* (1995) Structure-activity relationships of chemokines. *J Leukoc Biol* **57**: 703–711.
40. Rajagopalan L, Rajarathnam K. (2006) Structural basis of chemokine receptor function — A model for binding affinity and ligand selectivity. *Biosci Rep* **26**: 325–339.
41. Oppenheim JJ, Biragyn A, Kwak LW, Yang D. (2003) Roles of antimicrobial peptides such as defensins in innate and adaptive immunity. *Ann Rheum Dis* **62**(Suppl 2): ii17–ii21.
42. Yang D, Biragyn A, Hoover DM *et al.* (2004) Multiple roles of antimicrobial defensins, cathelicidins, and eosinophil-derived neurotoxin in host defense. *Annu Rev Immunol* **22**: 181–215.

43. Oppenheim JJ, Yang D. (2005) Alarmins: Chemotactic activators of immune responses. *Curr Opin Immunol* **17**: 359–365.
44. Yang D, Oppenheim JJ. (2004) Antimicrobial proteins act as "alarmins" in joint immune defense. *Arthritis Rheum* **50**: 3401–3403.
45. Degryse B, de Virgilio M. (2003) The nuclear protein HMGB1, a new kind of chemokine? *FEBS Lett* **553**: 11–17.
46. Rajarathnam K, Sykes BD, Kay CM *et al.* (1994) Neutrophil activation by monomeric interleukin-8. *Science* **264**: 90–92.
47. Nasser MW, Raghuwanshi SK, Grant DJ *et al.* (2009) Differential activation and regulation of CXCR1 and CXCR2 by CXCL8 monomer and dimer. *J Immunol* **183**: 3425–3432.
48. Das ST, Rajagopalan L, Guerrero-Plata A *et al.* (2010) Monomeric and dimeric CXCL8 are both essential for *in vivo* neutrophil recruitment. *PLoS One* **5**: e11754.
49. Weber C, Koenen RR. (2006) Fine-tuning leukocyte responses: Towards a chemokine 'interactome'. *Trends Immunol* **27**: 268–273.
50. Veldkamp CT, Peterson FC, Pelzek AJ, Volkman BF. (2005) The monomer-dimer equilibrium of stromal cell-derived factor-1 (CXCL 12) is altered by pH, phosphate, sulfate, and heparin. *Protein Sci* **14**: 1071–1081.
51. Veldkamp CT, Ziarek JJ, Su J *et al.* (2009) Monomeric structure of the cardioprotective chemokine SDF-1/CXCL12. *Protein Sci* **18**: 1359–1369.
52. Murphy JW, Cho Y, Sachpatzidis A *et al.* (2007) Structural and functional basis of CXCL12 (stromal cell-derived factor-1 alpha) binding to heparin. *J Biol Chem* **282**: 10018–10027.
53. Ravindran A, Joseph PR, Rajarathnam K. (2009) Structural basis for differential binding of the interleukin-8 monomer and dimer to the CXCR1 N-domain: Role of coupled interactions and dynamics. *Biochemistry* **48**: 8795–8805.
54. Koenen RR, von Hundelshausen P, Nesmelova IV *et al.* (2009) Disrupting functional interactions between platelet chemokines inhibits atherosclerosis in hyperlipidemic mice. *Nat Med* **15**: 97–103.
55. Handel TM, Domaille PJ. (1996) Heteronuclear (1H, 13C, 15N) NMR assignments and solution structure of the monocyte chemoattractant protein-1 (MCP-1) dimer. *Biochemistry* **35**: 6569–6584.
56. Wu B, Chien EY, Mol CD *et al.* (2010) Structures of the CXCR4 chemokine GPCR with small-molecule and cyclic peptide antagonists. *Science* **330**: 1066–1071.
57. Trettel F, Di Bartolomeo S, Lauro C *et al.* (2003) Ligand-independent CXCR2 dimerization. *J Biol Chem* **278**: 40980–40988.

58. Wilson S, Wilkinson G, Milligan G. (2005) The CXCR1 and CXCR2 receptors form constitutive homo- and heterodimers selectively and with equal apparent affinities. *J Biol Chem* **280**: 28663–28674.
59. Weiser WY, Temple DM, Witek-Gianotti JS *et al.* (1989) Molecular cloning of cDNA encoding a human macrophage migration inhibtion factor. *Proc Natl Acad Sci USA* **86**: 7522–7526.
60. Mühlhahn P, Bernhagen J, Czisch M *et al.* (1996) NMR characterization of structure, backbone dynamics and glutathione binding of the human macrophage migration inhibitory factor (MIF). *Protein Sci* **5**: 2095–2103.
61. Sun HW, Bernhagen J, Bucala R, Lolis E. (1996) Crystal structure at 2.6-A resolution of human macrophage migration inhibitory factor. *Proc Natl Acad Sci USA* **93**: 5191–5196.
62. Sun HW, Swope M, Cinquina C *et al.* (1996) The subunit structure of human macrophage migration inhibitory factor: Evidence for a trimer. *Protein Eng* **9**: 631–635.
63. Sugimoto H, Suzuki M, Nakagawa A *et al.* (1996) Crystal structure of macrophage migration inhibitory factor from human lymphocyte at 2.1 Å resolution. *FEBS Lett* **389**: 145–148.
64. Calandra T, Roger T. (2003) Macrophage migration inhibitory factor: A regulator of innate immunity. *Nat Rev Immunol* **3**: 791–800.
65. Mischke R, Kleemann R, Brunner H, Bernhagen J. (1998) Cross-linking and mutational analysis of the oligomerization state of the cytokine macrophage migration inhibitory factor (MIF). *FEBS Lett* **427**: 85–90.
66. Philo JS, Yang TH, LaBarre M. (2004) Re-examining the oligomerization state of macrophage migration inhibitory factor (MIF) in solution. *Biophys Chem* **108**: 77–87.
67. El-Turk F, Cascella M, Ouertatani-Sakouhi H *et al.* (2008) The conformational flexibility of the carboxy terminal residues 105–114 is a key modulator of the catalytic activity and stability of macrophage migration inhibitory factor. *Biochemistry* **47**: 10740–10756.
68. Kato Y, Muto T, Tomura T *et al.* (1996) The crystal structure of human glycosylation-inhibiting factor is a trimeric barrel with three 6-stranded b-sheets. *Proc Natl Acad Sci USA* **93**: 3007–3010.
69. Lolis E, Bucala R. (2003) Therapeutic approaches to innate immunity: Severe sepsis and septic shock. *Nat Rev Drug Discov* **2**: 635–645.
70. Thiele M, Bernhagen J. (2005) Link between macrophage migration inhibitory factor and cellular redox regulation. *Antioxid Redox Signal* **7**: 1234–1248.
71. Kleemann R, Kapurniotu A, Frank RW *et al.* (1998) Disulfide analysis reveals a role for macrophage migration inhibitory factor (MIF) as thiol-protein oxidoreductase. *J Mol Biol* **280**: 85–102.

72. Kleemann R, Hausser A, Geiger G *et al.* (2000) Intracellular action of the cytokine MIF to modulate AP-1 activity and the cell cycle through Jab1. *Nature* **408**: 211–216.
73. Burger-Kentischer A, Goebel H, Seiler R *et al.* (2002) Expression of macrophage migration inhibitory factor in different stages of human atherosclerosis. *Circulation* **105**: 1561–1566.
74. Simons D, Grieb G, Hristov M *et al.* (2010) Hypoxia-induced endothelial secretion of macrophage migration inhibitory factor and role in endothelial progenitor cell recruitment. *J Cell Mol Med* **15**(3): 668–678.
75. Miller EJ, Li J, Leng L *et al.* (2008) Macrophage migration inhibitory factor stimulates AMP-activated protein kinase in the ischaemic heart. *Nature* **451**: 578–582.
76. Mitchell RA. (2004) Mechanisms and effectors of MIF-dependent promotion of tumourigenesis. *Cell Signal* **16**: 13–19.
77. Verjans E, Noetzel E, Bektas N *et al.* (2009) Dual role of macrophage migration inhibitory factor (MIF) in human breast cancer. *BMC Cancer* **9**: 230.
78. Flieger O, Engling A, Bucala R *et al.* (2003) Regulated secretion of macrophage migration inhibitory factor is mediated by a non-classical pathway involving an ABC transporter. *FEBS Lett* **551**: 78–86.
79. Simons D, Grieb G, Hristov M *et al.* (2011) Hypoxia-induced endothelial secretion of macrophage migration inhibitory factor and role in endothelial progenitor cell recruitment. *J Cell Mol Med* **15**: 668–678.
80. Merk M, Baugh J, Zierow S *et al.* (2009) The Golgi-associated protein p115 mediates the secretion of macrophage migration inhibitory factor. *J Immunol* **182**: 6896–6906.
81. Lue H, Thiele M, Franz J *et al.* (2007) Macrophage migration inhibitory factor (MIF) promotes cell survival by activation of the Akt pathway and role for CSN5/JAB1 in the control of autocrine MIF activity. *Oncogene* **26**: 5046–5059.
82. Yang D, Chen Q, Yang H *et al.* (2007) High mobility group box-1 protein induces the migration and activation of human dendritic cells and acts as an alarmin. *J Leukoc Biol* **81**: 59–66.
83. Frye BC, Halfter S, Djudjaj S *et al.* (2009) Y-box protein-1 is actively secreted through a non-classical pathway and acts as an extracellular mitogen. *EMBO Rep* **10**: 783–789.
84. Pekkari K, Holmgren A. (2004) Truncated thioredoxin: Physiological functions and mechanism. *Antioxid Redox Signal* **6**: 53–61.
85. Wakasugi K, Schimmel P. (1999) Two distinct cytokines released from a human aminoacyl-tRNA synthetase. *Science* **284**: 147–151.
86. Wakasugi K, Slike BM, Hood J *et al.* (2002) Induction of angiogenesis by a fragment of human tyrosyl-tRNA synthetase. *J Biol Chem* **277**: 20124–20126.

87. Hoover DM, Boulegue C, Yang D *et al.* (2002) The structure of human macrophage inflammatory protein-3alpha/CCL20. Linking antimicrobial and CC chemokine receptor-6-binding activities with human beta-defensins. *J Biol Chem* **277**: 37647–37654.
88. Yang D, Chertov O, Bykovskaia SN *et al.* (1999) Beta-defensins: Linking innate and adaptive immunity through dendritic and T cell CCR6. *Science* **286**: 525–528.
89. Yang D, Chertov O, Bykovskaia SN *et al.* (1999) Beta-defensins: Linking innate and adaptive immunity through dendritic and T cell CCR6. *Science* **286**: 525–528.
90. Aliberti J, Valenzuela JG, Carruthers VB *et al.* (2003) Molecular mimicry of a CCR5 binding-domain in the microbial activation of dendritic cells. *Nat Immunol* **4**: 485–490.
91. Alcami A. (2003) Viral mimicry of cytokines, chemokines and their receptors. *Nat Rev Immunol* **3**: 36–50.
92. Zernecke A, Bot I, Djalali-Talab Y *et al.* (2008) Protective role of CXC receptor 4/CXC ligand 12 unveils the importance of neutrophils in atherosclerosis. *Circ Res* **102**: 209–217.
93. Frascaroli G, Varani S, Blankenhorn N *et al.* (2009) Human cytomegalovirus paralyzes macrophage motility through down-regulation of chemokine receptors, reorganization of the cytoskeleton, and release of macrophage migration inhibitory factor. *J Immunol* **182**: 477–488.
94. Cho Y, Crichlow GV, Vermeire JJ *et al.* (2010) Allosteric inhibition of macrophage migration inhibitory factor revealed by ibudilast. *Proc Natl Acad Sci USA* **107**: 11313–11318.
95. Ren Y, Tsui HT, Poon RT *et al.* (2003) Macrophage migration inhibitory factor: Roles in regulating tumor cell migration and expression of angiogenic factors in hepatocellular carcinoma. *Int J Cancer* **107**: 22–29.
96. Brandau S, Jakob M, Hemeda H *et al.* (2010) Tissue-resident mesenchymal stem cells attract peripheral blood neutrophils and enhance their inflammatory activity in response to microbial challenge. *J Leukoc Biol* **88**: 1005–1015.
97. Kamir D, Zierow S, Leng L *et al.* (2008) A *Leishmania* ortholog of macrophage migration inhibitory factor modulates host macrophage responses. *J Immunol* **180**: 8250–8261.
98. Hermanowski-Vosatka A, Mundt SS, Ayala JM *et al.* (1999) Enzymatically inactive macrophage migration inhibitory factor inhibits monocyte chemotaxis and random migration. *Biochemistry* **38**: 12841–12849.
99. Kraemer S, Lue H, Zernecke A *et al.* (2010) MIF-chemokine receptor interactions in atherogenesis are dependent on an N-loop-based 2-site binding mechanism. *FASEB J* **25**(3): 894–906.

100. Weber C, Kraemer S, Drechsler M *et al.* (2008) Structural determinants of MIF functions in CXCR2-mediated inflammatory and atherogenic leukocyte recruitment. *Proc Natl Acad Sci USA* **105**: 16278–16283.
101. Tarnowski M, Grymula K, Liu R *et al.* (2010) Macrophage migration inhibitory factor is secreted by rhabdomyosarcoma cells, modulates tumor metastasis by binding to CXCR4 and CXCR7 receptors and inhibits recruitment of cancer-associated fibroblasts. *Mol Cancer Res* **8**: 1328–1343.
102. Binsky I, Haran M, Starlets D *et al.* (2007) IL-8 secreted in a macrophage migration-inhibitory factor- and CD74-dependent manner regulates B cell chronic lymphocytic leukemia survival. *Proc Natl Acad Sci USA* **104**: 13408–13413.
103. Ren Y, Chan HM, Fan J *et al.* (2006) Inhibition of tumor growth and metastasis *in vitro* and *in vivo* by targeting macrophage migration inhibitory factor in human neuroblastoma. *Oncogene* **25**: 3501–3508.
104. Fan H, Hall P, Santos LL *et al.* (2011) Macrophage migration inhibitory factor and CD74 regulate macrophage chemotactic responses via MAPK and Rho GTPase. *J Immunol* **186**(8): 4915–4924.
105. Lue H, Dewor M, Leng L *et al.* (2011) Activation of the JNK signalling pathway by macrophage migration inhibitory factor (MIF) and dependence on CXCR4 and CD74. *Cell Signal* **23**: 135–144.
106. Hoi AY, Hickey MJ, Hall P *et al.* (2006) Macrophage migration inhibitory factor deficiency attenuates macrophage recruitment, glomerulonephritis, and lethality in MRL/lpr mice. *J Immunol* **177**: 5687–5696.
107. Santos LL, Fan H, Hall P *et al.* (2011) Macrophage migration inhibitory factor regulates neutrophil chemotactic responses in inflammatory arthritis in mice. *Arthritis Rheum* **63**: 960–970.
108. Gregory JL, Hall P, Leech M *et al.* (2009) Independent roles of macrophage migration inhibitory factor and endogenous, but not exogenous glucocorticoids in regulating leukocyte trafficking. *Microcirculation* **16**: 735–748.
109. Leng L, Metz CN, Fang Y *et al.* (2003) MIF signal transduction initiated by binding to CD74. *J Exp Med* **197**: 1467–1476.
110. Borghese F, Clanchy FL. (2011) CD74: An emerging opportunity as a therapeutic target in cancer and autoimmune disease. *Expert Opin Ther Targets* **15**: 237–251.
111. Meyer-Siegler KL, Vera PL. (2005) Substance P induced changes in CD74 and CD44 in the rat bladder. *J Urol* **173**: 615–620.
112. Meyer-Siegler KL, Iczkowski KA, Leng L *et al.* (2006) Inhibition of macrophage migration inhibitory factor or its receptor (CD74) attenuates growth and invasion of DU-145 prostate cancer cells. *J Immunol* **177**: 8730–8739.
113. Matza D, Kerem A, Shachar I. (2003) Invariant chain, a chain of command. *Trends Immunol* **24**: 264–268.

114. Gore Y, Starlets D, Maharshak N *et al.* (2008) Macrophage migration inhibitory factor induces B cell survival by activation of a CD74-CD44 receptor complex. *J Biol Chem* **283**: 2784–2792.
115. Shachar I, Flavell RA. (1996) Requirement for invariant chain in B cell maturation and function. *Science* **274**: 106–108.
116. Leng L, Metz CN, Fang Y *et al.* (2003) MIF signal transduction initiated by binding to CD74. *J Exp Med* **197**: 1467–1476.
117. Shi X, Leng L, Wang T *et al.* (2006) CD44 is the signaling component of the macrophage migration inhibitory factor-CD74 receptor complex. *Immunity* **25**: 595–606.
118. Lantner F, Starlets D, Gore Y *et al.* (2007) CD74 induces TAp63 expression leading to B-cell survival. *Blood* **110**: 4303–4311.
119. Starlets D, Gore Y, Binsky I *et al.* (2006) Cell-surface CD74 initiates a signaling cascade leading to cell proliferation and survival. *Blood* **107**: 4807–4816.
120. Becker-Herman S, Arie G, Medvedovsky H *et al.* (2005) CD74 is a member of the regulated intramembrane proteolysis-processed protein family. *Mol Biol Cell* **16**: 5061–5069.
121. Matza D, Kerem A, Medvedovsky H *et al.* (2002) Invariant chain-induced B cell differentiation requires intramembrane proteolytic release of the cytosolic domain. *Immunity* **17**: 549–560.
122. Balabanian K, Lagane B, Infantino S *et al.* (2005) The chemokine SDF-1/CXCL12 binds to and signals through the orphan receptor RDC1 in T lymphocytes. *J Biol Chem* **280**: 35760–35766.
123. Burns JM, Summers BC, Wang Y *et al.* (2006) A novel chemokine receptor for SDF-1 and I-TAC involved in cell survival, cell adhesion, and tumor development. *J Exp Med* **203**: 2201–2213.
124. Thelen M, Thelen S. (2008) CXCR7, CXCR4 and CXCL12: An eccentric trio? *J Neuroimmunol* **198**: 9–13.
125. Vera PL, Iczkowski KA, Wang X, Meyer-Siegler KL. (2008) Cyclophosphamide-induced cystitis increases bladder CXCR4 expression and CXCR4-macrophage migration inhibitory factor association. *PLoS One* **3**: e3898.
126. Levoye A, Balabanian K, Baleux F *et al.* (2009) CXCR7 heterodimerizes with CXCR4 and regulates CXCL12-mediated G protein signaling. *Blood* **113**: 6085–6093.
127. Dessein AF, Stechly L, Jonckheere N *et al.* (2010) Autocrine induction of invasive and metastatic phenotypes by the MIF-CXCR4 axis in drug-resistant human colon cancer cells. *Cancer Res* **70**: 4644–4654.
128. Boisvert WA, Rose DM, Johnson KA *et al.* (2006) Up-regulated expression of the CXCR2 ligand KC/GRO-alpha in atherosclerotic lesions plays a central

role in macrophage accumulation and lesion progression. *Am J Pathol* **168**: 1385–1395.

129. Zernecke A, Bot I, Talab YD *et al.* (2007) Protective role of CXC receptor 4/CXC ligand 12 unveils the importance of neutrophils in atherosclerosis. *Circ Res* **102**: 209–217.
130. Laudanna C, Alon R. (2006) Right on the spot. Chemokine triggering of integrin-mediated arrest of rolling leukocytes. *Thromb Haemost* **95**: 5–11.
131. Huo Y, Weber C, Forlow SB *et al.* (2001) The chemokine KC, but not monocyte chemoattractant protein-1, triggers monocyte arrest on early atherosclerotic endothelium. *J Clin Invest* **108**: 1307–1314.
132. Huo Y, Schober A, Forlow SB *et al.* (2003) Circulating activated platelets exacerbate atherosclerosis in mice deficient in apolipoprotein E. *Nat Med* **9**: 61–67.
133. Weber C, Schober A, Zernecke A. (2004) Chemokines: Key regulators of mononuclear cell recruitment in atherosclerotic vascular disease. *Arterioscler Thromb Vasc Biol* **24**: 1997–2008.
134. Ceradini DJ, Kulkarni AR, Callaghan MJ *et al.* (2004) Progenitor cell trafficking is regulated by hypoxic gradients through HIF-1 induction of SDF-1. *Nat Med* **10**: 858–864.
135. Clore GM, Appella E, Yamada M *et al.* (1990) Three-dimensional structure of interleukin 8 in solution. *Biochemistry* **29**: 1689–1696.
136. Crump MP, Gong JH, Loetscher P *et al.* (1997) Solution structure and basis for functional activity of stromal cell-derived factor-1: Dissociation of CXCR4 activation from binding and inhibition of HIV-1. *EMBO J* **16**: 6996–7007.
137. Katancik JA, Sharma A, de Nardin E. (2000) Interleukin 8, neutrophil-activating peptide-2 and GRO-alpha bind to and elicit cell activation via specific and different amino acid residues of CXCR2. *Cytokine* **12**: 1480–1488.
138. Rajagopalan L, Rajarathnam K. (2004) Ligand selectivity and affinity of chemokine receptor CXCR1. Role of N-terminal domain. *J Biol Chem* **279**: 30000–30008.
139. Hebert CA, Vitangcol RV, Baker JB. (1991) Scanning mutagenesis of interleukin-8 identifies a cluster of residues required for receptor binding. *J Biol Chem* **266**: 18989–18994.
140. Proudfoot AE. (2002) Chemokine receptors: Multifaceted therapeutic targets. *Nat Rev Immunol* **2**: 106–115.
141. Thelen M, Baggiolini M. (2001) Is dimerization of chemokine receptors functionally relevant? *Sci STKE* **2001**: pe34.
142. Thelen M, Munoz LM, Rodriguez-Frade JM, Mellado M. (2010) Chemokine receptor oligomerization: Functional considerations. *Curr Opin Pharmacol* **10**: 38–43.

143. Sohy D, Parmentier M, Springael JY. (2007) Allosteric transinhibition by specific antagonists in CCR2/CXCR4 heterodimers. *J Biol Chem* **282**: 30062–30069.
144. Schwartz V, Lue H, Kraemer S *et al.* (2009) A functional heteromeric MIF receptor formed by CD74 and CXCR4. *FEBS Lett* **583**: 2749–2757.
145. Roscic-Mrkic B, Fischer M, Leemann C *et al.* (2003) RANTES (CCL5) uses the proteoglycan CD44 as an auxiliary receptor to mediate cellular activation signals and HIV-1 enhancement. *Blood* **102**: 1169–1177.
146. Gordin M, Tesio M, Cohen S *et al.* (2010) c-Met and its ligand hepatocyte growth factor/scatter factor regulate mature B cell survival in a pathway induced by CD74. *J Immunol* **185**: 2020–2031.
147. Koenen RR, Weber C. (2010) Therapeutic targeting of chemokine interactions in atherosclerosis. *Nat Rev Drug Discov* **9**: 141–153.

I-3

CD74, the Natural Receptor for MIF, Regulates Cell Survival in Health and Disease

Idit Shachar*, Maya Gordin, Sivan Cohen, Inbal Binsky, Ayelet Marom and Shirly Becker-Herman

1. Introduction

This Chapter deals with the MIF receptor, CD74 (invariant chain, Ii). CD74 structure and mechanism of action will be discussed with an emphasis on their role in the regulation of B cell survival in health and disease.

2. CD74 (Invariant Chain, Ii)

The CD74 gene is located on human chromosome 5 (q32) and was first identified in 1979 by Jones *et al.*[1] However, it was not until 1989 that CD74 was shown to have a role in antigen presentation.[2] CD74 is nonpolymorphic type II integral membrane protein, which exists in different isoforms defined by its primary amino acid sequence. There are four isoforms of CD74 in humans: p33, p35, p41 and p43. CD74 p33 and p41 are distinguished by alternative splicing of the CD74 transcript, where the p41 isoform contains an extra exon (exon 6b). These two isoforms yield two additional protein products due to an N-terminal cytoplasmic extension of 16 residues, which results from an alternative translation initiation site. The major human p33 isoform has an N-terminal cytosolic tail of 30 amino acids, a transmembrane (TM) domain consisting of amino acid 31–56, and a C-terminal 160 residue luminal domain.[3]

The murine CD74 gene encodes two polypeptide chains, one of a relative molecular mass of 31 kDa (p31) (Fig. 1) and another less abundant 41-kDa

*Corresponding author: Department of Immunology, Weizmann Institute of Science, Rehovot, Israel. Email: idit.shachar@weizman.ac.il

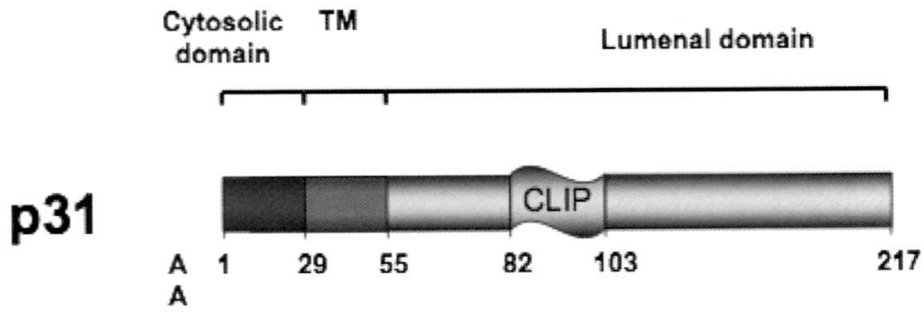

Fig. 1. Schematic presentation of the p31 murine CD74 molecule.

species (p41).[4,5] Exon 6b is alternatively spliced into the mRNA coding for the p41 isoform.[6]

3. CD74 Functions as a Chaperone for MHC Class II

Since its discovery in the late 1970s, CD74 has been associated with an increasing number of functions related to major histocompatibility complex class II (MHC II)-restricted antigen presentation and beyond.[3] MHC class II molecules are heterodimeric complexes that present foreign antigenic peptides on the cell surface of antigen-presenting cells (APCs) to CD4+ T cells.[7–9]

MHC class II synthesis and assembly begins in the endoplasmic reticulum (ER) with the noncovalent association of the MHC α- and β-chains with trimers of CD74. Three MHC class II αβ dimers bind sequentially to a trimer of CD74 to form a nonameric complex $(\alpha\beta CD74)_3$, which then exits the ER.[10] After being transported to the *trans*-Golgi, the αβCD74 complex is diverted from the secretory pathway to the endocytic system and ultimately to acidic endosome/lysosome-like structures called MHC class II compartments (MIIC or CIIV). Surface expression of newly synthesized CD74 followed by its rapid internalization to the endosomal pathway has also has been known for many years. Experiments investigating cell surface CD74 are complicated by the fact that CD74 on the cell surface is characterized by a very rapid turnover.[11–13] It was shown that cell surface CD74 is modified by the addition of chondroitin sulfate (CD74-CS) at amino acid position 201, and this form of CD74 is associated with MHC class II on the surface of antigen presenting cells.[14–16]

The N-terminal cytoplasmic tail of CD74 contains two extensively characterized dileucine-based endosomal targeting motifs.[17–19] These motifs mediate its internalization from the plasma membrane and from the trans-Golgi network. In the endocytic compartments, CD74 is gradually proteolytically processed, leaving only a small fragment, the class II-associated Ii chain peptide (CLIP), bound to the released αβ dimers. The final step for MHC class

II expression requires interaction of αβCLIP complexes with another class II-related αβ dimer, called HLA-DM in the human system, and H2-M in mice. Binding of this molecule drives out the residual CLIP, rendering the αβ dimers ultimately competent to bind antigenic peptides, which are mainly derived from internalized antigens and are also delivered to the endocytic pathway.[20,21] The peptide-loaded class II molecules then leave this compartment, by an unknown route, to be expressed on the cell surface and surveyed by CD4+ T cells.

Thus, CD74 was thought to function mainly as an MHC class II chaperone, which promotes ER exit of MHC class II molecules, directs them to endocytic compartments, prevents peptide binding in the ER, and contributes to peptide editing in the MHC class II compartment.

4. CD74 as a Signaling Molecule

4.1 *CD74 is a cell surface receptor that induces signaling cascades*

It has been known for many years that a fraction of the CD74 protein is expressed on the cell surface. The picture that emerges from studies from several laboratories is that in addition to its role in antigen presentation, the CD74 molecule serves as a receptor in many types of cells and can initiate various signaling cascades. CD74 was reported to be a high-affinity binding protein for the cytokine macrophage migration inhibitory factor (MIF) providing further evidence for a role in signal transduction pathways.[22] *Helicobacter pylori* was shown to bind to CD74 on gastric epithelial cells and thereby to stimulate interleukin-8 production.[23] In addition, it was shown that in mice lacking CD74, there is an accumulation of transitional I (TI) B cells in the periphery (Fig. 2), characterized by low expression levels of IgD and CD23 and poor response to T-independent antigens,[24–27] while the mature population responsible for the humoral immune response is missing.[24,28]

4.2 *CD74 is subjected to regulated intramembrane proteolysis (RIP)*

To dissect the chaperonin activity from the role of CD74 in shaping the B cell repertoire, the region of CD74 controlling the B cell peripheral repertoire was identified. To this end, transgenic mice were generated expressing truncated CD74, lacking the lumenal domain (and containing only amino acids 1–82). This truncated CD74 form, when expressed in transgenic mice, is

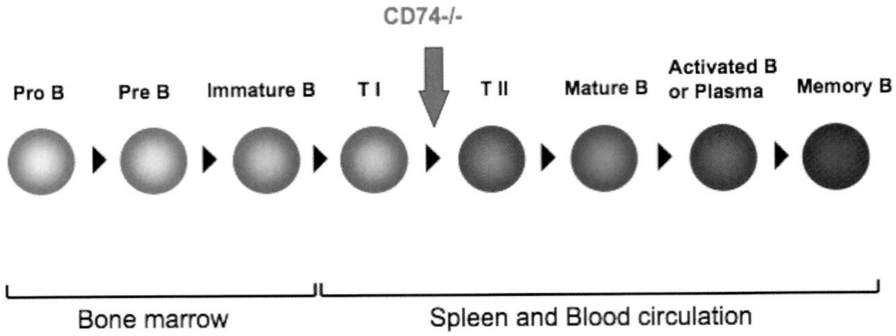

Fig. 2. Multistep differentiation of B cells in the bone marrow and spleen. In the absence of CD74, there is an accumulation of transitional I B cells (TI).

unable to allow MHC class II cell surface expression and formation of MHC class II compact dimers.[28] However, this segment is sufficient to retrieve the mature B cells. These studies showed that the CD74 N-terminal domain is directly involved in maintenance of peripheral B cells.[28]

Further studies demonstrated that CD74 functions as a signaling molecule that induces a specific gene expression program. Essential elements in this process include the p65 component of NF-κB and its co-activator, $TAF_{II}105$,[29] a subunit of the basal transcription factor, TFIID, which is highly expressed in B cells.[30] These studies suggested that CD74 initiates a signaling cascade that is transmitted to the nucleus and activates protein kinases, leading to modulation of either the p65 activation domain or of its co-activators.

To follow the mechanism by which full length or truncated (aa 1–82) CD74 transmits the signal to NF-κB in the nucleus, the localization of the CD74 cytosolic fragment (CD74 intracellular domain, CD74-ICD) was analyzed. These studies showed that the CD74-ICD is released into the cytoplasm in transfected 293 cells and in primary B cells. This cleavage and release of the CD74-ICD fragment is essential for NF-κB activation and B cell maintenance.[31] Previously, Lipp and Dobberstein described a cleavage site in the transmembrane region of human CD74 at amino acid 42.[32] The significance of their observation became fully apparent only when CD74 amino acids 42–44, which are located in the center of the transmembrane domain, were mutated. These amino acids were found to be crucial in the CD74 intramembrane cleavage event, as their mutation completely blocked the proteolytic release of the CD74 cytosolic fragment. Blocking this cleavage event largely inhibits NF-κB activation and maintenance of the mature B cell population. Moreover, the cytosolic domain of CD74 (1–42 fragment) by itself was shown to be capable of inducing NF-κB activation.[31]

Quite a few regulatory proteins, including transcription factors, are normally kept in a dormant state to be activated following internal or environmental cues. Previously, a novel strategy for the release of transcription factors by proteolytic cleavage from integral membrane proteins was described. These transcription factors are initially synthesized in an inactive form, while "nesting" in integral membrane precursor proteins. Following a cleavage event, these now-active factors are released from the membrane and can migrate into the nucleus to drive regulated gene transcription. This mechanism, "regulated intramembrane proteolysis (RIP)," is known for its conservation along evolution, and its ability to control diverse biological processes in prokaryotes and eukaryotes in response to a variety of signals.[33–35]

Some common features of proteins processed by RIP have shown remarkable similarities to the behavior of CD74. In general, in most RIP cases reported, cleavage proceeds through a two-step sequential proteolytic process. The first step involves the cleavage of the extracytoplasmic segment to shorten the ectodomain to less than 30 aa. This seems to be a requirement for the second proteolytic event, which occurs at the transmembrane domain. It appears that the initial shedding prepares the substrate for intramembrane proteolysis, such that the transmembrane region becomes accessible to the second protease, which then releases the product from the lipid bilayer towards the cytosol. This mechanism, when properly regulated, guarantees controlled proteolysis in the plane of the membrane and prevents random degradation of membrane proteins.[33–35]

It was shown that arrival to the endocytic compartments and a series of proteolytic cleavages in these compartments are necessary for the release of CD74-ICD from the membrane.[36] Once the CD74 lumenal domain has been removed, CD74 can serve as a substrate for I-CLiP cleavage, a process that occurs exclusively in the endocytic compartments. Thus, transport to the endocytic compartment is essential for both CD74 processing and for its intramembrane cleavage. The removal of the CD74 lumenal domain is thus the first step required for the release of the N-terminal cytosolic fragment leading to NF-κB activation.

In most RIP proteins, the released cytosolic fragments translocate to the nucleus to elicit the appropriate biological responses. Immunostaining and biochemical fractionation of cells transfected with the CD74 1–42 construct revealed that this fragment can be detected in the various fractions, membrane, cytosolic and nuclear, confirming that CD74-ICD is released from the membrane to the cytosol, and that a portion of this proteolytic product translocates to the nucleus.[36]

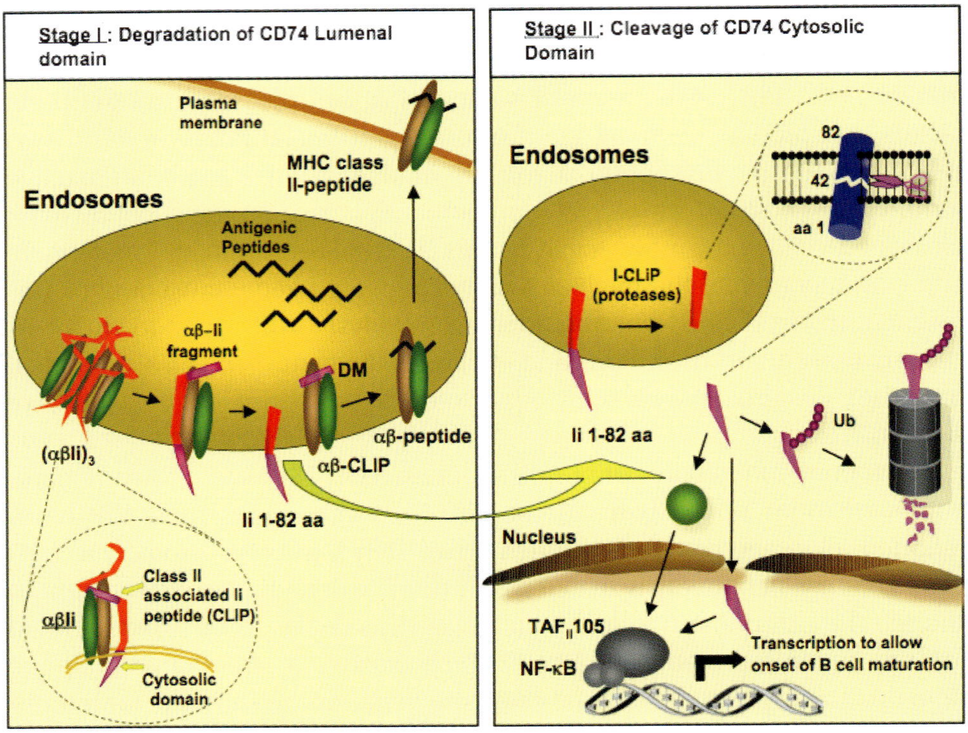

Fig. 3. Model for proteolytic cleavages of CD74. The 1–82 aa CD74 fragment is formed by a series of cleavage events during the MHC class II biosynthetic and expression pathway (stage I). Upon receiving a survival signal, B cells release the 1–42 aa cytosolic domain by cleavage within the transmembrane region (stage II). The released cytosolic fragment then induces NF-κB activity either directly or by activating another protein or pathway that can induce NF-κB; the fragment is then degraded, probably by the proteasome-ubiquitin pathway. Abbreviations: CLIP, class II-associated Ii chain peptide; I-CliP, intramembrane-cleaving proteases; TAF, TAB-associated factor; ub, ubiquitin.

To conclude, these studies add CD74 to the growing family of RIP proteins and show that the roles of CD74 as a chaperone and as a signaling molecule are intertwined in a tightly regulated pathway (Fig. 3).

5. MIF Regulates B Cell Survival in a CD74-Dependent Manner

B cells develop in the bone marrow (BM) and, through a process that involves rearrangement and expression of Ig genes, produce an antigen-specific receptor, which is first manifested in the immature stage. Immature

B cells emerge from the BM to the periphery and migrate into the spleen for their final maturation step.[37] However, up to a quarter of the mature IgD^+ IgM^+ B pool can subsequently be found recirculating in the steady state through the BM.[38] During their development, B cells encounter various checkpoints that control cell survival. Under steady state conditions, the number and distribution of B cells is under homeostatic control due to a balance between survival and apoptosis.

Regulation of mature B cell survival involves multiple mechanisms. The B cell receptor (BCR) provides survival signals essential for maintaining the mature B cell pool. Deletion of the *Igh* gene[39] or conditional deletion of either *Igh* or the signal-transducing *CD79α* (*Igα*) genes[40] leads to a loss of B cells. In addition, survival of mature naive B cells depends on signals delivered by the ligand-receptor pair, BAFF and BAFF-receptor (BAFF-R).[41,42] Mice that lack BAFF expression or that are subjected to treatments designed to block the action of BAFF fail to produce or maintain a mature B cell pool.[43–45]

This chapter discusses the role of MIF and CD74 in the regulation of peripheral B cell survival.

5.1 *A complex of CD74 and CD44 is essential for initiating the signaling cascade induced by MIF*

It was suggested that activation of cell surface CD74 in B cells might induce a signaling cascade resulting in gene transcription. To demonstrate such a pathway, cell surface CD74 was stimulated in B cells with an activating anti-CD74 antibody. CD74 stimulation was shown to induce a signaling pathway that involves Syk tyrosine kinase and the PI3K/Akt pathway. In addition, stimulation of CD74 induces CD74 intramembrane cleavage and the release of CD74-ICD. CD74-ICD translocates to the nucleus, where it regulates transcription of genes that control B cell proliferation and survival.[46] The trafficking of cell surface CD74 in this cascade is yet not clear. It can be suggested that cell surface activated CD74 is endocytosed to the endocytic compartments where it is cleaved. Alternatively, stimulation of cell surface CD74 initiates a cascade that induces the processing of a pool of CD74 that exists in the endocytic compartment.

Further studies identified the physiological ligand of CD74 in various cell types. In macrophages, MIF binds to the CD74 extracellular domain, a process that results in initiation of a signaling pathway.[22] An additional cell surface molecule, CD44, was described as an integral component of the CD74 receptor complex.[47,48] While CD74 is sufficient for the binding of MIF to the cell surface, CD44 was found to be necessary for MIF signal transduction in these cells.[48]

In B cells, CD74 forms a similar complex with CD44 that is essential for the MIF-induced signaling cascade.[49] The complex is formed even in the absence of MIF, since no change is detected in the formation of the complex or in CD44 expression following MIF stimulation. Nevertheless, formation of the CD74/CD44 complex was shown to be crucial for the signaling cascade induced by MIF.[49] MIF was found to regulate cell entry into the S-phase in a CD74- and CD44-dependent fashion by elevating cyclin E levels, resulting in cell proliferation. In addition, this cascade elevates Bcl-2 expression supporting cell survival.[49] Thus, MIF binding to both CD74 and CD44 initiates a survival pathway, resulting in proliferation of the mature B cell population and their rescue from death.

CD74 was also found to be expressed on normal colon epithelial cells (CECs). Similar to B cells, stimulation of CD74 on CECs by MIF, which is expressed throughout the human gastrointestinal tract, induces a signaling cascade leading to upregulation of Bcl-2 and cyclin E expression, resulting in a significant increased survival of these cells in a CD44 dependent manner.[50]

5.2 *c-Met and its ligand HGF regulate mature B cell survival in a pathway induced by MIF binding to CD74/CD44*

Cell surface receptor CD44 has been implicated in the regulation of the activation of the tyrosine kinase receptor, c-Met,[51–53] although the precise mechanism of this interaction is unknown.

c-Met is a unique disulfide-linked αβ heterodimeric receptor tyrosine kinase with a versatile role in regulating numerous biological functions in response to its natural ligand, hepatocyte growth factor/scatter factor (HGF). HGF is a multifunctional cytokine with a domain structure and proteolytic mechanism of activation similar to that of the serine protease plasminogen. Activation of the HGF/c-Met signaling pathway, which requires phosphorylation of various specific tyrosine residues on c-Met itself, leads to cellular responses, including increased motility, proliferation, morphogenesis, and cell survival.[54–60]

Murine splenic B cells express both c-Met and its ligand, HGF. Immunoprecipitation studies support the formation of a CD74/CD44/c-Met complex. The interaction between CD74 and c-Met is not observed in CD44-deficient B cells, suggesting that CD44 is essential for the formation of the CD74/CD44/c-Met complex. In addition, stimulation of the CD74/CD44 complex on B cells augments both c-Met cell surface expression and HGF secretion.[61]

Following MIF stimulation, c-Met engages with CD74 and CD44 on the cell membrane and, together with HGF, triggers an additional signaling pathway, which is necessary to initiate the MIF-induced survival signaling cascade.[61] The HGF-induced survival pathway controls proliferation and survival of peripheral B cell subsets. HGF elevates the survival of the mature population in the spleen, whereas there is no change in the cell death of the immature population. In addition, *in vivo* studies showed that activation of c-Met by HGF induces cell division and suppression of apoptosis, which enlarges the mature B cell compartment, whereas its blockade results in mature B cell death. The CD74/CD44 cell surface complex regulates the HGF/c-Met-induced survival cascade, since exogenous HGF is able to bypass the absent survival signal and rescue the mature B cell population that is missing in cells lacking CD74. Moreover, blocking HGF or c-Met activity abolishes MIF-induced Syk phosphorylation and Bcl-2 elevation, thereby inhibiting cell survival.[61] The precise mechanism by which MIF activates c-Met is still unclear. However, since HGF was shown to be sufficient to support survival of mature B cells, and its blocking inhibits the MIF-induced survival pathway, it is believed that HGF is involved in the MIF-induced survival cascade.[61]

Thus, c-Met participates in controlling MIF-induced signaling by forming a survival complex together with CD74 and CD44 in B cells. These findings establish a key role for the HGF/c-Met pair in the regulation of B cell survival, demonstrating an additional level of control of the humoral immune response.

5.3 *MIF induces TAp63 expression, which regulates B cell survival in a CD74-dependent manner*

Stimulation by MIF was shown to induce the activation of the p65/RelA member of the NF-κB family, which in turn upregulates TAp63 transcription and expression.[62]

The *p63* gene exhibits a high sequence and structural homology to *p53*.[63] Like *p53*, the *p63* gene encodes an N-terminal transactivation domain, a core DNA binding domain, and a carboxy-terminal oligomerization domain. The *p63* gene contains two transcriptional start sites that are used to generate transcripts encoding proteins with or without an N-terminal transactivation domain. Proteins with the transactivation domain are termed TAp63, and proteins lacking the transactivation domain are known as DNp63. *p63* was shown to play a role in developmental regulation of limbs, skin, most epithelial tissues, and in epidermal differentiation.[64,65]

In most cell types studied thus far, TA isoforms of *p63* were shown to possess proapoptotic functions, whereas delta-N isoforms were often found to exert an opposing effect and thus display features consistent with a pro-oncogenic role.[66–68] This is also reflected in gene expression profiling analysis, where TAp63 was found to activate a variety of proapoptotic genes.[69] Surprisingly, in B cells, TAp63 can also exhibit a very different behavior, favoring cell survival rather than cell death. It thus appears that the outcome of *p63* induction is strongly dependent on the cellular context, and therefore can vary greatly among different cell types.

TAp63 binds to the Bcl-2 promoter and induces the transcription of Bcl-2 mRNA and production of the Bcl-2 antiapoptotic protein, which enhances cell survival. Thus, the MIF/CD74/NF-κB/TAp63 axis defines a novel antiapoptotic pathway in mature B cells, resulting in the shaping of both the B cell repertoire and the immune response.[62] Fig. 4 summarizes the MIF/CD74 induced survival cascade in normal B cells.

6. MIF Is Secreted from Perivascular Bone Marrow Dendritic Cell Clusters that Regulate B Cell Survival

The mammalian BM is the major site of adult hematopoiesis. Importantly, the recent advent of advanced imaging studies has led to the identification of unique niches that provide a highly specialized microenvironment for distinct developmental processes. These include anatomically defined niches for hematopoetic stem cells[70,71] and for B cell development.[72]

The BM harbors dendritic cells (CD11c^{high} DC, bmDC) that function as myeloid BM cells and display an activated phenotype (MHC II^{high}, CD80^{high}). Most intriguingly, multiphoton analysis of the cranial BM of mice that harbor GFP-labeled bmDC (CX$_3$CR1^{GFP} mice) revealed that these cells are concentrated into unique perivascular clusters that wrap a distinct set of sinusoids and venules.[73]

Conditional ablation of bmDC results in the specific loss of both endogenous and adoptively transferred mature B cells from the BM immune niches. This failure of bmDC-depleted BM to support B cell engraftment could be overcome by the overexpression of the antiapoptotic factor bcl2 in the mature B cells, suggesting that bmDC provide a unique survival factor. Studies with mixed BM chimeras subsequently showed that this factor is MIF. Thus, mature B cell maintenance requires MIF-producing bmDC.[73] Newly formed mature B cells emerge from the spleen and circulate in the body. In the BM, a survival signal induced by MIF and secreted from bmDC is essential for their maintenance (Fig. 5).

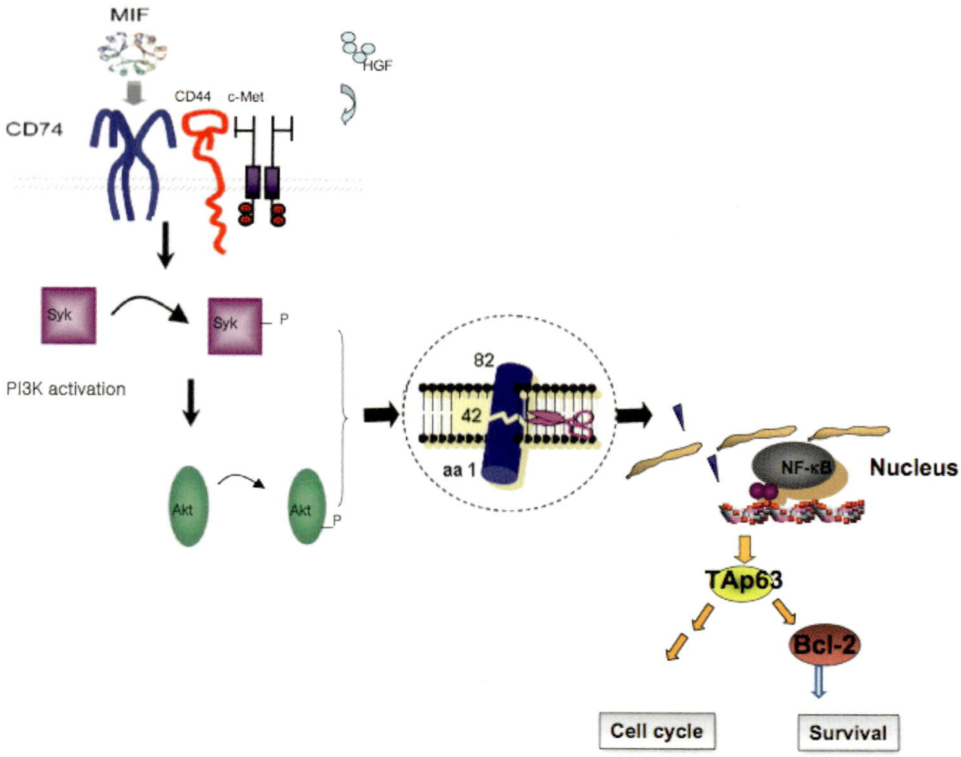

Fig. 4. Following MIF stimulation, c-Met engages with CD74 and CD44 on the cell membrane and, together with HGF, serves as an additional signaling pathway, which is necessary to initiate the MIF-induced survival signaling cascade. The CD74 stimulation by MIF activates the Syk and PI3K/Akt pathways, leading to intramembrane cleavage, CD7-ICD release, and NF-κB activation, enabling entry of the stimulated B cells into the S phase. In addition, NF-κB activation induces TAp63 expression. TAp63 binds to the Bcl-2 promoter and induces the transcription of Bcl-2 mRNA and production of the Bcl-2 antiapoptotic protein, which enhances cell survival.

7. MIF and CD74 in Tumors

Both MIF and CD74 have been associated with tumor progression. It was reported that MIF mRNA is overexpressed in various tumors, and MIF has also been associated with the growth of malignant cells. MIF was shown to be overexpressed in solid tumors,[74] and it is frequently overexpressed in primary breast cancer tissues, where it plays a role in tumor-stroma interactions of primary breast cancers.[75] MIF is also associated with tumor growth and tumor-associated angiogenesis in a murine colon cancer cell line.[76] In addition, anti-MIF Ig therapy has been shown to suppress tumor growth.[77] The overexpression of CD74, the cell surface receptor for MIF, was also

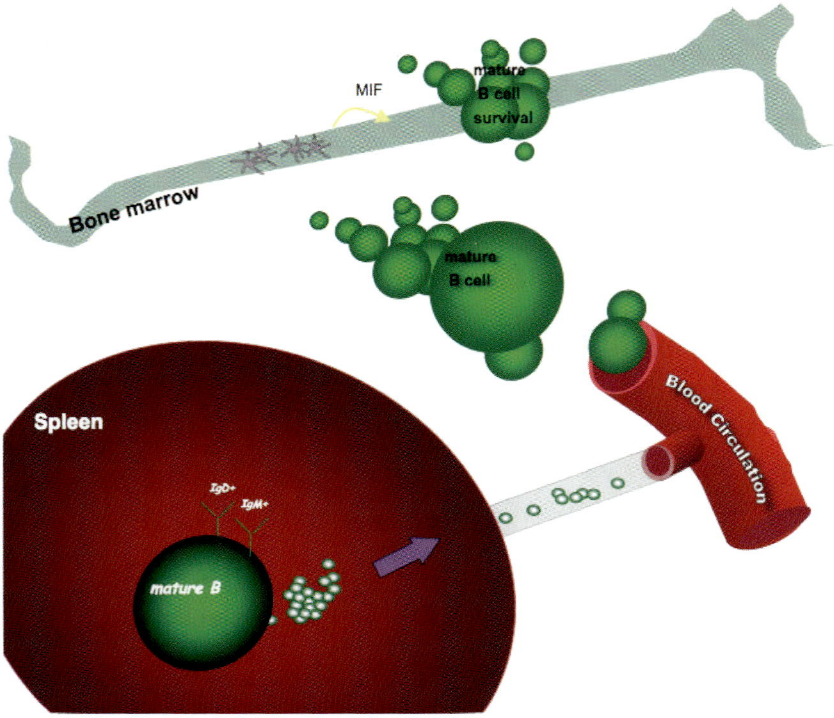

Fig. 5. Schematic illustration of the regulation of BM B cell survival by MIF secreted by BM dendritic cells. A survival signal induced by MIF and secreted from bmDC is essential for mature B cell maintenance.

observed in various cancers,[47,78–82] including CLL.[83,84] CD74 expression in many of these cancers has been suggested to serve as a prognostic factor, with higher relative expression of CD74 behaving as a marker of tumor progression.[85]

Moreover, a humanized anti-CD74 mAb (hLL1) was shown to have therapeutic activity in multiple myeloma, perhaps by inhibiting the high level of CD74 expression in this plasma cell malignancy.[86]

7.1 *MIF and CD74 in CLL survival*

Chronic lymphocytic leukemia (CLL), the most common leukemia in the Western world, is characterized by the progressive accumulation of small, mature CD5+ B lymphocytes in the peripheral blood, lymphoid organs, and bone marrow. CLL cells display features that are consistent with a defect in programmed cell death, and exhibit prolonged *in vivo* survival.[87] In the bone marrow and lymph nodes, CLL cells interact with a variety of accessory and bystander cells, which support CLL cell survival and are believed to promote resistance to therapy.

CD74 and its ligand, MIF, play a pivotal role in the regulation of CLL cell survival.[88] CLL cells markedly upregulate both expression of their cell surface CD74 and their MIF production. Stimulation of CD74 with the MIF ligand (as well as with an agonistic antibody) initiates a signaling cascade leading to IL-8 transcription and secretion in all CLL cells, regardless of the clinical status of the patients. Secreted IL-8 induces the transcription and translation of the antiapoptotic protein, Bcl-2, and thus regulates an antiapoptotic pathway, while no effect on proliferation is observed. Blocking of CD74 (by hLL1), or of MIF or IL-8 results in dramatic downregulation of Bcl-2 expression and augmentation of apoptosis.

IL-8 is a member of the CXC chemokine family that plays an important role in autoimmune, inflammatory, and infectious diseases.[89–91] Because of its potent proinflammatory properties, IL-8 is tightly regulated, and its expression is low or undetectable in normal tissues. However, it is now known that IL-8 also possesses tumorigenic and proangiogenic properties.[92] The signaling cascade induced by CD74 that activates NF-κB results in increased IL-8 expression and in an autocrine/paracrine survival response. Thus, in CLL, MIF and CD74 induce an important survival mechanism, which appears to be an early event in the pathogenesis of the disease.[88]

In addition, blocking CD74 by hLLI specifically downregulates TAp63 mRNA levels, demonstrating that blocking CD74 activity inhibits TAp63 expression, which was shown to regulate normal B cell survival in a CD74-dependent manner.[62] Moreover, downregulation of p63 in CLL cells specifically inhibits the MIF-induced elevation of Bcl-2 mRNA levels, showing that the MIF/CD74-induced survival cascade is mediated through TAp63.[93]

7.2 *MIF and CD74 regulate CLL homing to the BM*

The BM stroma plays an essential role in B lymphopoiesis and can provide survival niches for both normal and leukemic mature B cells. The adhesion of CLL cells to BM stromal cells or to the BM vasculature has been shown to rescue these lymphocytes from apoptosis and to extend their life span.[73,94,95] The increased accumulation of CLL cells in the BM during disease progression suggests a change in the migratory and homing pattern of the cells. Advanced stage CLL cells express higher levels of the VLA-4 integrin compared to early stage cells.[93,96–98]

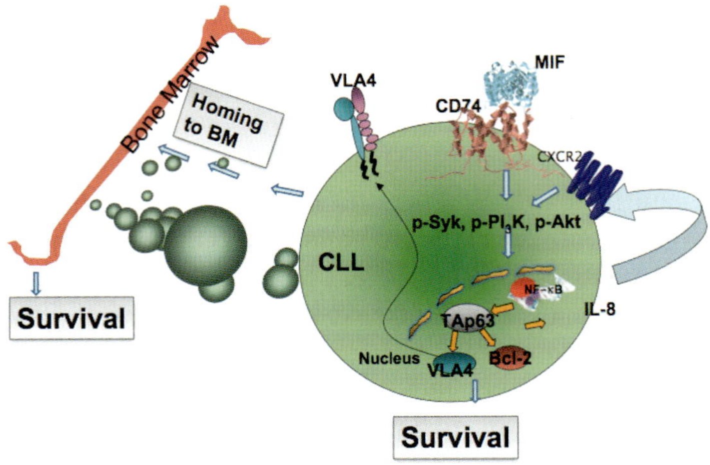

Fig. 6. Schematic representation of the dual role of the MIF-induced regulation in CLL survival affecting cell survival both directly (by elevating Bcl-2 expression) and indirectly (by elevating VLA-4 expression, promoting homing to the BM).

MIF and CD74 were demonstrated to play a significant role in the regulation of VLA-4 expression, and therefore to affect homing and survival of CLL cells.[93] MIF is secreted from all types of cells; hence, CLL cells are stimulated by this chemokine in all compartments. MIF stimulation elevates VLA-4 cell surface expression levels during advanced stage disease. It is likely that only after their progress to the advanced disease stage do the cells increase their VLA-4 expression to levels that support their homing to the BM, though the mechanism of VLA-4 regulation is not known. Thus, homing to the BM requires threshold levels of VLA-4 expression that enable retention and survival of CLL in BM, an environment that is enriched with the VLA-4 ligands, VCAM-1 and fibronectin, and further supports their retention and survival. It is possible that CLL exposure to systemic MIF redirects circulating CLL cells back to the BM, where they may encounter more MIF, and additionally elevate their VLA-4 expression and retention on stromal VLA-4 ligands. This may create a cycle that can promote disease-associated BM failure. Fig. 6 summarizes the role of MIF/CD74 induced cascade in the direct and indirect regulation of CLL survival.

These results suggest that blocking MIF expression or its receptor, such as with an antagonistic anti-CD74 antibody, might inhibit survival of CLL cells and their homing to the BM, which results in an additional way to regulate their survival. Thus, novel therapeutic strategies aimed at blocking the MIF/CD74 pathway could lead to better and more targeted eradication of the disease due to decreased cell survival and/or alteration of disease progression by decreasing BM homing.

References

1. Jones PP, Murphy DB, Hewgill D, Mcdevitt HO. (1979) Detection of a common polypeptide-chain in I-a and I-E sub-region immunoprecipitates. *Mol Immunol* **16**: 51–60.
2. Stockinger B, Pessara U, Lin RH *et al.* (1989). A role of Ia-associated invariant chains in antigen processing and presentation. *Cell* **56**: 683–689.
3. Landsverk OJB, Bakke O, Gregers TF. (2009) MHC II and the endocytic pathway: Regulation by invariant chain. *Scand J Immunol* **70**: 184–193.
4. Yamamoto K, Koch N, Steinmetz M, Hammerling GJ. (1985) One gene encodes 2 distinct Ia-associated invariant chains. *J Immunol* **134**: 3461–3467.
5. Strubin M, Long EO, Mach B. (1986) Two forms of the Ia antigen-associated invariant chain result from alternative initiations at two in-phase AUGs. *Cell* **47**: 619–625.
6. Koch N, Lauer W, Habicht J, Dobberstein B. (1987) Primary structure of the gene for the murine Ia antigen-associated invariant chains (Ii) — An alternatively spliced exon encodes a cysteine-rich domain highly homologous to a repetitive sequence of thyroglobulin. *EMBO J* **6**: 1677–1683.
7. Unanue ER. (1984) Antigen-presenting function of the macrophage. *Annu Rev Immunol* **2**: 395–428.
8. Long EO. (1989) Intracellular traffic and antigen processing. *Immunol Today* **10**: 232–234.
9. Harding CV, Unanue ER. (1990) Cellular mechanisms of antigen processing and the function of class I and II major histocompatibility complex molecules. *Cell Regul* **1**: 499–509.
10. Cresswell P. (1994) Assembly, transport, and function of MHC class II molecules. *Annu Rev Immunol* **12**: 259–293.
11. Freisewinkel IM, Schenck K, Koch N. (1993) The segment of invariant chain that is critical for association with major histocompatibility complex class II molecules contains the sequence of a peptide eluted from class II polypeptides. *Proc Natl Acad Sci USA* **90**: 9703–9706.
12. Ghosh P, Amaya M, Mellins E, Wiley DC. (1995) The structure of an intermediate in class II MHC maturation: CLIP bound to HLA-DR3. *Nature* **378**: 457–462.
13. Roche PA, Teletski CL, Stang E *et al.* (1993) Cell surface HLA-DR-invariant chain complexes are targeted to endosomes by rapid internalization. *Proc Natl Acad Sci USA* **90**: 8581–8585.

14. Sant AJ, Cullen SE, Schwartz BD. (1985) Biosynthetic relationships of the chondroitin sulfate proteoglycan with Ia and invariant chain glycoproteins. *J Immunol* **135**: 416–422.
15. Sant AJ, Cullen SE, Giacoletto KS, Schwartz BD. (1985) Invariant chain is the core protein of the Ia-associated chondroitin sulfate proteoglycan. *J Exp Med* **162**: 1916–1934.
16. Miller J, Hatch JA, Simonis S, Cullen SE. (1988) Identification of the glycosaminoglycan-attachment site of mouse invariant-chain proteoglycan core protein by site-directed mutagenesis. *Proc Natl Acad Sci USA* **85**: 1359–1363.
17. Lotteau V, Teyton L, Peleraux A *et al.* (1990) Intracellular transport of class II MHC molecules directed by invariant chain. *Nature* **348**: 600–605.
18. Odorizzi CG, Trowbridge IS, Xue L *et al.* (1994) Sorting signals in the MHC class II invariant chain cytoplasmic tail and transmembrane region determine trafficking to an endocytic processing compartment. *J Cell Biol* **126**: 317–330.
19. Pond L, Kuhn LA, Teyton L *et al.* (1995) A role for acidic residues in di-leucine motif-based targeting to the endocytic pathway. *J Biol Chem* **270**: 19989–19997.
20. Roche PA, Cresswell P. (1991) Proteolysis of the class II-associated invariant chain generates a peptide binding site in intracellular HLA-DR molecules. *Proc Natl Acad Sci USA* **88**: 3150–3154.
21. Neefjes JJ, Stollorz V, Peters PJ *et al.* (1990) The biosynthetic pathway of MHC class II but not class I molecules intersects the endocytic route. *Cell* **61**: 171–183.
22. Leng L, Metz CN, Fang Y *et al.* (2003) MIF signal transduction initiated by binding to CD74. *J Exp Med* **197**: 1467–1476.
23. Beswick EJ, Bland DA, Suarez G *et al.* (2005) *Helicobacter pylori* binds to CD74 on gastric epithelial cells and stimulates interleukin-8 production. *Infect Immun* **73**: 2736–2743.
24. Shachar I, Flavell RA. (1996) Requirement for invariant chain in B cell maturation and function. *Science* **274**: 106–108.
25. Kenty G, Martin WD, Van Kaer L, Bikoff EK. (1998) MHC class II expression in double mutant mice lacking invariant chain and DM functions. *J Immunol* **160**: 606–614.
26. Kenty G, Bikoff EK. (1999) BALB/C invariant chain mutant mice display relatively efficient maturation of CD4+ T cell in the periphery and secondary proliferative responses elicited upon peptide challenge. *J Immunol* **163**: 232–242.
27. Rajagopalan G, Smart MK, Krco CJ, David CS. (2002) Expression and function of transgenic HLA-DQ molecules and lymphocyte development in mice lacking invariant chain. *J Immunol* **169**: 1774–1783.
28. Matza D, Lantner D, Bogoch Y *et al.* (2002) Invariant chain induces B cell maturation in a process which is independent of its chaperonic activity. *Proc Natl Acad Sci USA* **99**: 3018–3023.

29. Matza D, Wolstein O, Dikstein R, Shachar I. (2001) Invariant chain induces B cell maturation by activating TAFII105-NF-kB dependent transcription program. *J Biol Chem* **276**: 27203–27206.
30. Dikstein R, Zhou S, Tjian R. (1996) Human TAFII 105 is a cell type-specific TFIID subunit related to hTAFII130. *Cell* **87**: 137–146.
31. Matza D, Kerem A, Lantner F, Shachar I. (2002) Invariant chain induced B cell differentiation requires intramembrane-proteolytic release of the cytosolic domain. *Immunity* **17**: 549–560.
32. Lipp J, Dobberstein B. (1986) The membrane-spanning segment of invariant chain (I gamma) contains a potentially cleavable signal sequence. *Cell* **46**: 1103–1112.
33. Brown MS, Ye J, Rawson RB, Goldstein JL. (2000) Regulated intramembrane proteolysis: A control mechanism conserved from bacteria to humans. *Cell* **100**: 391–398.
34. Urban S, Freeman M. (2002) Intramembrane proteolysis controls diverse signalling pathways throughout evolution. *Curr Opin Genet Dev* **12**: 512–518.
35. Hoppe T, Rape M, Jentsch S. (2001) Membrane-bound transcription factors: Regulated release by RIP or RUP. *Curr Opin Cell Biol* **13**: 344–348.
36. Becker-Herman S, Arie G, Medvedovsky H *et al.* (2005) CD74 is a member of the regulated intramembrane proteolysis (RIP) processed protein family. *Mol Biol Cell* **16**: 5061–5069.
37. Hardy RR, Carmack CE, Shinton SA *et al.* (1991) Resolution and characterization of pro-B and pre-pro-B cell stages in normal mouse bone marrow. *J Exp Med* **173**: 1213–1225.
38. Cariappa A, Mazo IB, Chase C *et al.* (2005) Perisinusoidal B cells in the bone marrow participate in T-independent responses to blood-borne microbes. *Immunity* **23**: 397–407.
39. Lam KP, Kuhn R, Rajewsky K. (1997) *In vivo* ablation of surface immunoglobulin on mature B cells by inducible gene targeting results in rapid cell death. *Cell* **90**: 1073–1083.
40. Kraus M, Alimzhanov MB, Rajewsky N, Rajewsky K. (2004) Survival of resting mature B lymphocytes depends on BCR signaling via the Ig alpha/beta heterodimer. *Cell* **117**: 787–800.
41. Crowley JE, Treml LS, Stadanlick JE *et al.* (2005) Homeostatic niche specification among naive and activated B cells: A growing role for the BLyS family of receptors and ligands. *Semin Immunol* **17**: 193–199.
42. Woodland RT, Schmidt MR, Thompson CB. (2006) BLyS and B cell homeostasis. *Semin Immunol* **18**: 318–326.
43. Schiemann B, Gommerman JL, Vora K *et al.* (2001) An essential role for BAFF in the normal development of B cells through a BCMA-independent pathway. *Science* **293**: 2111–2114.

44. Schneider P, Takatsuka H, Wilson A *et al.* (2001) Maturation of marginal zone and follicular B cells requires B cell activating factor of the tumor necrosis factor family and is independent of B cell maturation antigen. *J Exp Med* **194**: 1691–1697.
45. Thompson JS, Schneider P, Kalled SL *et al.* (2000) BAFF binds to the tumor necrosis factor receptor-like molecule B cell maturation antigen and is important for maintaining the peripheral B cell population. *J Exp Med* **192**: 129–135.
46. Starlets D, Gore Y, Binsky I *et al.* (2006) Cell surface CD74 initiates a signaling cascade leading to cell proliferation and survival. *Blood* **107**: 4807–4816.
47. Meyer-Siegler KL, Leifheit EC, Vera PL. (2004) Inhibition of macrophage migration inhibitory factor decreases proliferation and cytokine expression in bladder cancer cells. *BMC Cancer* **4**: 34–45.
48. Shi X, Leng L, Wang T *et al.* (2006) CD44 is the signaling component of the macrophage migration inhibitory factor-CD74 receptor complex. *Immunity* **25**: 595–606.
49. Gore Y, Starlets D, Maharshak N *et al.* (2008) Macrophage migration inhibitory factor (MIF) induces B cell survival by activation of a CD74/CD44 receptor complex. *J Biol Chem* **283**: 2784–2792.
50. Maharshak N, Cohen S, Lantner F *et al.* (2010) CD74 is a survival receptor on colon epithelial cells. *World J Gastroenterol* **16**: 3258–3266.
51. van der Voort R, Taher TE, Wielenga VJ *et al.* (1999) Heparan sulfate-modified CD44 promotes hepatocyte growth factor/scatter factor-induced signal transduction through the receptor tyrosine kinase c-Met. *J Biol Chem* **274**: 6499–6506.
52. Orian-Rousseau V, Chen L, Sleeman JP *et al.* (2002) CD44 is required for two consecutive steps in HGF/c-Met signaling. *Genes Dev* **16**: 3074–3086.
53. Orian-Rousseau V, Morrison H, Matzke A *et al.* (2007) Hepatocyte growth factor-induced Ras activation requires ERM proteins linked to both CD44v6 and F-actin. *Mol Biol Cell* **18**: 76–83.
54. Birchmeier C, Birchmeier W, Gherardi E, Vande Woude GF. (2003) Met, metastasis, motility and more. *Nat Rev Mol Cell Biol* **4**: 915–925.
55. Bertotti A, Comoglio PM. (2003) Tyrosine kinase signal specificity: Lessons from the HGF receptor. *Trends Biochem Sci* **28**: 527–533.
56. Zhang YW, Vande Woude GF. (2003) HGF/SF-met signaling in the control of branching morphogenesis and invasion. *J Cell Biochem* **88**: 408–417.
57. Corso S, Comoglio PM, Giordano S. (2005) Cancer therapy: Can the challenge be MET? *Trends Mol Med* **11**: 284–292.
58. Schmidt C, Bladt F, Goedecke S *et al.* (1995) Scatter factor/hepatocyte growth-factor is essential for liver development. *Nature* **373**: 699–702.
59. Uehara Y, Minowa O, Mori C *et al.* (1995) Placental defect and embryonic lethality in mice lacking hepatocyte growth factor/scatter factor. *Nature* **373**: 702–705.

60. Tulasne D, Foveau B. (2008) The shadow of death on the MET tyrosine kinase receptor. *Cell Death Differ* **15**: 427–434.
61. Gordin M, Tesio M, Cohen S *et al.* (2010) c-Met and its ligand hepatocyte growth factor/scatter factor regulate mature B cell survival in a pathway induced by CD74. *J Immunol* **185**: 2020–2031.
62. Lantner F, Starlets D, Gore Y *et al.* (2007) CD74 induces TAp63 expression leading to B cell survival. *Blood* **110**: 4303–4311.
63. Yang A, Kaghad M, Wang Y *et al.* (1998) p63, a p53 homolog at 3q27–29, encodes multiple products with transactivating, death-inducing, and dominant-negative activities. *Mol Cell* **2**: 305–316.
64. Yang A, Schweitzer R, Sun D *et al.* (1999) p63 is essential for regenerative proliferation in limb, craniofacial and epithelial development. *Nature* **398**: 714–718.
65. Mills AA, Zheng B, Wang XJ *et al.* (1999) p63 is a p53 homologue required for limb and epidermal morphogenesis. *Nature* **398**: 708–713.
66. Moll UM, Slade N. (2004) p63 and p73: Roles in development and tumor formation. *Mol Cancer Res* **2**: 371–386.
67. Gressner O, Schilling T, Lorenz K *et al.* (2005) TAp63alpha induces apoptosis by activating signaling via death receptors and mitochondria. *EMBO J* **24**: 2458–2471.
68. Jacobs WB, Govoni G, Ho D *et al.* (2005) p63 is an essential proapoptotic protein during neural development. *Neuron* **48**: 743–756.
69. Wu G, Nomoto S, Hoque MO *et al.* (2003) DeltaNp63alpha and TAp63alpha regulate transcription of genes with distinct biological functions in cancer and development. *Cancer Res* **63**: 2351–2357.
70. Zhang J, Niu C, Ye L *et al.* (2003) Identification of the haematopoietic stem cell niche and control of the niche size. *Nature* **425**: 836–841.
71. Calvi LM, Adams GB, Weibrecht KW *et al.* (2003) Osteoblastic cells regulate the haematopoietic stem cell niche. *Nature* **425**: 841–846.
72. Tokoyoda K, Egawa T, Sugiyama T *et al.* (2004) Cellular niches controlling B lymphocyte behavior within bone marrow during development. *Immunity* **20**: 707–718.
73. Sapoznikov A, Pewzner-Jung Y, Kalchenko V *et al.* (2008) Perivascular clusters of dendritic cells provide critical survival signals to B cells in bone marrow niches. *Nat Immunol* **9**: 388–395.
74. Meyer-Siegler K, Hudson PB. (1996) Enhanced expression of macrophage migration inhibitory factor in prostatic adenocarcinoma metastases. *Urology* **48**: 448–452.
75. Bando H, Matsumoto G, Bando M *et al.* (2002) Expression of macrophage migration inhibitory factor in human breast cancer: Association with nodal spread. *Jpn J Cancer Res* **93**: 389–396.

76. Nishihira J, Ishibashi T, Fukushima T *et al.* (2003) Macrophage migration inhibitory factor (MIF): Its potential role in tumor growth and tumor-associated angiogenesis. *Ann N Y Acad Sci* **995**: 171–182.
77. Chesney J, Metz C, Bacher M *et al.* (1999) An essential role for macrophage migration inhibitory factor (MIF) in angiogenesis and the growth of a murine lymphoma. *Mol Med* **5**: 181–191.
78. Ishigami S, Natsugoe S, Tokuda K *et al.* (2001) Invariant chain expression in gastric cancer. *Cancer Lett* **168**: 87–91.
79. Young AN, Amin MB, Moreno CS *et al.* (2001) Expression profiling of renal epithelial neoplasms: A method for tumor classification and discovery of diagnostic molecular markers. *Am J Pathol* **158**: 1639–1651.
80. Ioachim HL, Pambuccian SE, Hekimgil M *et al.* (1996) Lymphoid monoclonal antibodies reactive with lung tumors. Diagnostic applications. *Am J Surg Pathol* **20**: 64–71.
81. Datta MW, Shahsafaei A, Nadler LM *et al.* (2000) Expression of MHC class II-associated invariant chain (Ii;CD74) in thymic epithelial neoplasms. *Appl Immunohistochem Mol Morphol* **8**: 210–215.
82. Lazova R, Moynes R, May D, Scott G. (1997) LN-2 (CD74). A marker to distinguish atypical fibroxanthoma from malignant fibrous histiocytoma. *Cancer* **79**: 2115–2124.
83. Narni F, Kudo J, Mars W *et al.* (1986) HLA-DR-associated invariant chain is highly expressed in chronic lymphocytic leukemia. *Blood* **68**: 372–377.
84. Veenstra H, Jacobs P, Dowdle EB. (1996) Abnormal association between invariant chain and HLA class II alpha and beta chains in chronic lymphocytic leukemia. *Cell Immunol* **171**: 68–73.
85. Mizue Y, Nishihira J, Miyazaki T *et al.* (2000) Quantitation of macrophage migration inhibitory factor (MIF) using the one-step sandwich enzyme immunosorbent assay: Elevated serum MIF concentrations in patients with autoimmune diseases and identification of MIF in erythrocytes. *Int J Mol Med* **5**: 397–403.
86. Stein R, Qu Z, Cardillo TM *et al.* (2004) Antiproliferative activity of a humanized anti-CD74 monoclonal antibody, hLL1, on B-cell malignancies. *Blood* **104**: 3705–3711.
87. Caligaris-Cappio F, Hamblin TJ. (1999) B-cell chronic lymphocytic leukemia: A bird of a different feather. *J Clin Oncol* **17**: 399–408.
88. Binsky I, Haran M, Starlets D *et al.* (2007) IL-8 secreted in a macrophage migration-inhibitory factor- and CD74-dependent manner regulates B cell chronic lymphocytic leukemia survival. *Proc Natl Acad Sci USA* **104**: 13408–13413.
89. Harada A, Sekido N, Akahoshi T *et al.* (1994) Essential involvement of interleukin-8 (IL-8) in acute inflammation. *J Leukoc Biol* **56**: 559–564.

90. Koch AE, Polverini PJ, Kunkel SL *et al.* (1992) Interleukin-8 as a macrophage-derived mediator of angiogenesis. *Science* **258**: 1798–1801.
91. Smyth MJ, Zachariae CO, Norihisa Y *et al.* (1991) IL-8 gene expression and production in human peripheral blood lymphocyte subsets. *J Immunol* **146**: 3815–3823.
92. Brat DJ, Bellail AC, Van Meir EG. (2005) The role of interleukin-8 and its receptors in gliomagenesis and tumoral angiogenesis. *Neuro-oncol* **7**: 122–133.
93. Binsky I, Lantner F, Grabovsky V *et al.* (2010) TAp63 regulates VLA-4 expression and CLL cell migration to the BM in a CD74 dependent manner. *J Immunol* **184**: 4761–4769.
94. Ghia P, Granziero L, Chilosi M, Caligaris-Cappio F. (2002) Chronic B cell malignancies and bone marrow microenvironment. *Semin Cancer Biol* **12**: 149–155.
95. Chappell CP, Clark EA. (2008) Survival niches: B cells get MIFed as well as BAFFled by dendritic cells. *Immunol Cell Biol* **86**: 487–488.
96. Shanafelt TD, Geyer SM, Bone ND, *et al.* (2008) CD49d expression is an independent predictor of overall survival in patients with chronic lymphocytic leukaemia: A prognostic parameter with therapeutic potential. *British Journal of Haematology* **140**: 537–546.
97. Gattei V, Bulian P, Del Principe MI, *et al.* (2008) Relevance of CD49d protein expression as overall survival and progressive disease prognosticator in chronic lymphocytic leukemia. *Blood* **111**: 865–873.
98. Rossi D, Zucchetto A, Rossi FM, *et al.* (2008) CD49d expression is an independent risk factor of progressive disease in early stage chronic lymphocytic leukemia. *Haematologica-the Hematology Journal* **93**: 1575–1579.

I-4

Towards the MIF Interactome

Jörg Klug* and Andreas Meinhardt

1. Introduction

Mapping the interactome of a single protein, a protein complex, and ultimately a whole cell is crucial for a complete understanding of cellular function and represents a major challenge in current biological research. This task is intriguingly complex because protein interactions are often transient, occur only in a specific cellular context or at particular times during development. Adding another layer of complexity, protein-protein interactions identified *in vitro* may not occur *in vivo*, nor may they have a biological function. Nevertheless, characterization of a protein's interactome still represents a valuable first step in elucidating protein function. Thus, after cloning and initial characterization of MIF's enzymatic functions and 3D structure,[1–3] many laboratories attempted to identify MIF interacting proteins (MIPs) to delineate the molecular mechanisms by which MIF is exerting its various functions (see Table 1). To date, a substantial number of interacting proteins have been identified, reflecting the multitude of cellular processes that involve MIF. Besides furthering our understanding of the biological/cellular role of MIF, the identification of MIF interactions relevant for disease induction or progression could help to demarcate novel targets for therapeutic intervention. With MIF levels in body fluids and tissues determined by ELISA being largely ignored, one has to consider that the interaction between MIF and other proteins may mask epitopes, thereby leading to an underestimation of total MIF content.

The aim of this chapter is to provide an overview of currently known proteins that have been shown to interact directly with MIF, and to describe the

*Corresponding author: Department of Anatomy and Cell Biology, Justus-Liebig-University of Giessen, Aulweg 123, 35385 Giessen, Germany. Emails: joerg.klug@anatomie.med.uni-giessen.de, andreas.meinhardt@anatomie.med.uni-giessen.de

Table 1. List in chronological order of only those proteins for which a direct protein-protein interaction with MIF has been reported. Abbreviations used: Y2H = yeast two-hybrid screen, Co-IP = co-immunoprecipitation, TPOR = thiol-protein oxidoreductase, HUVEC = human umbilical vein endothelial cells, BPAEC = bovine pulmonary artery endothelial cells, SPR = surface plasmon resonance, in = intracellular interaction, ex = extracellular interaction, rec = recombinant, wt = wild type, DBD = DNA binding domain. CD44 and p97/VCP have been omitted, as they do not directly interact with MIF. Bold lettering indicates MIPs that both directly interact with MIF but also indirectly to another MIP.

Interactor	Symbol	Source	Method of Identification	Function	K_d	Function of Interaction	Loc.	Ref.
Jun activation domain-binding protein 1 (**JAB1**) or COP9 signalosome subunit 5 (**CSN5**)	COPS5	Fetal brain cDNA library Bait: MIF	Y2H*	**Signalosome** subunit with MPN domain,[18] transcription co-activator,[10] cell-cycle regulator,[19] also interacts with **thioredoxin**[20]		Inhibition of general signalosome and JAB1/ CSN5-specific functions	in	21
COP9 signalosome subunit 6 (**CSN6**)	COPS6	HEK 293neo	Endogenous Co-IP	**Signalosome** subunit with MPN domain, cleaved during apoptosis[22]		Tighter binding to signalosome? Role in apoptosis?	in	23
Peroxiredoxin 1, proliferation-associated gene protein (PAG)	PRDX1	HeLa cDNA library Bait: MIF HEK 293T	Y2H Co-IP of overexpressed proteins	Antioxidant enzyme controlling cellular peroxide levels[24]		Inhibition of MIFs tautomerase activity. Oxidative inhibition of glutamine synthetase by PRDX1 is blocked.	in	25

(*Continued*)

Table 1. (*Continued*)

Interactor	Symbol	Source	Method of Identification	Function	K_d	Function of Interaction	Loc.	Ref.
Insulin	INS	*In vitro* assays	ELISA, Enzymatic activity (TOPOR)	Binding of insulin to the insulin receptor stimulates glucose uptake		Role in disulfide reduction for antigen presentation	ex	26
HBsAg (14–33)	—	*In vitro* assays	ELISA CD spectroscopy	HLA class II processed peptide		Role in antigen presentation	ex	26
HLA-DP2 β	HLA-DPB1	*In vitro* assays NB-1	Immunoaffinity purification	HLA class II peptide		Role in antigen presentation	ex	26
Bcl-2/adenovirus E1B 19-kDa interacting protein 2-like (**BNIPL**)	BNIPL	Human lung cDNA library	Y2H	Apoptosis-associated Interacts with **GFER**		Inhibition of cell proliferation	in	27
Growth factor, augmenter of liver regeneration (**GFER**) Hepatopoietin (HPO)	GFER	HepG2 HEK 293T	Targeted Y2H Co-IP of overexpressed proteins Co-localization of overexpressed proteins	Proliferation of hepatoma cells, liver regeneration, TPOR activity (oxidation), forms homo- and heterodimers. Interacts with **BNIPL** and **JAB1/CSN5**		Stimulates GFER promoter acitivity and proliferation of hepatoma cells	in	28

(*Continued*)

Table 1. (*Continued*)

Interactor	Symbol	Source	Method of Identification	Function	K_d	Function of Interaction	Loc.	Ref.
CD74	CD74	THP-1 rec wt MIF and soluble CD74	Transfection of THP-1 cDNA expression library into COS-7 cells followed by cell surface MIF-binding analyses SPR	MHC class II invariant chain, trimer, type II transmembrane protein Interacts also with **CXCR2**	9.0 nM	MIF binding induces serine phosphorylation of CD74/CD44 receptor complex by Src followed by activation of MAP kinases. Cell proliferation. PGE 2 production. Protection from p53-mediated apoptosis.	ex	29,30
Endothelial myosin light chain kinase (MLCK)	MYLK	Primary HUVEC cDNA library Bait: MLCK1 (aa 1-922)	Y2H *In vitro* binding to GST-MLCK after TNF α and thrombin treatment of BPAECs Co-localization of phospho-MLCK and tagged MIF	Phosphorylation of myosin light chain upon Ca^{2+}-binding to associated calmodulin,[31] involved in endothelial barrier function		Role in regulating the endothelial barrier? Role in angiogenesis?	in	32
α1-inhibitor-3 (a1-I3) Murinoglobulin 1	Mug1	Rat bladder intraluminal fluid	Native PAGE and non-reducing SDS-PAGE Endogenous Co-IP	Acute-phase protein, serine-type endopeptidase inhibitor, rat specific protein		Release of MIF/ α 1-I3	ex	33, 34

(*Continued*)

Table 1. (*Continued*)

Interactor	Symbol	Source	Method of Identification	Function	K_d	Function of Interaction	Loc.	Ref.
Bcl2-like 11 BimL	BCL2L11	human fetal brain cDNA library Bait: BimL	Y2H	Apoptotic activator MIF also interacts with isoforms BimEL and BimS		Inhibition of Bim-mediated apoptosis	in	35
		HEK293T	Co-IP of overexpressed proteins					
		K562	Endogenous Co-IP *In vitro* pull down assays					
p53	TP53	293T MCF7	Co-IP of overexpressed MIF by p53 antibody	Interaction depends on Cys^{242} (and Cys^{238}) in the DBD of p53 and Cys^{81} of MIF and reducing conditions.		Suppression of p53-mediated transcription. Stabilization of the p53/Mdm2 complex (ternary complex formation). Increased p53 ubiquitination and p53 instability. Inhibition of p53 nuclear translocation. Inhibition of p53-induced apoptosis and cell cylcle arrest.	in	36
		HCT116	Endogenous Co-IP					

(*Continued*)

Table 1. (*Continued*)

Interactor	Symbol	Source	Method of Identification	Function	K_d	Function of Interaction	Loc.	Ref.
Protein (NM23A) expressed in non-metastatic cells NM23-H1	NME1	HEK293 HEK293T, HEK293, MCF	Co-IP of overexpressed proteins Endogenous Co-IP	Possesses nucleoside-diphosphate kinase, serine/threonine-specific protein kinase, geranyl and farnesyl pyrophosphate kinase, histidine protein kinase and 3′-5′ exonuclease activities. Interaction depends on Cys^{145} of NM23A and Cys^{60} of MIF as well as reducing conditions.		Intracellular negative regulator of MIF: (i) Inhibition of MIF-induced proliferation of NIH3T3 cells (ii) Inhibition of MIF-induced MAP kinase activation	in	37

(*Continued*)

Table 1. (*Continued*)

Interactor	Symbol	Source	Method of Identification	Function	K_d	Function of Interaction	Loc.	Ref.
Chemokine (C-X-C motif) receptor 2 (**CXCR2**)	CXCR2	HEK293-CXCR2 transfectant RAW264.7-CXCR2 HEK293-CXCR2 transfectant CD74 transient transfection	Receptor-binding competition with iodinated CXCL8 Pulldown of biotinylated MIF with streptavidin beads Co-IP of CXCR2/CD74	G-protein coupled receptor for chemokine ligands CXCL1 and CXCL8 mediating neutrophil migration to sites of inflammation Also interacts with **CD74**	1.4 nM	CXCR2 internalization. Inhibition of MIF-triggered monocyte arrest after blocking CXCR2. Stimulation of leukocyte chemotaxis. Stimulation of calcium influx in primary human neutrophils. Integrin aLβ2-dependent arrest of MonoMac6 cells on CHO/ICAM-1 cells.	ex	38
Chemokine (C-X-C motif) receptor 4 (CXCR4)	CXCR4	Jurkat	Receptor-binding competition with iodinated CXCL12	Receptor for the chemokine CXCL12/SDF-1 that transduces a signal by increasing intracellular calcium levels and enhanicng activation of MAP kinases	1.5 nM	CXCR4 internalization. Inhibition of MIF-triggered T-cell arrest after blocking CXCR4 in HAoECs. Stimulation of leukocyte chemotaxis.	ex	38

(*Continued*)

Table 1. (*Continued*)

Interactor	Symbol	Source	Method of Identification	Function	K_d	Function of Interaction	Loc.	Ref.
Ribosomal protein S19	RPS19	NIH 3T3	Endogenous Co-IP Endogenous biotin-tagging of MIF *In vitro* GST pulldowns SPR	Structural ribosomal protein of small subunit Extraribosomal: RPS19 dimer functions as extracellular monocyte-selective chemoattractant factor[39]	1.0 μM	RPS19 could work as a native extracellular inhibitor of MIF function (inhibits binding to CD74 and mononuclear cell arrest)	in/ex	6
Heat shock 70-kDa protein 5; glucose-regulated protein 78-kDa (GRP78); BiP	HSPA5	Human bladder homogenate	Endogenous Co-IP	Luminal ER chaperone, also expressed on cell membrane, Ca^{2+}-binding protein[40]		Eliciting an inflammatory response after substance P-induced surface expression of HSPA5	in	41
Thioredoxin	TXN	ATL-2 lysate and supernatant Recombinant proteins	Endogenous Co-IP SPR	Also interacts with **JAB1/CSN5**[20]		Surface thioredoxin augments internalization of MIF into cells	in/ex	42
USO1 vesicle docking protein homolog (yeast) p115	USO1	Human pituitary cDNA library Bait: MIF	Y2H Co-IP with overexpressed proteins *In vitro* competition binding assays	Transport from ER to cis/medial Golgi compartments requires USO1		Mediates secretion of MIF	in/ex	43

functional consequences associated with complex formation on cellular processes. While we have tried to provide a comprehensive list of identified interactors (see Table 1), the MIPs we selected for a more detailed description are focusing on proteins involved in modulating the ubiquitin proteasome system, on RPS19 and on MIPs involved in redox activities. More MIPs are considered in greater detail in other chapters.

2. MIF Affects Degradation of Regulatory Proteins by Modulating the Ubiquitin Proteasome System

Jürgen Bernhagen's group was the first to perform a yeast two-hybrid screen in order to find intracellular MIPs motivated by ample evidence for intracellular functions of MIF.[2–6] From a fetal brain cDNA library, they isolated three types of interacting clones: two were identified as MIF itself and the third as JAB1/CSN5 (see Table 1). MIF self-interaction can be easily explained as it oligomerizes in crystals as well as in solution, mainly forming homotrimers.[7] JAB1/CSN5 was originally discovered as a transcriptional co-activator[8] and identified as subunit 5 of the COP9 signalosome (CSN)[9] two years later. When analyzing the consequences of MIF binding to JAB1/CSN5, Bernhagen's group focused on described co-activator functions of JAB1 and showed that JAB1 and stimulus-enhanced AP1-acitivity, activation of JNK, increase of phospho-cJun levels, and JAB1-dependent cell-cycle regulation through stabilization of $p27^{Kip1}$ were inhibited by MIF.

Within the past decade, JAB1/CSN5 has come more into focus as the fifth subunit of the CSN complex, and it is conceivable now that at least some of its co-activator functions can be explained by functions of the CSN holocomplex.[10] The CSN is a highly conserved protein complex in higher eukaryotes consisting of eight proteins that was originally described as a repressor of light-dependent growth in *Arabidopsis*.[11] Its two subunits, CSN5 and CSN6, possess a MOV34/PAD N-terminal (MPN) domain. The MPN domain of CSN5 contains a JAMM metalloprotease motif that is responsible for the removal of the ubiquitin-like modifier Nedd8 from Nedd8-modified proteins, often referred to as deneddylase activity.[12] Only the CSN holocomplex is deneddylase competent, but not its monomeric CSN5 subunit that is providing the enzymatic activity. A number of neddylated proteins have been identified,[13] with all cullins, the scaffold subunits of cullin RING ligases (CRLs), considered as the most important ones. CRL complexes comprise the largest class of ubiquitin ligases that are activated by neddylation. Although CSN-mediated deneddylation of cullins abolishes CRL activity *in vitro*, deneddylation seems to be required for CRL activity *in vivo*. Therefore,

Petroski and Deshaies suggested that deneddylated cullin-RING protein complexes could be sequestered by cullin-associated and neddylation-dissociated protein-1 (CAND1, also called TIP120A). CAND1 can, in turn, be displaced by neddylation of the cullin, enabling the assembly of SKP1-F box substrate-recognition modules that ubiquitylate the substrate. Ubiquitylation is stopped by displacement of SKP1 and F-box protein by CAND1 again so that another cycle of assembly can be started if required.[14]

In MIF-deficient cells, a large proportion of cullin 1 is deneddylated and remains sequestered by CAND1 due to a more active CSN deneddylase that is unrestricted by binding of MIF to JAB1/CSN5.[15] Moreover, loss of MIF uncouples Chk1/Chk2 cell cycle checkpoints from proteasomal degradation of key cell cycle regulators like Cdc25A, cyclin A2, E2F1, and DP1. Therefore, Nemajerova *et al.* suggested that regulation of JAB1/CSN5 by MIF is required to sustain a CAND1 free CRL pool that can interact with Chk1 instead.[16] Recently, Schmidt *et al.* modified this model, suggesting that in the absence of substrate CRL complexes undergo continuous and rapid adapter exchange in a CAND1 cycle.[17] After substrate-induced neddylation of the cullin, CRLs transition into the CSN cycle, where the CSN deneddyates the CRL once substrate has been forwarded to the proteasome. Following deneddylation the CRLs toggle back into the CAND1 cycle for maintenance and loading with the proper SKP1/F-box module. In this model, MIF interacting with JAB1/CSN5 would lead to a block in the CSN cycle so that specific CRLs with neddylated cullins accumulate leading to an increase of specific ubiquitylated substrates like $p27^{Kip1}$ that are destabilized as a result. Because the CSN controls important cellular regulators, including proto-oncogenes such as c-Jun and β-catenin or the tumor suppressors $p27^{Kip1}$ and p53, known to be frequently malfunctioning in cancer, MIF also impinges on the cell cycle and cancer development as it regulates the CSN.

3. Ribosomal Protein S19 as Endogenous Inhibitor of Extracellular MIF

MIF is an established mediator crucial for immune function and inflammatory-based disease. MIF functions within the cytokine cascade to control the initiation and progression of an inflammatory response, but persistent high levels of MIF increase the likelihood of systemic inflammation or sepsis.[44] However, it is presently not known how MIF activity is controlled or terminated. In a search for endogenous MIF inhibitors, the ribosomal protein S19 (RPS19) was identified by endogenous co-immunoprecipitation with an anti-MIF antibody in NIH 3T3 cells.[45] Interaction of MIF and RPS19 was verified using an undirected *in vivo* biotinylation tagging approach for the identification of MIF binding proteins and surface plasmon resonance. The dissociation constant of the protein complex ($K_D = 1.3 \times 10^{-6}$ M) in the micromolar range is well within the range commonly observed for biologically relevant interactions, such as in intracellular signal transduction cascades. Interestingly, RPS19, which is part of the small subunit of the ribosome, is released from apoptotic cells into the extracellular fluid and was also found in serum.[39,46] Low concentrations of RPS19 (0.5 μM) inhibited the binding of MIF to its receptor CD74 and significantly compromised CXCR2-dependent MIF-initiated adhesion of monocytes to endothelial cells under flow conditions. Recently, the chemokines CXCR2 and CXCR4 have been identified as noncognate receptors for MIF.[47] Taken together with the kinetic data of complex formation,[45] which are strongly indicative of a fast and reversible adjustment of the equilibrium, this indicates that RPS19 serves to limit the bioavailability of extracellular MIF for receptor binding. However, it is tempting to speculate that MIF binding to RPS19 may also influence RPS19 actions. Notably, in apoptotic cells, RPS19 can be released as a dimer cross-linked by a transglutaminase.[46] The RPS19 dimer, like MIF itself, exerts a strong chemotactic stimulus on monocytes by mimicking the complement factor C5a and binding as a ligand to the C5a receptor (CD88).[48] Thus, one may hypothesize that MIF's proinflammatory activity can be controlled by monomeric RPS19, and *vice versa*, the chemotactic function of RPS19 on monocyte dimers can be limited by interaction with MIF. Therefore, RPS19 monomers may counteract excessive levels of MIF at sites of inflammation, thereby decreasing the likelihood of unregulated inflammatory-associated diseases such as septic shock or autoimmune reactions.

MIF mutational analysis employing the C60S mutant, which abolishes the thiol-protein oxidoreductase activity of MIF based on the Cys57-Ala-Leu-Cys60 (CALC) motif,[49] revealed that C60 is required for interaction with RPS19. However, in contrast to other molecules that interact with MIF such as Nm23-H1, insulin, and HPO, binding does not appear to be dependent on redox status or a transitory disulfide linkage with partner proteins in the complex formation, as RPS19 does not harbor any cysteine residues. Thus, RPS19 does not fall in the subgroup of binding partners that link MIF to cellular redox activity.

4. Interactors Involved in MIF Redox Activity

The connection between MIF and cellular redox regulation has been provided by numerous studies. To summarise, the links between MIF and redox activities are based on three lines of evidence, namely (i) MIF's own thiol-protein oxidoreductase activity initially identified by conserved sequence motifs also found in the catalytic center of other redox enzymes, (ii) the involvement of cysteine residues in the physical interaction of MIF with binding partners presumably engaged in the temporary formation of disulfide bonds, and (iii) interaction of MIF with binding partners with an established role in cellular redox reactions. The focus of this chapter is only on the role of MIF interacting partners in MIF's redox regulation. For a comprehensive review on all aspects of the role of MIF in cellular redox reactions, we refer the reader to the work of Thiele and Bernhagen.[50]

Shown to catalyze the reduction of insulin and 2-hydroxyethyldisulfide in a glutathione- and lipoamide-dependent manner,[51] MIF's thiol-protein oxidoreductase (TPOR) activity has been found to be dependent on the presence of the cysteines in the motif Cys57-Ala-Leu-Cys60.[52,53] This CXXC motif is a common feature of all TPOR enzymes and can be found in thioredoxin, glutaredoxin, and protein disulfide isomerase. Similar to the other TPORs, MIF catalyzes the reduction of protein disulfides in a cysteine based dithiol/disulfide reaction. Although not unanimously clarified, at least part of the immunologic activity of MIF has been related to this enzymatic activity. This has been demonstrated by the substitution of either Cys 57 or Cys 60 by serine, which results in the partial or complete inactivation of some immunological actions of MIF, such as macrophage activating properties.[54–57]

Recently, a role for MIF in oxidative stress in cardiac myocytes was suggested. Oxidative stress is important in the pathogenesis of ischemic/reperfusion injury in myocardium, apoptosis, and hypertrophy in cardiomyocytes.[58,59]

Inflammation in cardiovascular disease is associated with increased production and secretion of proinflammatory cytokines such as IL-1, IL-6, and TNF-α, which are released following acute coronary ischemia. Stress and damage is further substantiated by the generation of reactive oxygen species (ROS) during the ischemic period and reperfusion after ischemia.[58,60,61] Consistent with previously published data by Kleemann *et al.*,[51] RAW 264.7 macrophages and COS-1 cells were also able to induce substantial amounts of MIF from isolated rat cardiomyocytes following treatment with H_2O_2. This upregulation could be reversed following the addition of the H_2O_2 degrading enzyme catalase. Intracellularly, the H_2O_2 effect on MIF expression was mediated by protein kinases such as protein kinase C and tyrosine protein kinases, both known to be involved in redox-sensitive signal transduction in the myocardium. Interestingly, TNF-α was ineffective in eliciting an MIF release despite observations made in numerous other cell types. This suggests a role for MIF as a redox-sensitive mediator at least for this cell type.

Protein kinase C activation by H_2O_2 was not dependent on calcium influx as shown by calcium chelation indicating the involvement of an atypical protein kinase C isoform (ζ, λ).[62] This atypical protein kinase C appears to be involved in the regulation of MIF secretion rather than synthesis.[62]

A number of studies have expanded on the role of MIF's participation in cellular redox reactions by modulating redox activities of binding partners via direct association, as detailed below.

4.1. *Insulin*

The majority of published literature to date demonstrates that MIF's thiol-protein oxidoreductase (TPOR) enzymatic activity is predominantly a disulfide reductase.[50,51,63] Although the redox potential of full-length MIF has not been determined, an estimate is provided by the redox potential of a short MIF peptide spanning amino acids 50–65, giving an E'_0 value of −258 mV.[53] More specifically, the reducing activity of MIF is supported by the reduction of insulin disulfides linking the A and B chain with glutathione and dihydrolipoamide, a therapeutic antioxidant with vitamin E- and GSH-like antioxidant activity functioning as co-substrate in this reaction.[51] Furthermore, MIF was also shown to reduce low molecular weight disulfides such as 2-hydroxyethyldisulfide. Nonetheless, in comparison with thioredoxin and glutaredoxin, two other members of the TPOR family, MIF's insulin reducing activity is relatively low, and it remains unclear what biological role this could play. Although MIF has been shown to co-localize with insulin in secretory vesicles of pancreatic β-cells,[64] there is a lack of evidence proving that physical interaction of MIF is involved in the mechanism for insulin secretion. Nonetheless, a general regulatory role of MIF in controlling glucose homeostasis and insulin sensitivity in the development of type I and II diabetes is supported by numerous studies.[65–68]

4.2. *Hepatopoietin (HPO)*

HPO, or growth factor, augmenter of liver regeneration (GFER), was initially identified as a MIF binding protein using the yeast two-hybrid system. The interaction was verified by co-immunoprecipitation of overexpressed proteins in animal cells.[69] HPO is a liver specific hepatic mitogen that stimulates liver regeneration following damage. HPO, which is similar in structure and function to MIF, has also been shown to rescue acute hepatic failure by activating proliferation of hepatoma cells.[70] HPO can mediate its action both extracellularly via a specific surface receptor and intracellularly by stimulating the

AP-1 pathway through Jun activation domain binding protein (JAB1, also termed COP9 signalosome subunit 5, or CSN5). In yeast, the affinity of MIF for HPO was shown to be higher than for JAB1/CSN5, which is also indicated by reduced binding of MIF to JAB1/CSN5 in co-expression assays with increasing amounts of HPO in HepG2 cells.[69] This study further showed that ectopically expressed MIF did not compete for binding of HPO to JAB1/CSN5. HPO, like MIF, is characterized by the conserved CXXC-motif of thiol-protein oxidoreductases and serves as a flavin-linked sulfhydryl oxidase that catalyzes the oxidation of sulfhydryl groups to disulfide bonds in substrate proteins.[63,71] It has been suggested that HPO binding may have an inhibitory effect on MIF's oxidoreductase activity, which is important for interaction with JAB1/CSN5.[69] In contrast, the augmentation of JAB1/CSN5 action on AP-1 by HPO can be decreased by ectopic MIF expression, indicating a modulation of HPO redox activities by MIF. Given previous findings, it appears that MIF and HPO may also compete for binding to specific cell surface receptors, thus modulating the amount of available bioactive ligands.

4.3. *Proliferation-associated gene (PAG) protein*

Like many other MIF interacting proteins, the thiol-specific antioxidant protein PAG was identified by yeast two-hybrid analysis.[72] *In vitro* ectopic expression studies with MIF and PAG resulted in the formation of MIF-PAG complexes in HEK293T cells.[72] A role in cellular redox regulation is indicated by the involvement of disulfide bonds in MIF-PAG binding. Association of the protein complex was substantially impaired using the PAG C173S mutant.

Specifically, the interaction was strongly dependent on redox status. Reducing agents such as DTT and β-mercaptoethanol substantially decreased the amount of co-precipitated MIF in co-expression experiments, whilst the oxidant H_2O_2 had no effect. A direct binding of PAG to the N-terminus of MIF is suggested by the PAG-dependent inhibition of MIF's D-dopachrome tautomerase activity. The N-terminal proline has been established as critical for this enzymatic activity, and modifications of this residue or those surrounding it negatively affect the D-dopachrome tautomerase activity.[55,73] In contrast, the antioxidant activity of PAG is suppressed by binding of MIF. Therefore, MIF can influence the redox status of cells by modifying PAG, which belongs to the constitutively expressed peroxiredoxin family of proteins, comprised of low-efficiency peroxidases that serve as antioxidants in the cell.[74] The peroxidative activity is based on conserved cysteines, not prosthetic groups or heteroatoms. Peroxiredoxins also inhibit bacterial antioxidant defense during infection as part of the innate immune response initiated by the host. The mutual modulation of

MIF and PAG reflects a noteworthy overlap in key cellular functions such as proliferation, differentiation, tumor promotion, and apoptosis. In all of these instances, MIF and redox status may play a major role.[50,75–77]

4.4. *p53*

p53 acts as an important tumor suppressor protein, which is critical for the regulation of normal cell cycle activities. Interestingly, MIF treatment suppressed p53 activity as a transcriptional activator. The observation that MIF, which behaves as a proinflammatory cytokine, is capable of functionally inactivating tumor suppressor p53 may provide a link between inflammation and tumorigenesis.[78,79] Jung *et al.*[36] showed that MIF interacts with p53 in cellular extracts and *in vitro* and that binding requires the presence of cysteine residues in both partners. Mutational analysis demonstrated that C81 of MIF as well as C242 of p53 are essential for complex association. Other cysteine residues (C57 or C60 in MIF, and C238 in p53) show no or limited involvement. Similarly with PAG, Jung and colleagues also demonstrated that MIF-p53 binding is dependent on redox conditions, as the addition of reducing agents (β-mercaptoethanol, DDT) was effective in preventing interaction.[36] Ectopic expression of wild type MIF, but not the C81S mutant, suppressed p53 transcriptional activity as well as p53 mediated apoptosis and cell cycle arrest. This is consistent with other studies that have illustrated MIF to be a potent regulator of p53 dependent signaling.[79–81] Intracellularly, MIF inhibited nuclear transfer of p53 from the cytoplasm, which represents one likely mechanism as to how MIF may suppress activation of p53. Moreover, MIF stabilizes the complex between p53 and its negative regulator Mdm2 in MCF7 and HCT116 cells, which is dependent on the presence of C81 within MIF.[36] Mdm2 can inhibit p53 function in resting cells in two different ways. Mdm2 can either directly associate with p53, preventing p53 transactivation, or alternatively it can indirectly facilitate polyubiquitylation and the subsequent degradation of p53 in the proteasome.[82,83] Overall, these findings in addition to previous studies have important implications for our understanding of the role MIF plays in the association of chronic inflammation and tumor development and progression.[79,81,84]

4.5. *Protein (NM23A) expressed in nonmetastatic cells (NM23-H1)*

Shortly after the discovery of PAG and p53 as MIF interacting partners, Jung and Ha identified Nm23-H1 as another MIP.[37] The Nm23-H1 gene encodes a nucleoside diphosphate kinase-A, implicated in suppression of tumor metastasis in breast cancer. However, previous evidence also suggested that Nm23-H1/NDPK-A protein promotes metastasis in other types of tumors.[85] Beside its role in metastasis, Nm23-H1/NDPK-A is a multifunctional protein involved in proliferation, differentiation, endocytosis, and signal transduction. Similar to PAG, MIF binding to Nm23-H1 is also dependent on cysteine residues and redox conditions. This was demonstrated by dissociation of the protein complex following replacement of cysteine residues Cys145 (Nm23-H1) and Cys60 (MIF), while exchange of Cys4 and Cys109 (Nm23-H1) as well as Cys57 and Cys81 (MIF) had no effect. Although Cys209 is not involved in binding MIF, this residue is essential for the metastatic suppressor activity of Nm23-H1.[86] Like the interaction of MIF with p53 or PAG, the binding of MIF to Nm23-H1 was dependent on the redox status. Using binding assays, the addition of reducing agents such as β-mercaptoethanol and dithiothreitol significantly diminished the amount of bound MIF in co-immunoprecipitation experiments.[37] Expression of Nm23-H1 interfered with MIF mediated cell cycle regulation on multiple levels. Firstly, ectopic expression of Nm23-H1 facilitated the dissociation of MIF from the MIF-p53 complex, thereby inhibiting the suppressive effects of MIF on p53 mediated cell cycle arrest and apoptosis. Furthermore, physical interaction of Nm23-H1 and MIF negatively interfered with MIF-induced activation of mitogen activated kinases such as p44/p42, an extracellular signal regulated protein kinase, and PI3K/PDK1 in 3T3 fibroblasts.[37] Although the exact functions of Nm23 proteins are not well characterized, they are implicated in a broad range of cellular activities that partly overlap with MIF functions. Due to the wide range of overlap of Nm23-H1 and MIF functions, reciprocal effects of MIF on Nm23-H1 functions were difficult to assess.

4.6. *α1-inhibitor-3 (α1-I3)*

The α1-inhibitor-3 (α1-I3) is a 180-kDa protease found in plasma and other bodily fluids. It is a member of the α-macroglobulin proteinase inhibitor family found to be complexed with MIF in the intraluminal fluid of the bladder in rats.[33] Little is known about the function of this interaction; however, consistent with previous findings, the interaction of MIF and α1-I3 is dependent on redox conditions. Reducing agents have been shown to disrupt MIF binding to α1-I3, again suggesting the involvement of disulfide bonds in complex formation.[33] During Substance P induced inflammation, MIF/α1-I3 complexes are released from the bladder into the lumen. Since α1-I3 is a specific rodent protein, these findings cannot be transferred to other species.

5 Concluding Remarks

Table 1 demonstrates that co-immunoprecipitation and the yeast two-hybrid screen are the two methods most often employed to identify MIPs. Surprisingly, beside MIF itself, each report could identify only one single new MIP and none could confirm a previously identified interacting protein. Although this could easily be explained by the use of different cDNA libraries, it is still not easy to reason why MIF as almost ubiquitous bait protein does not consistently associate with other previously identified MIPs, like JAB1/CSN5, p53 or peroxiredoxin 1, that are also ubiquitously expressed.

This was not observed in a screen employing a MIF protein as bait that is equipped with a biotinylation and a calmodulin binding-peptide tag both separated by a cleavage site for the TEV protease, which allows a systematic approach to define the MIF interactome[87] (see Fig. 1). The tagged MIF construct was stably transfected into NIH 3T3 cells and cell lysates affinity-purified on streptavidin agarose.[6] Bound proteins were separated by SDS-PAGE and identified by mass spectrometry. In our initial experiments, we could simultaneously identify numerous new putative MIF interactors including RPS19[6] (see above), in addition to a number of established interactors (e.g., peroxiredoxin 1, thioredoxin 1, and MIF itself, *unpublished data*). However, as the spectrum of MIPs is influenced by the tag selected and binding conditions used, a holistic picture of the MIF interactome is far from being attained.

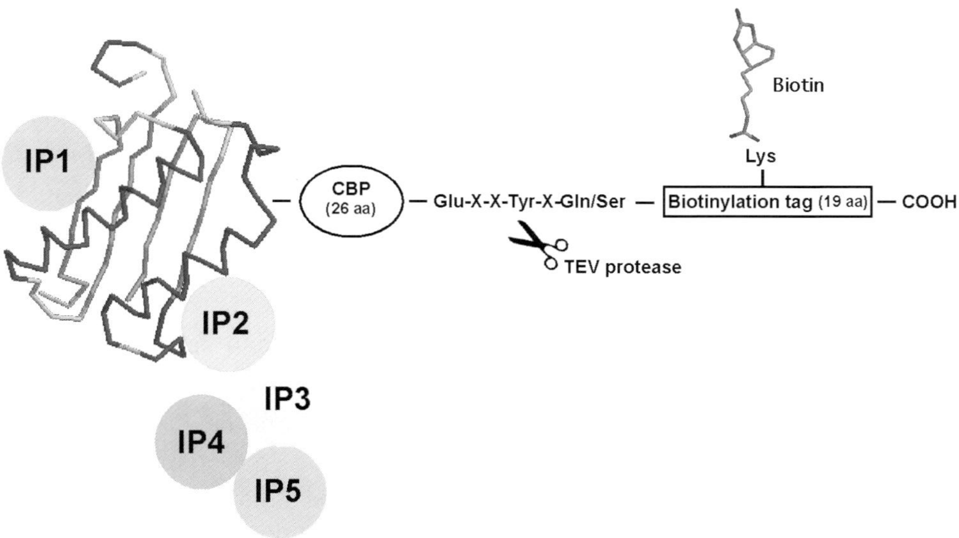

Fig. 1. Sketch of tagged MIF used for affinity purification of directly (IP1 and IP2) or indirectly associated binding proteins (IP3–5). CBP = calmodulin binding-peptide.

Acknowledgments

The authors would like to thank Dr. Tali Lang for critically reading the manuscript and language editing. Their work on MIF is supported by grants from the Deutsche Forschungsgemeinschaft (DFG, Me 1323/2–3 and 2–4), funds from the Universitätsklinikum Giessen-Marburg GmbH, the Deutsche Akademische Austauschdienst (DAAD), and the Von-Behring-Roentgen-Foundation.

References

1. Suzuki M *et al.* (1996) Crystal structure of the macrophage migration inhibitory factor from rat liver. *Nat Struct Biol* **3**(3): 259–266.
2. Kleemann R *et al.* (1998) Disulfide analysis reveals a role for macrophage migration inhibitory factor (MIF) as thiol-protein oxidoreductase. *J Mol Biol* **280**(1): 85–102.
3. Rosengren E *et al.* (1996) The immunoregulatory mediator macrophage migration inhibitory factor (MIF) catalyzes a tautomerization reaction. *Mol Med* **2**(1): 143–149.
4. Bernhagen J *et al.* (1993) MIF is a pituitary-derived cytokine that potentiates lethal endotoxaemia. *Nature* **365**(6448): 756–759.
5. Bernhagen J, Calandra T, Bucala R. (1998) Regulation of the immune response by macrophage migration inhibitory factor: Biological and structural features. *J Mol Med* **76**(3–4): 151–161.
6. Filip AM *et al.* (2009) Ribosomal protein S19 interacts with macrophage migration inhibitory factor and attenuates its pro-inflammatory function. *J Biol Chem* **284**(12): 7977–7985.
7. El-Turk F *et al.* (2008) The conformational flexibility of the carboxy terminal residues 105–114 is a key modulator of the catalytic activity and stability of macrophage migration inhibitory factor. *Biochemistry* **47**(40): 10740–10756.
8. Claret FX, Hibi M, Dhut S *et al.* (1996) A new group of conserved coactivators that increase the specificity of AP-1 transcription factors. *Nature* **383**(6599): 453–457.
9. Seeger M *et al.* (1998) A novel protein complex involved in signal transduction possessing similarities to 26S proteasome subunits. *FASEB J* **12**(6): 469–478.
10. Chamovitz DA. (2009) Revisiting the COP9 signalosome as a transcriptional regulator. *EMBO Rep* **10**(4): 352–358.
11. Wei N, Chamovitz DA, Deng XW. (1994) Arabidopsis COP9 is a component of a novel signaling complex mediating light control of development. *Cell* **78**(1): 117–124.
12. Cope GA *et al.* (2002) Role of predicted metalloprotease motif of Jab1/Csn5 in cleavage of Nedd8 from Cul1. *Science* **298**(5593): 608–611.

13. Rabut G, Peter M. (2008) Function and regulation of protein neddylation. 'Protein Modifications: Beyond the Usual Suspects' Review Series. *EMBO Rep* **9**: 969–976.
14. Petroski MD, Deshaies RJ. (2005) Function and regulation of cullin-RING ubiquitin ligases. *Nat Rev Mol Cell Biol* **6**(1): 9–20.
15. Nemajerova A, Mena P, Fingerle-Rowson G *et al.* (2007) Impaired DNA damage checkpoint response in MIF-deficient mice. *EMBO J* **26**(4): 987–997.
16. Nemajerova A, Moll UM, Petrenko O, Fingerle-Rowson G. (2007) Macrophage migration inhibitory factor coordinates DNA damage response with the proteasomal control of the cell cycle. *Cell Cycle* **6**(9): 1030–1034.
17. Schmidt MW, McQuary PR, Wee S *et al.* (2009) F-box-directed CRL complex assembly and regulation by the CSN and CAND1. *Mol Cell* **35**(5): 586–597.
18. Wei N, Serino G, Deng XW (2008) The COP9 signalosome: More than a protease. *Trends Biochem Sci* **33**(12): 592–600.
19. Tomoda K *et al.* (2002) The cytoplasmic shuttling and subsequent degradation of p27Kip1 mediated by Jab1/CSN5 and the COP9 signalosome complex. *J Biol Chem* **277**(3): 2302–2310.
20. Hwang CY *et al.* (2004) Thioredoxin modulates activator protein 1 (AP-1) activity and p27Kip1 degradation through direct interaction with Jab1. *Oncogene* **23**(55): 8868–8875.
21. Kleemann R *et al.* (2000) Intracellular action of the cytokine MIF to modulate AP-1 activity and the cell cycle through Jab1. *Nature* **408**(6809): 211–216.
22. da Silva Correia J, Miranda Y, Leonard N, Ulevitch RJ. (2007) The subunit CSN6 of the COP9 signalosome is cleaved during apoptosis. *J Biol Chem* **282**(17): 12557–12565.
23. Burger-Kentischer A *et al.* (2005) Binding of JAB1/CSN5 to MIF is mediated by the MPN domain but is independent of the JAMM motif. *FEBS Lett* **579**(7): 1693–1701.
24. Wood ZA, Schroder E, Robin Harris J, Poole LB. (2003) Structure, mechanism and regulation of peroxiredoxins. *Trends Biochem Sci* **28**(1): 32–40.
25. Jung H, Kim T, Chae HZ *et al.* (2001) Regulation of macrophage migration inhibitory factor and thiol-specific antioxidant protein PAG by direct interaction. *J Biol Chem* **276**(18): 15504–15510.
26. Potolicchio I, Santambrogio L, Strominger JL. (2003) Molecular interaction and enzymatic activity of macrophage migration inhibitory factor with immunorelevant peptides. *J Biol Chem* **278**(33): 30889–30895.
27. Shen L *et al.* (2003) The apoptosis-associated protein BNIPL interacts with two cell proliferation-related proteins, MIF and GFER. *FEBS Lett* **540**(1–3): 86–90.
28. Li Y, Lu C, Xing G *et al.* (2004) Macrophage migration inhibitory factor directly interacts with hepatopoietin and regulates the proliferation of hepatoma cell. *Exp Cell Res* **300**(2): 379–387.

29. Leng L *et al.* (2003) MIF signal transduction initiated by binding to CD74. *J Exp Med* **197**(11): 1467–1476.
30. Shi X *et al.* (2006) CD44 is the signaling component of the macrophage migration inhibitory factor-CD74 receptor complex. *Immunity* **25**(4): 595–606.
31. Shen Q, Rigor RR, Pivetti CD *et al.* (2010) Myosin light chain kinase in microvascular endothelial barrier function. *Cardiovasc Res* **87**(2): 272–280.
32. Wadgaonkar R *et al.* (2005) Intracellular interaction of myosin light chain kinase with macrophage migration inhibition factor (MIF) in endothelium. *J Cell Biochem* **95**(4): 849–858.
33. Vera PL, Iczkowski KA, Leng L *et al.* (2005) Macrophage migration inhibitory factor is released as a complex with alpha1-inhibitor-3 in the intraluminal fluid during bladder inflammation in the rat. *J Urol* **174**(1): 338–343.
34. Vera PL, Meyer-Siegler KL. (2006) Substance P induces localization of MIF/alpha1-inhibitor-3 complexes to umbrella cells via paracellular transit through the urothelium in the rat bladder. *BMC Urol* **6**: 24.
35. Liu L *et al.* (2008) Macrophage migration inhibitory factor (MIF) interacts with Bim and inhibits Bim-mediated apoptosis. *Mol Cells* **26**(2): 193–199.
36. Jung H, Seong HA, Ha H. (2008) Critical role of cysteine residue 81 of macrophage migration inhibitory factor (MIF) in MIF-induced inhibition of p53 activity. *J Biol Chem* **283**(29): 20383–20396.
37. Jung H, Seong HA, Ha H (2008) Direct interaction between NM23-H1 and macrophage migration inhibitory factor (MIF) is critical for alleviation of MIF-mediated suppression of p53 activity. *J Biol Chem* **283**(47): 32669–32679.
38. Bernhagen J *et al.* (2007) MIF is a noncognate ligand of CXC chemokine receptors in inflammatory and atherogenic cell recruitment. *Nat Med* **13**(5): 587–596.
39. Semba U *et al.* (2010) A plasma protein indistinguishable from ribosomal protein S19: Conversion to a monocyte chemotactic factor by a factor XIIIa-catalyzed reaction on activated platelet membrane phosphatidylserine in association with blood coagulation. *Am J Pathol* **176**(3): 1542–1551.
40. Lee AS. (2001) The glucose-regulated proteins: Stress induction and clinical applications. *Trends Biochem Sci* **26**(8): 504–510.
41. Vera PL, Wang X, Bucala RJ, Meyer-Siegler KL. (2009) Intraluminal blockade of cell-surface CD74 and glucose regulated protein 78 prevents substance P-induced bladder inflammatory changes in the rat. *PLoS One* **4**(6): e5835.
42. Son A *et al.* (2009) Direct association of thioredoxin-1 (TRX) with macrophage migration inhibitory factor (MIF): Regulatory role of TRX on MIF internalization and signaling. *Antioxid Redox Signal* **11**(10): 2595–2605.
43. Merk M *et al.* (2009) The Golgi-associated protein p115 mediates the secretion of macrophage migration inhibitory factor. *J Immunol* **182**(11): 6896–6906.

44. Baugh JA, Bucala R. (2002) Macrophage migration inhibitory factor. *Crit Care Med* **30**(Suppl 1): S27–S35.
45. Filip AM *et al.* (2009) Ribosomal protein S19 interacts with macrophage migration inhibitory factor and attenuates its pro-inflammatory function. *J Biol Chem* **284**(12): 7977–7985.
46. Nishiura H *et al.* (1996) Monocyte chemotactic factor in rheumatoid arthritis synovial tissue. Probably a cross-linked derivative of S19 ribosomal protein. *J Biol Chem* **271**(2): 878–882.
47. Bernhagen J *et al.* (2007) MIF is a noncognate ligand of CXC chemokine receptors in inflammatory and atherogenic cell recruitment. *Nat Med* **13**(5): 587–596.
48. Nishiura H, Shibuya Y, Yamamoto T. (1998) S19 ribosomal protein cross-linked dimer causes monocyte-predominant infiltration by means of molecular mimicry to complement C5a. *Lab Invest* **78**(12): 1615–1623.
49. Kleemann R *et al.* (2000) Dissection of the enzymatic and immunologic functions of macrophage migration inhibitory factor. Full immunologic activity of N-terminally truncated mutants. *Eur J Biochem* **267**(24): 7183–7193.
50. Thiele M, Bernhagen J. (2005) Link between macrophage migration inhibitory factor and cellular redox regulation. *Antioxid Redox Signal* **7**(9–10): 1234–1248.
51. Kleemann R, Mischke R, Kapurniotu A *et al.* (1998) Specific reduction of insulin disulfides by macrophage migration inhibitory factor (MIF) with glutathione and dihydrolipoamide: Potential role in cellular redox processes. *FEBS Lett* **430**(3): 191–196.
52. Kleemann R, Kapurniotu A, Mischke R *et al.* (1999) Characterization of catalytic center mutants of macrophage migration inhibitory factor (MIF) and comparison to Cys81Ser MIF. *Eur J Biochem* **261**(3): 753–766.
53. Nguyen MT *et al.* (2003) A sixteen residue peptide fragment of macrophage migration inhibitory factor, MIF(50–65), exhibits redox activity and has MIF-like biological functions. *J Biol Chem* **278**: 33654–33671.
54. Swope M, Sun HW, Blake PR, Lolis E. (1998) Direct link between cytokine activity and a catalytic site for macrophage migration inhibitory factor. *EMBO J* **17**(13): 3534–3541.
55. Bendrat K *et al.* (1997) Biochemical and mutational investigations of the enzymatic activity of macrophage migration inhibitory factor. *Biochemistry* **36**(49): 15356–15362.
56. Hermanowski-Vosatka A *et al.* (1999) Enzymatically inactive macrophage migration inhibitory factor inhibits monocyte chemotaxis and random migration. *Biochemistry* **38**(39): 12841–12849.
57. Jung H, Kim T, Chae HZ *et al.* (2001) Regulation of macrophage migration inhibitory factor and thiol-specific antioxidant protein PAG by direct interaction. *J Biol Chem* **276**(18): 15504–15510.

58. Aikawa R *et al.* (2002) Reactive oxygen species induce cardiomyocyte apoptosis partly through TNF-alpha. *Cytokine* **18**(4): 179–183.
59. Takano H, Hasegawa H, Nagai T, Komuro I. (2003) Implication of cardiac remodeling in heart failure: Mechanisms and therapeutic strategies. *Intern Med* **42**(6): 465–469.
60. Takano H *et al.* (2003) Oxidative stress-induced signal transduction pathways in cardiac myocytes: Involvement of ROS in heart diseases. *Antioxid Redox Signal* **5**(6): 789–794.
61. Aikawa R *et al.* (2001) Reactive oxygen species in mechanical stress-induced cardiac hypertrophy. *Biochem Biophys Res Commun* **289**(4): 901–907.
62. Fukuzawa J *et al.* (2002) Contribution of macrophage migration inhibitory factor to extracellular signal-regulated kinase activation by oxidative stress in cardiomyocytes. *J Biol Chem* **277**(28): 24889–24895.
63. Kleemann R *et al.* (1998) Disulfide analysis reveals a role for macrophage migration inhibitory factor (MIF) as thiol-protein oxidoreductase. *J Mol Biol* **280**(1): 85–102.
64. Waeber G *et al.* (1997) Insulin secretion is regulated by the glucose-dependent production of islet beta cell macrophage migration inhibitory factor. *Proc Natl Acad Sci USA* **94**(9): 4782–4787.
65. Verschuren L *et al.* (2009) MIF deficiency reduces chronic inflammation in white adipose tissue and impairs the development of insulin resistance, glucose intolerance, and associated atherosclerotic disease. *Circ Res* **105**(1): 99–107.
66. Serre-Beinier V *et al.* (2010) Macrophage migration inhibitory factor deficiency leads to age-dependent impairment of glucose homeostasis in mice. *J Endocrinol* **206**(3): 297–306.
67. Toso C, Emamaullee JA, Merani S, Shapiro AM. (2008) The role of macrophage migration inhibitory factor on glucose metabolism and diabetes. *Diabetologia* **51**(11): 1937–1946.
68. Waeber G *et al.* (1997) Insulin secretion is regulated by the glucose-dependent production of islet beta cell macrophage migration inhibitory factor. *Proc Natl Acad Sci USA* **94**(9): 4782–4787.
69. Li Y, Lu C, Xing G *et al.* (2004) Macrophage migration inhibitory factor directly interacts with hepatopoietin and regulates the proliferation of hepatoma cell. *Exp Cell Res* **300**(2): 379–387.
70. Gatzidou E, Kouraklis G, Theocharis S. (2006) Insights on augmenter of liver regeneration cloning and function. *World J Gastroenterol* **12**(31): 4951–4958.
71. Lisowsky T, Lee JE, Polimeno L *et al.* (2001) Mammalian augmenter of liver regeneration protein is a sulfhydryl oxidase. *Dig Liver Dis* **33**(2): 173–180.
72. Jung H, Kim T, Chae HZ *et al.* (2001) Regulation of macrophage migration inhibitory factor and thiol-specific antioxidant protein PAG by direct interaction. *J Biol Chem* **276**(18): 15504–15510.

73. Nishihira J *et al.* (1998) Molecular cloning of human D-dopachrome tautomerase cDNA: N-terminal proline is essential for enzyme activation. *Biochem Biophys Res Commun* **243**(2): 538–544.
74. Robinson MW, Hutchinson AT, Dalton JP, Donnelly S. (2010) Peroxiredoxin: A central player in immune modulation. *Parasite Immunol* **32**(5): 305–313.
75. Bach JP *et al.* (2009) The role of macrophage inhibitory factor in tumorigenesis and central nervous system tumors. *Cancer* **115**(10): 2031–2040.
76. Lolis E, Bucala R. (2003) Macrophage migration inhibitory factor. *Expert Opin Ther Targets* **7**(2): 153–164.
77. Kudrin A, Ray D. (2008) Cunning factor: Macrophage migration inhibitory factor as a redox-regulated target. *Immunol Cell Biol* **86**(3): 232–238.
78. Bucala R, Donnelly SC. (2007) Macrophage migration inhibitory factor: A probable link between inflammation and cancer. *Immunity* **26**(3): 281–285.
79. Hudson JD *et al.* (1999) A proinflammatory cytokine inhibits p53 tumor suppressor activity. *J Exp Med* **190**(10): 1375–1382.
80. Petrenko O, Fingerle-Rowson G, Peng T *et al.* (2003) Macrophage migration inhibitory factor deficiency is associated with altered cell growth and reduced susceptibility to Ras-mediated transformation. *J Biol Chem* **278**(13): 11078–11085.
81. Mitchell RA *et al.* (2002) Macrophage migration inhibitory factor (MIF) sustains macrophage proinflammatory function by inhibiting p53: Regulatory role in the innate immune response. *Proc Natl Acad Sci USA* **99**(1): 345–350.
82. Haupt Y, Maya R, Kazaz A, Oren M. (1997) Mdm2 promotes the rapid degradation of p53. *Nature* **387**(6630): 296–299.
83. Levine AJ. (1997) p53, the cellular gatekeeper for growth and division. *Cell* **88**(3): 323–331.
84. Fingerle-Rowson G *et al.* (2003) The p53-dependent effects of macrophage migration inhibitory factor revealed by gene targeting. *Proc Natl Acad Sci USA* **100**(16): 9354–9359.
85. Almgren MA, Henriksson KC, Fujimoto J, Chang CL. (2004) Nucleoside diphosphate kinase A/nm23-H1 promotes metastasis of NB69-derived human neuroblastoma. *Mol Cancer Res* **2**(7): 387–394.
86. Lee E *et al.* (2009) Multiple functions of Nm23-H1 are regulated by oxido-reduction system. *PLoS One* **4**(11): e7949.
87. Rodriguez P *et al.* (2006) Isolation of transcription factor complexes by *in vivo* biotinylation tagging and direct binding to streptavidin beads. *Method Mol Biol* **338**: 305–323.

I-5

Structural Studies of Small Molecule Inhibitors of MIF

Yoonsang Cho* and Elias J. Lolis*,†

1. Introduction

For almost the past two decades, macrophage migration inhibitory factor (MIF) has been extensively studied, yet its mechanism of action is not fully understood. MIF is associated with a number of diseases but most prominently with those that involve inflammation, the innate immune response, autoimmune diseases, and cancer. A number of methodologies such as anti-MIF antibodies, knockout mice, and small molecule inhibitors have been successful in demonstrating the importance of MIF in these diseases. For example, initial experiments with polyclonal antibodies resulted in survival in mice injected with lipopolysaccharide (LPS) relative to control mice that were injected with normal rabbit serum.[1] Similar experiments using knockout mice,[2] monoclonal antibodies,[3–5] and small molecule inhibitors[6,7] demonstrate the importance of MIF in various inflammatory diseases. Similar to cytokines, chemokines, and other secreted agents, the effect of MIF on specific infectious agents is microbe-dependent, having both an appropriate innate response[8,9] and an inappropriate response that is damaging to the host.[2] A few preclinical studies have been performed on mouse models of sepsis,[6] lupus,[10] and cancer[11–13] with the small molecule inhibitor (S,R)-3-(4-hydrophenyl)-4,5-dihydro-5-isoxazole acetic acid methyl ester (ISO-1). In all published cases, the mice treated with ISO-1 produced statistically significant effects relative to control mice.

Various receptors for MIF have been identified, leading to several mechanisms for different biological activities.[14,15] For example, the interaction of

*Department of Pharmacology, Yale University School of Medicine, Sterling Hall of Medicine B345, 333 Cedar Street, New Haven, CT 06520, USA.
†Corresponding author: Email: elias.lolis@yale.edu

MIF and CD74 leads to the recruitment of CD44 at the cell surface to form a trimeric MIF/CD74/CD44 complex inducing phosphorylation of the cytoplasmic domains of both CD74 and CD44 leading to ERK-1/2 activation in an Srk-dependent manner.[16] MIF is also a noncognate agonist of CXCR4 and CXCR2, with considerable effects on atherosclerosis and plaque formation.[14]

Of value with respect to MIF-mediated disease are genetic polymorphisms that are important in the severity of various diseases. The first identification of a polymorphism was an SNP (G-to-C) mutation at −173 of the *MIF* gene, creating a binding site for activator protein 4 transcription factor and leading to an increased risk of systemic-onset juvenile idiopathic arthritis (JIA) compared to a control population.[17] The second identified polymorphism involved four alleles comprising 5-, 6-, 7- or 8-CATT tetranucleotide repeats at position −794 of the *MIF* gene. These different amounts of the tetranucleotide affect the promoter activity for MIF. The 5-repeat of CATT-tetranucleotide is associated with low expression of *MIF*, with increased repeats associated with higher *MIF* expression. In general, the 5-repeat tetranucleotide allele also correlates with the low severity of various inflammatory diseases (e.g., rheumatoid arthritis[18]), although there are also a few studies that dispute the correlation with arthritis but note higher MIF levels with higher numbers of CATT repeats associated with greater joint damage.[19] If these polymorphisms are confirmed in disease, as appears to be case by the preponderance of studies, they will be equally important in future clinical studies to stratify patients receiving a small molecule inhibitor of MIF.

Given the importance of MIF as a drug target, it is relevant to consider the unique properties of MIF for drug discovery, inhibitory mechanisms, and potential toxicity. First, it appears that all cells constitutively express MIF. This is in contrast to most cytokines and chemokines that are expressed and secreted when cells are activated. Second, the MIF cDNA does not contain a signal sequence. MIF is present in the cytosol and is exported from activated cells by a p115-mediated process.[20] The third unique property for this cytokine is the existence of a small molecule binding pocket similar to the enzymatic site of two microbial isomerases.[21,22] A known, *bona fide* physio-logical substrate has yet to be identified, but two pseudo-substrates, D-dopachrome and hydroxyphenylpyruvate (HPP), are used to measure tautomerase activity.[23,24] The relationship among the small-molecule binding site, potential enzymatic activity, the binding of MIF to its receptors, and the contribution of each of these in disease is still under investigation. While ISO-1 and

(*E*)-4-hydroxybenzaldehyde-*O*-cyclohexanecarbonyloxime (OXIM-11), one of the inhibitors described in this chapter, have been used in preclinical studies of inflammatory, infectious, autoimmune, vascular diseases, and cancer,[6,7,10,13,25,26] many of the compounds are being used to decipher the relationships of these unique properties to gain a better understanding of the mechanism of action of MIF.

2. Small Molecule Structures

Crystal structures of wildtype and mutant MIF have provided molecular insight into the protein and contributed to the development of potent small molecule inhibitors.[27–29] MIF is a 12-kDa protein that forms a three-fold trimer. The crystal structure reveals three binding sites for a substrate at the three interfaces formed by residues from two neighboring subunits within the trimer.[30] The complex between MIF and substrate, HPP, defined the substrate binding sites, the interacting residues, and a potential mechanism for enzymatic activity using Pro-1 as a catalytic base.[31] Each active site is about $11 \times 4 \times 5.5$ in Ångströms. The three-dimensional structure of MIF-HPP allowed medicinal chemists to develop various competitive inhibitors, including the widely used ISO-1. The emergence of ISO-1 ignited the discovery of more potent tautomerase inhibitors as well as biologically inhibitory small molecules against MIF (Table 1). The coordinates for currently available MIF-inhibitor complexes are listed in Table 2.

Not all the inhibitors have similar effects on both the enzymatic and biological activity of MIF. One of the inhibitors identified from virtual docking is N-benzyl-benzoxazol-2-ones. This compound inhibited both HPP tautomerization and CD74 binding to similar extent, with IC_{50}s of 0.5 μM and 1.5 μM, respectively.[32] In contrast, one of the chemical analogues, compd5, had a tautomerase inhibition IC_{50} of 0.01 μM but a maximum inhibition of 40% for MIF-CD74 binding. This indicates the inhibition of enzymatic activity does not directly correlate to CD74 binding. The partial inhibitory activity on receptor binding reveals the region in or around the active site has an important role in receptor binding, but the catalytic activity does not. Unfortunately, these two inhibitors have yet to be co-crystallized with MIF.

Most of the current inhibitors bind to the active site although the hydrophobic surface consisting of Y36 and W108, and F113 at the rim of the active site was also recognized as a potential binding site.[33] These hydrophobic residues have recently been discovered as part of a novel allosteric

Table 1. MIF inhibitors.

Inhibited MIF activity	Interaction mode	Inhibition type	Inhibitor	ID method	IC_{50} (μM) or max percentage inhibition[b]	Affected MIF-mediated activity	Ref.
Catalytic	Noncovalent	Competitive	D-dopachrome analogues		900–8,000[a]	Tautomerase	22
			Ketone bodies		0.1–476.2	Ketonase	43
			ISO-1 analogues		0.55–26	Tautomerase	36
			Unsaturated cyclic ketones		0.79–127.42	Tautomerase	44
			Quinolinones	VS[c]		Tautomerase	34
			Various	VS	1.04–64.2	Tautomerase	45
	Covalent		2-OPB			Tautomerase	46
			Hydroxyquinoline analogues			Tautomerase	39
Biological	Noncovalent	Competitive	ISO-1		50%	Arachidonic acid release	29
					33%	Guillain–Barré syndrome, experimental allergic neuritis	47
					75%	Experimental autoimmune diabetes	48
					85%	LPS-induced sepsis	6
					90%	Prostate cancer growth	13
					30%	Cardiocirculatory depression	49

(*Continued*)

Table 1. *(Continued)*

Inhibited MIF activity	Interaction mode	Inhibition type	Inhibitor	ID method	IC_{50} (μM) or max percentage inhibition[b]	Affected MIF-mediated activity	Ref.
					73–90%	TLR-4 signaling	50
					30%	Cleavage of caspase 3, reduction of Bcl-xL/Bax	51
					30%	Neutrophil accumulation	52
					37–43%	Glucocorticoid overriding	53
					25%	Colorectal carcinoma development	12
					43–70%	Angiogenic factor production	54
					44%	B cell maturation	55
					48%	Glucose-induced apoptosis	56
					50–80%	NF-κB activation, IL-6, IL-8, and TNF-α production	57
					66–68%	Cyclophosphamide cystitis	58
					86%	Amyloid β-dependent neurotoxicity	59
					30–80%	Cytokine secretion, MMP expression	60
					50%	Pain behavior	61

(Continued)

Table 1. (*Continued*)

Inhibited MIF activity	Interaction mode	Inhibition type	Inhibitor	ID method	IC_{50} (μM) or max percentage inhibition[b]	Affected MIF-mediated activity	Ref.
					45–80%	Tumor growth	62, 63
					60%	Cell permeability	25
			Imine conjugates	Rational design	1.65	ERK activation, apoptosis inhibition	64
			Phenolic hydrazones	Rational design	~84%	TNF-α secretion, sepsis	65
			OXIM-11	OXIM analog	80–90%	Glucocorticoid overriding, sepsis	7
			ISO-F	Docking	70%	DSS-induced colitis	37
			Benzoxazol-2-ones		0.0075	CD74, ERK activation	32
			CPSI-1306		45%	Experimental autoimmune encephalomyelitis (EAE)	66
			Furans		0–100%	MIF secretion	67
			Indoles		10–100%	Proinflammatory cytokine induction	67

(*Continued*)

Table 1. (*Continued*)

Inhibited MIF activity	Interaction mode	Inhibition type	Inhibitor	ID method	IC_{50} (μM) or max percentage inhibition[b]	Affected MIF-mediated activity	Ref.
		Noncompetitive	Ibudilast	Phosphodiesterase inhibitor	0.1	PBMC migration	35
	Covalent		4-IPP	VS	50–75%	Lung adenocarcinoma cell migration, anchorage-independent growth	40
					~80%	p115-dependent secretion	20
					60%	Keratitis	68
			Novel class	Isothiocyanate analogues	29–100%	Glucocorticoid overriding, AKT phosphorylation, MIF trimerization	69

[a]Inhibition constant of tautomerase activity
[b]Range of percent inhibition at various MIF concentrations
[c]Virtual screen

Table 2. MIF-inhibitor complex crystal structures.

Compound (molecule ID)	Source	Binding site	Resolution (Å)	PDB ID	Ref.	Selection method
(E)-2-Fluoro-p-hydroxycinnamate (FHC)	Murine	Active	1.8	1MFI	70	Enol analog
Chromene (YZ9)	Human	Active	1.9	1GCZ	33	Docking
ISO-1 (HDI)	Human	Active	2.0	1LJT	29	Isoxazoline (rational design based on HPP)
OXIM-6 (OX3)	Human	Active	1.85	2OOH	7	Indole mimicry
OXIM-11 (OX5)	Human	Active	1.80	2OOZ		
Fluoro-OXIM-11 (OX4)	Human	Active	1.75	2OOW		
4-IPP (RW1)	Human	Covalent	1.8	3B9S	40	Virtual screen
PMSM (PMS)	Human	Covalent	1.55	3CE4	38	Serendipity
Bi-APAP (TYL)	Human	Partial active	1.95	3DJI		Dopachrome methyl ester structural analogue
Hydroxyquinoline (XV1)	Human	Covalent	1.86	3JSF	39	Virtual screen
Hydroxyquinoline (OIN)	Human	Covalent	1.58	3JSG		
Hydroxyquinoline 708 (ZIN)	Human	Covalent	1.86	3JTU		
Benzothiazole (ZEC)	Human	Active	1.90	3L5U	34	Virtual screen
Phenylchromenone (47X)	Human	Active	1.94	3L5R		
Imidazopyrimidinylphenyl (88X)	Human	Allosteric	1.86	3L5S		
Imidazopyridazinol (428)	Human	Allosteric	1.80	3L5P		
Thiophenepiperazinylquinolinone (956)	Human	Mixed	1.86	3L5T		
AV1013 (AVR)	Human	Allosteric	1.25	3IJG	35	Repositioning

binding site.[34,35] In this chapter, we will classify the currently available MIF small molecule ligands and their catalytically and biologically inhibitory activities.

3. Catalytic Site Inhibitors

MIF inhibitors that compete with MIF substrates such as dopachrome and HPP bind deep within the active site (Fig. 1). Inhibitors such as ISO-1 and OXIM-11 form hydrogen bonds with Asn97, which is deep in the active site pocket (Fig. 1, Table 3). Hydrogen bonds, as well as hydrophobic contacts, are made with Tyr95 from the neighboring subunit. Three inhibitors, ISO-1, (E)-2-fluoro-p-hydroxycinnamate, and bi-APAP, possess a carboxyl oxygen at the entry of the active site that interacts with Lys32 (Table 3).

The most extensively studied competitive MIF inhibitor, ISO-1, has a phenylimine scaffold and is stable in aqueous solution.[29] Following the co-crystal structure of ISO-1 and MIF, structural analogues were synthesized to improve the inhibitory potency for the catalytic and biological activities. One of the ISO-1 analogues, compound 17, revealed 20-fold

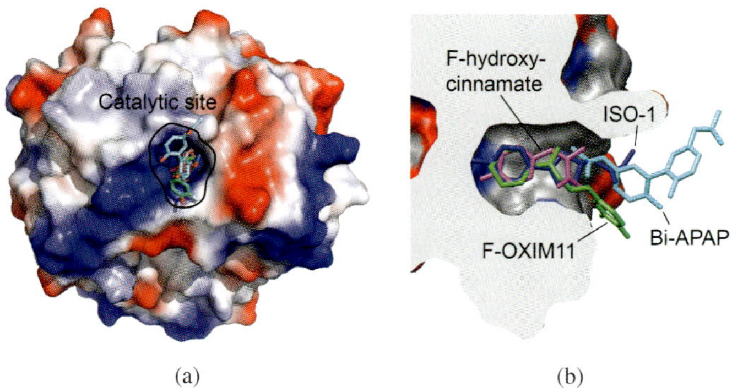

Fig. 1. Interactions of inhibitors with MIF. (a) Global view of catalytic site of the trimer superimposed with four inhibitors: ISO-1,[29] fluoro-OXIM-11,[7] fluoro-hydroxy-cinnamate,[70] and bi-APAP.[38] The electrostatic potential of the protein is colored red for negative, blue for positive, and white for hydrophobic. (b) A cutout section at approximately 90° rotation relative to (a) showing the position of all four inhibitors. Notice that bi-APAP is an acetaminophen dimer that does not enter into the active site.

increased inhibition in tautomerase inhibition assay.[36] A fluorinated ISO-1 analog (ISO-F) inhibits dopachrome tautomerase activity 10-fold greater than ISO-1 and also inhibits MIF-mediated murine colitis.[37] Derivatives of non-physiological MIF substrates with novel scaffolds have also been synthesized (e.g., indole-3-acrylic acid and 2,3-indolinedione) with single-digit mM inhibition constants.[22]

4. Covalent Inhibitors

Covalently modified MIF inhibitors have been discovered by either serendipity[38] or virtual docking of small molecule libraries.[39,40] The secondary amine of Pro-1 has a pKa of 5.6–6.0,[41,42] is unprotonated at physiological pH, and functions as a catalytic base during chemical reactions involving HPP and dopachrome or as a nucleophile making a covalent adduct in the MIF-phenylpyrimidine structure with 4-iodo-6-phenylpyrimidine (4-IPP) generating iodide as a leaving group.[40] Covalent adducts with Pro-1 inhibit tautomerase activity by continuously preventing access of a substrate to the active site. Secondly, in the event that a (physiologic) substrate could get into the active site in the presence of a covalent inhibitor, the lone pair electrons of Pro-1 necessary for enzymatic activity are now part of the covalent bond with the inhibitor and unable to serve as a catalytic base. As inhibitors of the active site, a covalent inhibitor such as 4-IPP also affects MIF-mediated lung adenocarcinoma cell migration and anchorage-independent growth.[40]

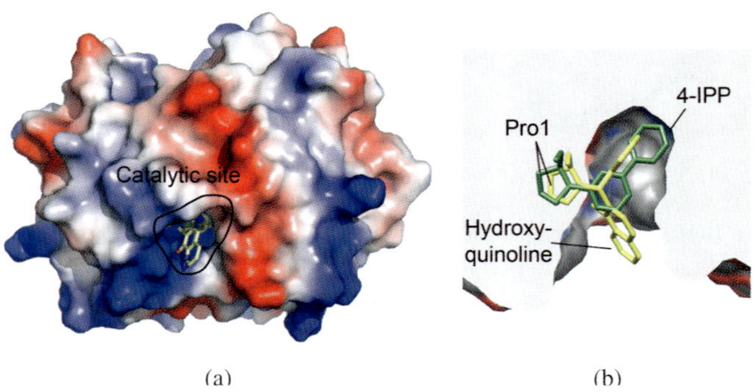

Fig. 2. Interactions of MIF with covalent inhibitors. (a) Global view of the catalytic site. (b) A cutout view of the Pro-1 covalently bound to hydroxyquinoline[39] and 4-IPP.[40]

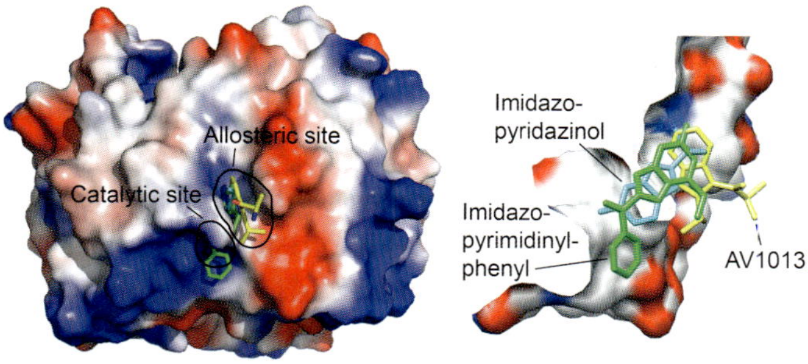

(a) (b)

Fig. 3. The relative orientation of the catalytic and allosteric sites. (a) Global view of the catalytic and allosteric sites. (b) The positions of AV1013,[35] imidazopyridazinol,[34] and imidazopyrimidinylphenyl[34] in the allosteric site.

5. Allosteric and Mixed Inhibitors

Noncompetitive (or allosteric) inhibitors were reported by two groups simultaneously (Fig. 3). McLean *et al.* crystallized inhibitors identified by virtual screening against MIF, which bound at a novel hydrophobic binding site formed by residues Y36, W108, and F113.[34] In another study, Cho *et al.* performed an MIF tautomerase inhibition assay with the phosphodiesterase (PDE) inhibitor, ibudilast(3-isobutyryl-2-isopropylpyrazolo-[1,5-a] pyridine), a drug used for treating asthma and post-stroke complications in Japan for two decades.[35] Ibudilast is currently in clinical trials for a number of central nervous system disorders, including neuropathic pain and multiple sclerosis, but its therapeutic function was not fully explained by inhibition of phosphodiesterases, as a chemically similar analogue, 2-amino-1-(2-isopropylpyrazolo[1,5-a]pyridin-3-yl)propan-1-one, retained therapeutic activity with decreased PDE inhibition. Ibudilast exhibited noncompetitive inhibition of MIF that was explained by its binding site, as determined by NMR and X-ray crystallography. The allosteric site is formed by residues Y36, W108, and F113, and the allosteric inhibitors form aromatic stacking interactions between W108 and the rotated phenol ring of Y36. These residues were identified as a hydrophobic surface adjacent to the active site that could be used to extend the size (and hydrophobicity) of an inhibitor to increase affinity. The difference between the two is described by Orita *et al.*[33] as a *static* hydrophobic site at the surface of the protein while McLean *et al.*[34] and Cho *et al.*[35] describe a *dynamic* set of residues that create a site for compounds (or moieties) to insert and induce allosteric effects at the active site. There is one inhibitor that binds both to the active and allosteric sites. The thiophenepiperinyl group of thiophenepiperazinylquinolinone binds to the active site, while the quinolinone group binds to the allosteric site with a very short linker involving atoms from both the piperinyl and quinolinone groups (Fig. 4).[34]

Most of the competitive, covalent, and allosteric interactions discussed in this chapter have been documented by X-ray structures in complex with MIF and illustrate the plasticity of this particular protein in binding small molecules in these different categories, with a large number of residues from both subunits (Fig. 5, Table 3). This bodes well for developing potent therapeutics and using the entire repertoire of inhibitors to study the mechanism of action and receptor binding of MIF.

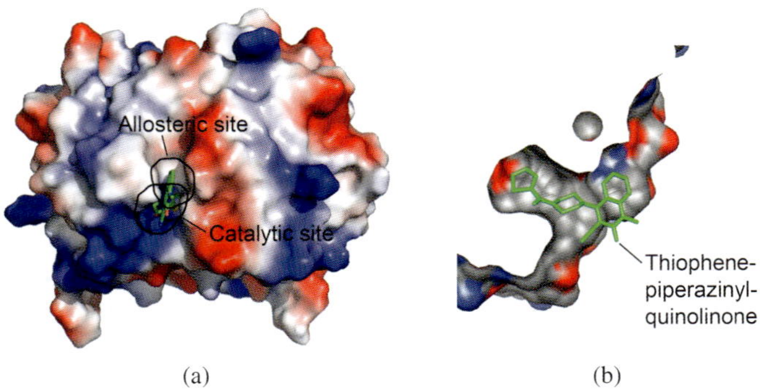

Fig. 4. Mixed inhibition. (a) Moieties of thiophenepiperazinylquinolinone binding to active and allosteric sites. (b) A cutout section at approximately 90° rotation relative to (a) showing the positions of the inhibitor.

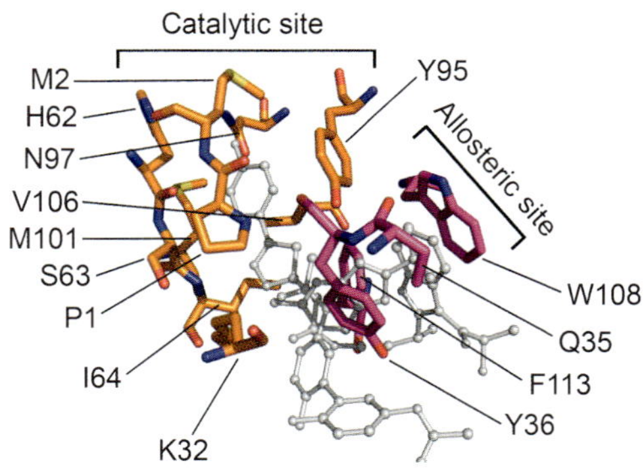

Fig. 5. The residues involved in both the catalytic (orange carbon atoms) and allosteric (purple carbon atoms) sites with catalytic and allosteric inhibitors (white).

Table 3. Residues interacting with inhibitors and types of interactions.

Residue	FHC	YZ9	HDI	OX3	OX5	RW1	PMS	TYL	XV1	428	AVR
P1	○	○	○	○		•	•		•		
M2	○	○	○	○	○	○	○		○		
K32	•	•	•	•				•	•		
Q35										○	
Y36			○	○	○			○	○	○	○
H62	○	○	○	○	○	○	○				
S63	○	○	○	○	○		○				
I64	○	•	○	•	○		•		•		
Y95	○	○	○	•	○	○	○		○		
N97			•	○	•	○					
M101			○	○							
V106	○	○		○	○	○	○		○		
W108										○	○
F113		○	○	○	○			○	○	○	○
(GOL)										•	

• = hydrogen bonding; ○ = hydrophobic interaction; GOL = glycerol

References

1. Bernhagen J, Calandra T, Mitchell RA *et al.* (1993) MIF is a pituitary-derived cytokine that potentiates lethal endotoxaemia. *Nature* **365**: 756–759.
2. Bozza M, Satoskar AR, Lin G *et al.* (1999) Targeted disruption of migration inhibitory factor gene reveals its critical role in sepsis. *J Exp Med* **189**: 341–346.
3. Lan HY, Bacher M, Yang N *et al.* (1997) The pathogenic role of macrophage migration inhibitory factor in immunologically induced kidney disease in the rat. *J Exp Med* **185**: 1455–1465.
4. Leech M, Metz C, Santos L *et al.* (1998) Involvement of macrophage migration inhibitory factor in the evolution of rat adjuvant arthritis. *Arthritis Rheum* **41**: 910–917.
5. Burger-Kentischer A, Gobel H, Kleemann R *et al.* (2006) Reduction of the aortic inflammatory response in spontaneous atherosclerosis by blockade of macrophage migration inhibitory factor (MIF). *Atherosclerosis* **184**: 28–38.
6. Al-Abed Y, Dabideen D, Aljabari B *et al.* (2005) ISO-1 binding to the tautomerase active site of MIF inhibits its pro-inflammatory activity and increases survival in severe sepsis. *J Biol Chem* **280**: 36541–36544.

7. Crichlow GV, Cheng KF, Dabideen D *et al.* (2007) Alternative chemical modifications reverse the binding orientation of a pharmacophore scaffold in the active site of macrophage migration inhibitory factor. *J Biol Chem* **282**: 23089–23095.
8. Koebernick H, Grode L, David JR *et al.* (2002) Macrophage migration inhibitory factor (MIF) plays a pivotal role in immunity against *Salmonella typhimurium*. *Proc Natl Acad Sci USA* **99**: 13681–13686.
9. Satoskar AR, Bozza M, Rodriguez-Sosa M *et al.* (2001) Migration-inhibitory factor gene-deficient mice are susceptible to cutaneous *Leishmania major* infection. *Infect Immun* **69**: 906–911.
10. Leng L, Chen L, Fan J *et al.* (2011) A small-molecule macrophage migration inhibitory factor antagonist protects against glomerulonephritis in lupus-prone NZB/NZW F1 and MRL/lpr mice. *J Immunol* **186**: 527–538.
11. Dessein AF, Stechly L, Jonckheere N *et al.* (2011) Autocrine induction of invasive and metastatic phenotypes by the MIF-CXCR4 axis in drug-resistant human colon cancer cells. *Cancer Res* **70**: 4644–4654.
12. He X-X, Chen K, Yang J *et al.* (2009) Macrophage migration inhibitory factor promotes colorectal cancer. *Mol Med* **15**: 1–10.
13. Meyer-Siegler KL, Iczkowski KA, Leng L *et al.* (2006) Inhibition of macrophage migration inhibitory factor or its receptor (CD74) attenuates growth and invasion of DU-145 prostate cancer cells. *J Immunol* **177**: 8730–8739.
14. Bernhagen J, Krohn R, Lue H *et al.* (2007) MIF is a noncognate ligand of CXC chemokine receptors in inflammatory and atherogenic cell recruitment. *Nat Med* **13**: 587–596.
15. Leng L, Metz CN, Fang Y *et al.* (2003) MIF signal transduction initiated by binding to CD74. *J Exp Med* **197**: 1467–1476.
16. Shi X, Leng L, Wang T *et al.* (2006) CD44 is the signaling component of the macrophage migration inhibitory factor-CD74 receptor complex. *Immunity* **25**: 595–606.
17. Donn RP, Shelley E, Ollier WE, Thomson W. (2001) A novel 5′-flanking region polymorphism of macrophage migration inhibitory factor is associated with systemic-onset juvenile idiopathic arthritis. *Arthritis Rheum* **44**: 1782–1785.
18. Baugh JA, Chitnis S, Donnelly SC *et al.* (2002) A functional promoter polymorphism in the macrophage migration inhibitory factor (MIF) gene associated with disease severity in rheumatoid arthritis. *Genes Immun* **3**: 170–176.
19. Radstake TR, Sweep FC, Welsing P *et al.* (2005) Correlation of rheumatoid arthritis severity with the genetic functional variants and circulating levels of macrophage migration inhibitory factor. *Arthritis Rheum* **52**: 3020–3029.
20. Merk M, Baugh J, Zierow S *et al.* (2009) The Golgi-associated protein p115 mediates the secretion of macrophage migration inhibitory factor. *J Immunol* **182**: 6896–6906.

21. Subramanya HS, Roper DI, Dauter Z *et al.* (1996) Enzymatic ketonization of 2-hydroxymuconate: Specificity and mechanism investigated by the crystal structures of two isomerases. *Biochemistry* **35**: 792–802.
22. Suzuki M, Sugimoto H, Tanaka I, Nishihira J. (1997) Substrate specificity for isomerase activity of macrophage migration inhibitory factor and its inhibition by indole derivatives. *J Biochem* **122**: 1040–1045.
23. Rosengren E, Aman P, Thelin S *et al.* (1997) The macrophage migration inhibitory factor MIF is a phenylpyruvate tautomerase. *FEBS Lett* **417**: 85–88.
24. Rosengren E, Bucala R, Aman P *et al.* (1996) The immunoregulatory mediator macrophage migration inhibitory factor (MIF) catalyzes a tautomerization reaction. *Mol Med* **2**: 143–149.
25. Chuang Y-C, Lei H-Y, Liu H-S *et al.* (2011) Macrophage migration inhibitory factor induced by dengue virus infection increases vascular permeability. *Cytokine* **54**: 222–231.
26. Inacio AR, Ruscher K, Leng L *et al.* (2011) Macrophage migration inhibitory factor promotes cell death and aggravates neurologic deficits after experimental stroke. *J Cereb Blood Flow Metab* **31**: 1093–1106.
27. Sun HW, Bernhagen J, Bucala R, Lolis E. (1996) Crystal structure at 2.6-Å resolution of human macrophage migration inhibitory factor. *Proc Natl Acad Sci USA* **93**: 5191–5196.
28. Stamps SL, Taylor AB, Wang SC *et al.* (2000) Mechanism of the phenylpyruvate tautomerase activity of macrophage migration inhibitory factor: Properties of the P1G, P1A, Y95F, and N97A mutants. *Biochemistry* **39**: 9671–9678.
29. Lubetsky JB, Dios A, Han J *et al.* (2002) The tautomerase active site of macrophage migration inhibitory factor is a potential target for discovery of novel anti-inflammatory agents. *J Biol Chem* **277**: 24976–24982.
30. Suzuki M, Sugimoto H, Nakagawa A *et al.* (1996) Crystal structure of the macrophage migration inhibitory factor from rat liver. *Nat Struct Biol* **3**: 259–266.
31. Lubetsky JB, Swope M, Dealwis C *et al.* (1999) Pro-1 of macrophage migration inhibitory factor functions as a catalytic base in the phenylpyruvate tautomerase activity. *Biochemistry* **38**: 7346–7354.
32. Hare AA, Leng L, Gandavadi S. (2010) Optimization of N-benzyl-benzoxazol-2-ones as receptor antagonists of macrophage migration inhibitory factor (MIF). *Bioorg Med Chem Lett* **20**: 5811–5814.
33. Orita M, Yamamoto S, Katayama N *et al.* (2001) Coumarin and chromen-4-one analogues as tautomerase inhibitors of macrophage migration inhibitory factor: Discovery and X-ray crystallography. *J Med Chem* **44**: 540–547.
34. McLean LR, Zhang Y, Li H *et al.* (2010) Fragment screening of inhibitors for MIF tautomerase reveals a cryptic surface binding site. *Bioorg Med Chem Lett* **20**: 1821–1824.

35. Cho Y, Crichlow GV, Vermeire JJ *et al.* (2010) Allosteric inhibition of macrophage migration inhibitory factor revealed by ibudilast. *Proc Natl Acad Sci USA* **107**: 11313–11318.
36. Cheng KF, Al-Abed Y. (2006) Critical modifications of the ISO-1 scaffold improve its potent inhibition of macrophage migration inhibitory factor (MIF) tautomerase activity. *Bioorg Med Chem Lett* **16**: 3376–3379.
37. Dagia NM, Kamath DV, Bhatt P *et al.* (2009) A fluorinated analog of ISO-1 blocks the recognition and biological function of MIF and is orally efficacious in a murine model of colitis. *Eur J Pharmacol* **607**: 201–212.
38. Crichlow GV, Lubetsky JB, Leng L *et al.* (2009) Structural and kinetic analyses of macrophage migration inhibitory factor active site interactions. *Biochemistry* **48**: 132–139.
39. McLean L, Zhang Y, Li H *et al.* (2009) Discovery of covalent inhibitors for MIF tautomerase via cocrystal structures with phantom hits from virtual screening. *Bioorg Med Chem Lett* **19**(23): 6717–6720.
40. Winner M, Meier J, Zierow S *et al.* (2008) A novel, macrophage migration inhibitory factor suicide substrate inhibits motility and growth of lung cancer cells, *Cancer Res* **68**: 7253–7257.
41. Stamps SL, Fitzgerald MC, Whitman CP. (1998) Characterization of the role of the amino-terminal proline in the enzymatic activity catalyzed by macrophage migration inhibitory factor. *Biochemistry* **37**: 10195–10202.
42. Swope M, Sun HW, Blake PR, Lolis E. (1998) Direct link between cytokine activity and a catalytic site for macrophage migration inhibitory factor. *EMBO J* **17**: 3534–3541.
43. Garai J, Lóránd T, Molnár V. (2005) Ketone bodies affect the enzymatic activity of macrophage migration inhibitory factor. *Life Sci* **77**: 1375–1380.
44. Garai J, Molnár V, Eros D *et al.* (2007) MIF tautomerase inhibitor potency of alpha, beta-unsaturated cyclic ketones. *Int Immunopharmacol* **7**: 1741–1746.
45. El Turk F, Fauvet B, Ouertatani-Sakouhi H *et al.* (2010) An integrative *in silico* methodology for the identification of modulators of macrophage migration inhibitory factor (MIF) tautomerase activity. *Bioorg Med Chem* **18**: 5425–5440.
46. Golubkov PA, Johnson WH, Czerwinski RM *et al.* (2006) Inactivation of the phenylpyruvate tautomerase activity of macrophage migration inhibitory factor by 2-oxo-4-phenyl-3-butynoate. *Bioorg Chem* **34**: 183–199.
47. Nicoletti F, Créange A, Orlikowski D *et al.* (2005) Macrophage migration inhibitory factor (MIF) seems crucially involved in Guillain–Barré syndrome and experimental allergic neuritis. *J Neuroimmunol* **168**: 168–174.
48. Cvetkovic I, Al-Abed Y, Miljkovic D *et al.* (2005) Critical role of macrophage migration inhibitory factor activity in experimental autoimmune diabetes. *Endocrinology* **146**: 2942–2951.

49. Sakuragi T, Lin X, Metz CN *et al.* (2007) Lung-derived macrophage migration inhibitory factor in sepsis induces cardio-circulatory depression. *Surg Infect (Larchmt)* **8**: 29–40.
50. West P, Parker L, Ward J, Sabroe I. (2008) Differential and cell-type specific regulation of responses to Toll-like receptor agonists by ISO-1. *Immunology* **125**(1): 101–110.
51. Dhanantwari P, Nadaraj S, Kenessey A *et al.* (2008) Macrophage migration inhibitory factor induces cardiomyocyte apoptosis. *Biochem Biophys Res Commun* **371**: 298–303.
52. Takahashi K, Koga K, Linge HM *et al.* (2009) Macrophage CD74 contributes to MIF-induced pulmonary inflammation. *Respir Res* **10**: 33.
53. Piette C, Deprez M, Roger T *et al.* (2009) The dexamethasone-induced inhibition of proliferation, migration, and invasion in glioma cell lines is antagonized by macrophage migration inhibitory factor (MIF) and can be enhanced by specific MIF inhibitors. *J Biol Chem* **284**: 32483–32492.
54. Veillat V, Carli C, Metz CN, *et al.* (2010) Macrophage migration inhibitory factor elicits an angiogenic phenotype in human ectopic endometrial cells and triggers the production of major angiogenic factors via CD44, CD74, and MAPK signaling pathways. *J Clin Endocrinol Metab* **95**: E403–E412.
55. Traggiai E, Casati A, Frascoli M *et al.* (2010) Selective preservation of bone marrow mature recirculating but not marginal zone B cells in murine models of chronic inflammation. *PLoS One* **5**: e11262.
56. Liang J-L, Xiao D-Z, Liu X-Y *et al.* (2010) High glucose induces apoptosis in AC16 human cardiomyocytes via macrophage migration inhibitory factor and c-Jun N-terminal kinase. *Clin Exp Pharmacol Physiol* **37**: 969–973.
57. Chuang C-C, Chuang Y-C, Chang W-T *et al.* (2010) Macrophage migration inhibitory factor regulates interleukin-6 production by facilitating nuclear factor-kappa B activation during *Vibrio vulnificus* infection. *BMC Immunol* **11**: 50.
58. Vera PL, Iczkowski KA, Howard DJ *et al.* (2010) Antagonism of macrophage migration inhibitory factor decreases cyclophosphamide cystitis in mice. *Neurourol Urodyn* **29**: 1451–1457.
59. Bacher M, Deuster O, Aljabari B *et al.* (2010) The role of macrophage migration inhibitory factor in Alzheimer's disease. *Mol Med* **16**: 116–121.
60. Assunção-Miranda I, Bozza MT, Da Poian AT. (2010) Pro-inflammatory response resulting from sindbis virus infection of human macrophages: Implications for the pathogenesis of viral arthritis. *J Med Virol* **82**: 164–174.
61. Wang F, Shen X, Guo X *et al.* (2010) Spinal macrophage migration inhibitory factor contributes to the pathogenesis of inflammatory hyperalgesia in rats. *Pain* **148**: 275–283.

62. Baron N, Deuster O, Noelker C *et al.* (2011) Role of macrophage migration inhibitory factor in primary glioblastoma multiforme cells. *J Neurosci Res* **89**: 711–717.
63. Schrader J, Deuster O, Rinn B *et al.* (2009) Restoration of contact inhibition in human glioblastoma cell lines after MIF knockdown. *BMC Cancer* **9**: 464.
64. Dios A, Mitchell RA, Aljabari B *et al.* (2002) Inhibition of MIF bioactivity by rational design of pharmacological inhibitors of MIF tautomerase activity. *J Med Chem* **45**: 2410–2416.
65. Dabideen DR, Cheng KF, Aljabari B *et al.* (2007) Phenolic hydrazones are potent inhibitors of macrophage migration inhibitory factor proinflammatory activity and survival improving agents in sepsis. *J Med Chem* **50**: 1993–1997.
66. Kithcart AP, Cox GM, Sielecki T *et al.* (2010) A small-molecule inhibitor of macrophage migration inhibitory factor for the treatment of inflammatory disease. *FASEB J* **24**(11): 4459–4466.
67. Balachandran S, Gadekar PK, Parkale S *et al.* (2011) Synthesis and biological activity of novel MIF antagonists. *Bioorg Med Chem Lett* **21**: 1508–1511.
68. Gadjeva M, Nagashima J, Zaidi T *et al.* (2010) Inhibition of macrophage migration inhibitory factor ameliorates ocular *Pseudomonas aeruginosa*-induced keratitis. *PLoS Pathog* **6**: e1000826.
69. Ouertatani-Sakouhi H, El-Turk F, Fauvet B *et al.* (2010) Identification and characterization of novel classes of macrophage migration inhibitory factor (MIF) inhibitors with distinct mechanisms of action. *J Biol Chem* **285**: 26581–26598.
70. Taylor AB, Johnson WH, Czerwinski RM *et al.* (1999) Crystal structure of macrophage migration inhibitory factor complexed with (E)-2-fluoro-p-hydroxycinnamate at 1.8 Å resolution: Implications for enzymatic catalysis and inhibition. *Biochemistry* **38**: 7444–7452.

PART II

Regulation of MIF Expression

II-1

Epigenetic Control of MIF Expression

Thierry Roger*,†, Jérôme Lugrin*, Xavier C. Ding*
and Thierry Calandra*

1. Introduction

Covalent modifications of DNA by methylation and of histones by acetylation are two main mechanisms by which epigenetics regulates gene expression in physiological and pathological situations. Inhibitors of DNA methyltransferases and histone deacetylases (HDACs) are potent anti-cancer drugs, displaying anti-inflammatory and immunomodulatory activities. MIF is a proinflammatory cytokine involved in the pathogenesis of inflammatory, autoimmune and infectious diseases. Recently, MIF has been shown to promote tumorigenesis, suggesting that epigenetic mechanisms participate in the control of MIF expression. Notably, the *MIF* gene lies in a CpG island, a DNA context prone to regulation by methylation. Yet the *MIF* promoter is hypomethylated in primary and tumor cells, and demethylating agents do not affect *MIF* expression. In contrast, HDAC inhibitors impair MIF mRNA and protein expression *in vitro* and *in vivo*. At the molecular level, HDAC inhibitors decrease the recruitment of the basal transcriptional machinery to the *MIF* promoter and thereby inhibits *MIF* transcription. These data indicate that HDACs are important regulators of *MIF* expression. Therefore, inhibition of *MIF* expression may contribute to the anti-cancer and anti-inflammatory activities of HDAC inhibitors.

2. Epigenetics

All cells from a given organism contain essentially the same DNA information. Developmental specification relies on qualitative and quantitative differences in gene expression that, for obvious reasons of parsimony, is

*Infectious Diseases Service, Department of Medicine, Centre Hospitalier Universitaire, Vaudois and University of Lausanne, CH-1011 Lausanne, Switzerland.
†Corresponding author: Email: thierry.roger@chuv.ch

primarily controlled at the level of transcription. In recent years, epigenetics, defined as all meiotically and mitotically heritable changes in gene expression that are not coded in the DNA sequence,[1] has profoundly transformed our vision of how gene expression is regulated.

Epigenetics comprises three main and inter-related mechanisms: DNA methylation, small interfering RNAs, and post-transcriptional modifications of histones.[2] Specific panels of epigenetic modifications shape the transcriptional program in a cell-specific manner. As such, the epigenome plays a central role in conserving cell characteristics by maintaining specific patterns of gene expression during somatic cell division.[3–6] Yet the epigenome is dynamic and flexible and accommodates transcriptional changes during development. Most importantly, epigenetic modifications have been directly linked to the dysregulated gene expression characterizing numerous human diseases.[7] Reflecting the great interest of biomedical research in epigenetics, ambitious projects and initiatives (NIH Roadmap Epigenomics Program, ENCODE project, AHEAD project, and the Epigenomics NCBI browser) have been developed to provide highly comprehensive epigenomic maps in human stem cells and in healthy and diseased tissues.[8,9] In this chapter, we will focus our attention on DNA methylation and histone acetylation as possible mechanisms involved in the control of *MIF* gene expression, as no MIF-specific micro-RNA (miRNA) has been identified thus far.

2.1 *DNA methylation*

DNA methylation is probably the most studied epigenetic modification in mammals. The reversible covalent modification of 5′-methyl cytosine residues mainly occurs in the context of CpG dinucleotides. DNA methylation is catalyzed by DNA methyltransferases (DNMTs). Around 3% of cytosines are methylated in the human genome. Repetitive genomic DNA sequences are heavily methylated, whereas CpG-rich regions of the genome, also known as CpG islands, are commonly unmethylated in normal cells. The CpG islands are not randomly distributed. About half of these are localized in the promoter region of genes that have a wide expression, such as *MIF*.

DNA methylation is usually associated with repressed gene expression. The methyl groups added by DNMTs protrude from cytosines and affect gene transcription through two main mechanisms. First, they inhibit the binding of transcription factors that positively regulate transcription. Second, they recruit methyl-CpG-binding proteins and other types of proteins that are involved in histone modification, chromatin compaction, and gene silencing.[10,11]

DNA hypomethylation was one of the very first epigenetic alterations reported in human cancer.[12] Interestingly, dysregulated DNA methylation has also been reported in neurological disorders and autoimmune diseases.[7] Yet very little is known about the mechanisms involved in DNA demethylation.[10,11] Although tumor cells have globally 20–60% less 5-methyl-cytosine methylation than normal cells, they also contain subsets of hypermethylated genes. Indeed, transcriptional silencing of tumor suppressor genes resulting from the hypermethylation of CpG island promoters is a common hallmark of tumor cells.[13,14] DNA demethylating agents have been developed to revert aberrant gene silencing in cancers. The nucleoside analog 5-azacytidine (5-aza-CR; azacitidine; Vidaza®) and its derivative 5-aza-2-deoxycytidine (5-aza-CdR; decitabine; Dacogen®) have been approved for the treatment of all subtypes of myelodysplastic syndrome.[15–17]

2.2 *Post-transcriptional modifications by histone deacetylases*

The nucleosome, the basic repeating unit of chromatin, is composed of a 147-bp section of DNA wrapped around a histone octamer composed of two copies of each of the four core histones, H2A, H2B, H3, and H4. Histones are subjected to post-transcriptional covalent modifications at amino-terminal tails through acetylation, ubiquitination, and sumoylation of lysine; methylation of arginine and lysine; and phosphorylation of serine and threonine. Histone acetylation usually associates with specific histone methylation marks.[7] Each modification affects the structure and the function of the chromatin. The open structure of transcriptionally active euchromatin is enriched in acetylated and trimethylated H3K4, H3K36, and H3K79 histones, whereas the transcriptionally inactive heterochromatin is characterized by hypoacetylated and highly methylated H3K9, H3K27, and H4K20 histones.

Histone acetylation is controlled by the antagonistic action of two enzyme families: histone acetyl transferases (HATs) and histone deacetylases (HDACs). HATs catalyze the transfer of an acetyl group from acetyl-coenzyme A to an amino group of lysine residues of histones. Conversely, HDACs catalyze the hydrolysis of acetamides, resulting in histone deacetylation. The name HDAC was coined because histones were the first substrates identified for lysine deacetylases.[18] However, HDACs deacetylate numerous non-histone proteins, such as tubulin, heat shock proteins, steroid receptors, and nuclear import and transcription regulators.[19,20] HDACs are classified into two main subfamilies: HDAC1–11 and the sirtuins.[21] In the following sections, we will use HDACs as a generic term referring to HDAC1–11.

Aberrant gene expression due to inactivation of HATs or overexpression of HDACs is common in cancer cells. Moreover, dysregulated recruitment of HDACs to promoters is associated with transcriptional repression, notably that of cell-cycle modifiers and tumor suppressor genes, thereby contributing to oncogenesis.[14,22–24] These observations strengthened the development of pharmacological inhibitors of HDACs as novel cancer therapeutics. Indeed, HDAC inhibitors counteract cancer development by blocking DNA synthesis and inducing growth arrest, differentiation, and apoptosis of tumor cells. They also reduce tumor angiogenesis, metastasis, and invasion. Suberoylanilide hydroxamic acid (SAHA; vorinostat; Zolinza®) and depsipeptide (romidepsin; Istodax®) have been approved for the treatment of cutaneous T cell lymphoma[17] and numerous additional HDAC inhibitors are now being tested in clinical trials.

Specific patterns of histone modifications by methylation and acetylation correlate with the expression of inflammatory and immune genes. HDAC inhibitors have recently been reported to exert potent anti-inflammatory and immunomodulatory activities *in vitro* and *in vivo*. HDAC inhibitors improved outcomes in models of inflammatory and autoimmune diseases, such as sepsis, rheumatoid arthritis, lupus, autoimmune encephalitis, multiple sclerosis, graft versus host disease, asthma, and colitis.[25–31] Therefore, HDAC inhibitors are attractive therapies not only for the treatment of oncologic disorders but also possibly for immune-related diseases.[32]

Numerous experimental, preclinical and clinical observations have ascribed a central role for MIF in the pathogenesis of inflammatory, autoimmune, and neoplastic diseases.[33–35] Considering that these diseases are characterized by dysregulation of epigenetic marks, we hypothesized that epigenetic mechanisms may participate in the control of *MIF* expression. Since epigenetic modifications profoundly affect gene transcription, we will first summarize our current knowledge about the various DNA binding sites and trans-acting transcription factors controlling *MIF* gene expression.

3. *MIF* Gene Structure, Expression and Transcriptional Regulation

3.1 *MIF gene structure and expression*

The first sequences of a human MIF cDNA and of the *MIF* gene were reported by Weiser *et al.* in 1989[36] and by Paralkar and Wistow in 1994,[37] respectively. A single *MIF* gene spans nucleotide positions 24236191 to 24237414 on chromosome 22 (22q11.2). This region of chromosome 22 is in syntenic

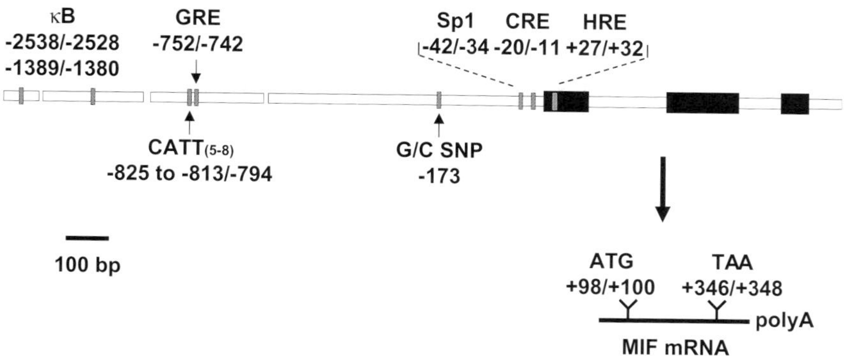

Fig. 1. Structure of the *MIF* gene and MIF mRNA. The three *MIF* exons are represented by black boxes. The κB, glucocorticoid response element (GRE), specificity protein 1 (Sp1), cAMP-responsive element (CRE), and hypoxia-inducible factor response element (HRE), which have been functionally characterized, are depicted by grey boxes. Their localization is relative to the transcriptional start site set at +1. Vertical arrows indicate the positions of the $CATT_{5-8}$ tetranucleotide microsatellite and the −173*G/C single nucleotide polymorphism (SNP). The translational start codon (ATG at +98/+100) and stop codon (TAA at 346+/+348) are pinpointed in the MIF mRNA.

conservation with part of mouse chromosome 10 containing the *Mif* gene.[38] Also located on chromosome 22 (22q11.23), D-dopachrome tautomerase is the only potential human *MIF* paralog.[39] The *MIF* gene is composed of three exons of 108, 173, and 67 bp interspaced by two introns of 189 and 95 bp (Fig. 1). A single RNA initiation start site located 97 bp upstream of the methionine codon is used to transcribed a 0.8 kb mRNA.[37] The 345-bp open reading frame of MIF mRNA encodes for a 115 amino acid nonglycosylated protein of 12.5 kDa.

Sequence analyses of the *MIF* gene revealed that it does not contain a TATA box but numerous CpG dinucleotides forming a CpG island. A CpG island is defined as a sequence of at least 200 bp with a G+C content of 50% or more and an observed to expected CpG dinucleotide ratio greater than 0.6. The *MIF* CpG island spans approximately 1.2 kb, starting 300 bp upstream of the transcriptional start site (Fig. 2). In agreement with the fact that broadly expressed genes are typically lying in CpG islands, MIF is constitutively expressed as a single mRNA species of 0.8 kb in virtually all organs and cell types (summarized in Ref. 34).

MIF gene expression increases through the action of cytokines [tumor necrosis factor, interferon-γ, interleukin(IL)-1, IL-2],[40–44] mitogens,[40–44] microbial products,[40,41,45–47] glucose,[48] low-density lipoproteins,[49,50] UV-B,[51]

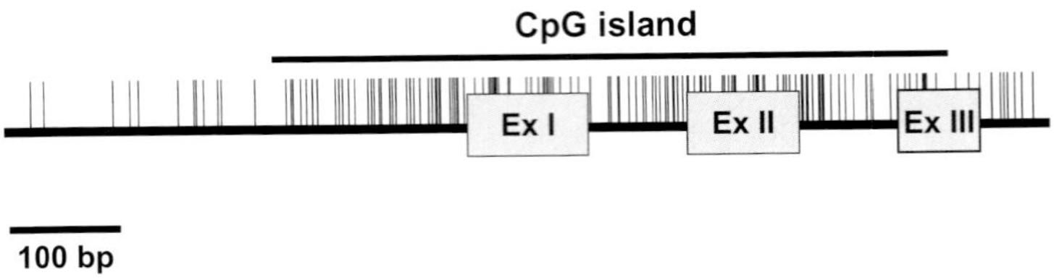

Fig. 2. The *MIF* gene is located in a cytosine guanine dinucleotide (CpG) island. *In silico* analysis of CpG sites (vertical lines) within the human *MIF* gene. *MIF* exons are depicted by grey boxes.

hypoxia,[52,53] and hormones [glucocorticoids, corticotropin-releasing factor (CRF), human chorionic gonadotropin, angiotensin].[54,55] Of great interest, MIF is commonly overexpressed in prostate, breast, colon, brain, skin, and lung cancers.[56–63] Altogether, the expression patterns of MIF — its well-characterized proinflammatory, proproliferative, prosurvival, and proangiogenic biological activities — point towards a crucial role of MIF in the pathogenesis of infectious, inflammatory, autoimmune, and neoplastic diseases.[33–35]

3.2 *MIF gene transcriptional regulation*

Despite the involvement of MIF in the pathogenesis of numerous diseases, few studies have analyzed the molecular mechanisms underlying the transcriptional regulation of *MIF*. The *MIF* promoter region contains putative DNA binding sites for transcription factors, such as activator protein-1 (AP-1), nuclear factor (NF)-κB, E-twenty six (Ets), GATA, cAMP-responsive element (CRE) binding protein (CREB), specificity protein 1 (Sp1), hypoxia-inducible factor (HIF), and a glucocorticoid receptor (GR). The first insights about *MIF* gene transcriptional regulation were obtained by the analysis of the activity of mouse *Mif* promoter reporter constructs in rat anterior pituitary cells. A CRE site in the vicinity of the transcriptional start site of the mouse *Mif* gene (located at position −48/41) was shown to mediate forskolin- and CRF-induced *Mif* promoter activation.[64] Subsequently, Baugh *et al.* reported that hypoxia and HIF-1α activate human *MIF* promoter activity through an HIF responsive element (HRE located at position +25). Conversely, CREB overexpression decreased *MIF* promoter activity under hypoxic conditions, whereas disruption of a proximal (−20/−11) CRE site increased HIF-1α-mediated *MIF* promoter activity.[52] The functional role of the HRE site in response to hypoxia was recently confirmed.[65] We have shown that CREB and Sp1 interact with proximal CRE (−20/−11) and Sp1 (−42/−34) sites in the human *MIF* promoter to positively regulate constitutive promoter transcriptional activity in human monocytic (THP-1), epithelial (HeLa and A549), and keratinocytic (HaCat) cell lines, and in peripheral blood mononuclear cells (PBMCs). The CRE and Sp1 sites also cooperate to mediate microbial product-induced *MIF* gene expression in monocytic cells.[66] The CRE site has also been reported to relay glucocorticoid-induced *MIF* gene expression in CEM-C7 T cells.[54] Altogether, these studies indicate that DNA regulatory elements surrounding the transcriptional start site play a central role in controlling *MIF* gene transcription. Yet several lines of evidence suggest that more distant regulatory elements also may have a functional role (Fig. 1). For example, disruption of a distal consensus GR element (GRE at –742) completely abolished glucocorticoid-inducible *MIF* promoter activity.[54] Moreover, NF-κB recruitment to putative κB sites at –2538 or –1389 trans-activates *MIF* promoter in response to IL-1β and TNF in endometrial cells.[42,43]

The *MIF* gene contains two major functional polymorphisms, a 5- to 8-CATT tetranucleotide repeat at –794[67] and a G/C single nucleotide polymorphism at –173[68] (Fig. 1). These polymorphisms have been reported to modulate *MIF* promoter activity, to correlate with MIF expression levels, and to be associated with the susceptibility to or the outcome of infectious, inflammatory, autoimmune, and neoplastic diseases, as discussed in other chapters of this book. The exact mechanisms whereby these polymorphisms affect *MIF* gene transcriptional activity remain poorly understood. The –173*C SNP creates a putative AP-4 DNA binding site.[68] HMG box-containing protein 1 (HBP1), a known negative regulator of tumorigenesis, has recently been proposed to inhibit *MIF* gene transcription in prostate cancer cells by interacting with a sequence (–811/–792) covering five CATT repeats.[69]

4. Epigenetic Control of MIF Expression

4.1 *The proximal MIF promoter is not methylated*

The human *MIF* promoter contains numerous CpG sites typically found in the proximal promoter of housekeeping genes (Fig. 1). Two CpG sites are part of the proximal Sp1 and CRE binding sequences implicated in basal *MIF* promoter activity.[66] To test whether CpG methylation affects *MIF* gene expression in THP-1, HeLa, A549, and HaCat cell lines, we sequenced the proximal *MIF* promoter region using sodium bisulfite-treated genomic DNA, in which 5-methyl cytosines are protected from bisulfite-induced conversion to uracils. These analyses focused on 34 CpG sites confined in a region extending from position –300 to +1. Only two methylated cytosines located

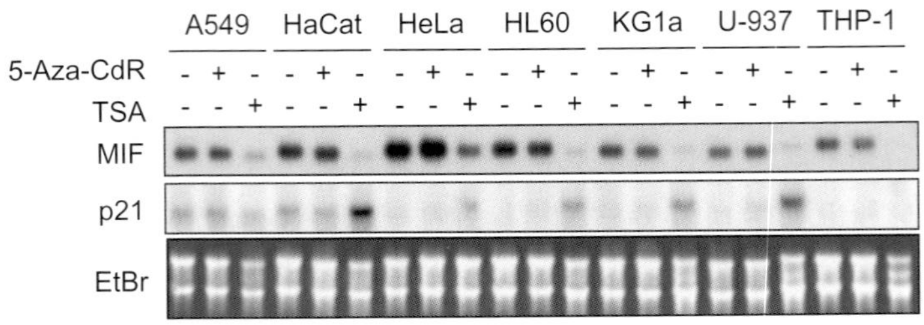

Fig. 3. Inhibition of histone deacetylases (HDACs), but not of DNA methyl transferases (DNMTs), inhibits MIF mRNA expression. A549, HaCat, HeLa, HL60, KG1a, U937, and THP1 cells were cultured for 18 hours with or without 5-aza-2′-deoxycytidine (5-aza-CdR at 5 μM), an inhibitor of DNA methyl transferases, or trichostatin A (TSA at 1 μM), an inhibitor of histone deacetylases (HDACs). MIF and p21 (CDKN1A) mRNA expression was analyzed by northern blotting. EtBr = ethidium bromide.

at –211 and –121 were detected in one out of five sequences in HaCat keratinocytes and THP-1 monocytic cells. Thus, the proximal *MIF* promoter is essentially not methylated in cell lines of different origins.[70] Given that hypomethylation of the *MIF* promoter could account for higher MIF mRNA expression in tumor cells compared to normal cells, we analyzed CpG methylation in primary cells by bisulfite DNA sequencing. Two methylated cytosines located at –65 and –28 were detected among 12 sequences (408 CpG sites analyzed) obtained from PBMCs isolated from three healthy subjects. Altogether, these data indicate that methylation of CpG sites within the proximal *MIF* promoter is a very rare event and does not account for increased MIF expression in tumor cell lines.

4.2 *Inhibition of DNA methyl transferases (DNMTs) does not affect MIF gene expression*

CpG island shores that refer to regions of lower CpG density close to CpG islands are subjected to methylation.[71] Moreover, methylation of CpG sites located in coding sequences or distant from transcriptional start sites have been reported to affect gene expression.[72,73] We thus explored whether methylation of CpG sites outside the proximal *MIF* promoter influences *MIF* gene expression. As illustrated in Fig. 3, treatment of epithelial (A549 and HeLa), keratinocytic (HaCat) and myeloid (HL-60, KG1a, U-937, THP-1) cell lines with 5-aza-CdR does not alter MIF mRNA expression. Similar results were obtained with bone marrow-derived macrophages.[66] This groundwork argues against a role for DNA methylation as an epigenetic mechanism affecting *MIF* gene expression in tumor cell lines.

4.3 *Inhibition of histone deacetylases (HDACs) impairs MIF gene expression*

Dysregulated HDAC activity contributes to abnormal gene expression in tumors. Given that MIF is overexpressed in tumor cells and that MIF levels correlate with tumor aggressiveness and metastatic potential,[59,74,75] HDACs were considered attractive candidate molecules regulating *MIF* gene expression.

Trichostatin A (TSA), a natural hydroxamic acid that has inspired the design of many synthetic HDAC inhibitors including SAHA,[76] is widely used to assess the role of HDACs *in vitro* and *in vivo*. Valproic acid (VPA) is a chemically unrelated HDAC inhibitor used to treat epileptic seizures and bipolar disorders. TSA, SAHA, and VPA powerfully inhibited MIF mRNA expression in a time- and dose-dependent manner in a panel of tumor cell lines (A549, HeLa, HaCat, HL-60, KG1a, U-937, and THP-1), in B16 melanoma and in primary macrophages (Fig. 3 and Ref. 70). The inhibitory effect of TSA on *MIF* expression was specific, as TSA reactivated the expression of the *CDKN1A* gene (encoding for the p21/WAF cell-cycle inhibitor). TSA reduced MIF protein expression in cell lines, whole blood, and in the circulation of mice injected with TSA.[70] Altogether, these data suggest that HDACs are important regulators of *MIF* gene expression.

Combinatorial epigenetic therapies associating DNMT inhibitors and HDAC inhibitors have been shown to exert additive and synergistic clinical effects in patients with hematologic malignancies.[77–79] In HeLa, HL60, and THP-1 cells, the association of 5-aza-CdR and TSA does not amplify TSA-mediated inhibition of MIF mRNA expression (Roger *et al., unpublished data*), in line with the observation that the MIF promoter is hypomethylated in tumor cells.

MIF is overexpressed in multiple types of tumors and it promotes malignancies by increasing survival, proliferation, and migration of tumor cells,[80–82] by promoting angiogenesis[74,83,84] and by altering antitumor adaptive immune responses.[75,85–87] This led us to speculate that MIF might be a common target of the anti-cancer activity of HDAC inhibitors. Similarly, considering that MIF is a central mediator of the pathogenesis of sepsis,[45,47,55,88] arthritis,[89] colitis,[90] and lupus,[91] we propose that the benefit afforded by HDAC inhibitors in inflammatory and autoimmune diseases may be related to the inhibition of MIF expression.

4.4 *Mechanisms by which HDAC inhibitors inhibit MIF gene expression*

Studies initiated to unravel the molecular mechanisms by which HDAC inhibition affects *MIF* gene expression revealed several unique features. Nuclear run-on assays demonstrated that TSA inhibits *MIF* gene transcription. Surprisingly, however, chromatin immunoprecipitation analyses showed that TSA deacetylates histones H3 and H4 associated with the proximal *MIF* promoter.[70] This observation was unexpected given that HDAC inhibitors increased overall histone acetylation. Yet genome-wide expression studies have established that HDAC inhibition impacts on a minority of the transcriptome (2–10%), with similar proportions of genes upregulated and downregulated. Moreover, local hypoacetylation following HDAC inhibition has previously been observed within the high-mobility-group A2 (*HMGA2*) and *BCL2* genes.[92,93]

Because TSA reduced the acetylation of MIF promoter-associated histones, we hypothesized that the MIF promoter was less accessible to the transcription machinery. Indeed, chromatin immunoprecipitation experiments confirmed that TSA impairs the binding of Sp1, CREB, and RNA polymerase II to the proximal MIF promoter. However, challenging the concept that hypoacetylated chromatin forms a compact structure less accessible to transcriptional regulators, accessibility studies revealed that the proximal region of the *MIF* promoter was accessible in cells cultured either with or without TSA.[70] Blocking protein synthesis with cycloheximide increased MIF mRNA expression early on without modifying the acetylation of histones associated with the *MIF* promoter in cells treated with TSA. Therefore, we assume that TSA repressed *MIF* gene transcription through *de novo* protein synthesis.

5. Conclusions

The biological activities ascribed to MIF over the last 20 years have established this cytokine as a central mediator of cell proliferation and survival, angiogenesis, and inflammatory and immune responses. In line with these observations, preclinical and clinical studies suggest that MIF represents a therapeutic target for the treatment of immune-related and neoplastic diseases. While our understanding of MIF biology has improved markedly in recent years, little is known about the mechanisms controlling *MIF* gene expression. Very few studies have characterized the DNA-binding elements and cognate transcription factors regulating basal and stimulus-induced *MIF* transcription. Much less is known about the role of epigenetics in regulating

MIF expression. Most recent data suggest that HDAC activity, but not DNA methylation, strongly impacts on *MIF* transcription. We speculate that the powerful inhibition of MIF expression by HDAC inhibitors contributes, at least in part, to the antitumorigenic and anti-inflammatory activities of these drugs. Further work will be required to more deeply decipher the genetic and epigenetic mechanisms controlling MIF expression in health and diseases. Besides increasing our knowledge on the biology of MIF, these studies may help to develop novel MIF-directed intervention strategies for diseases associated with dysregulated MIF expression.

Acknowledgments

The authors are supported by the Swiss National Science Foundation (310000–114073, 310030–118266 and 310030–132744), the Bristol-Myers Squibb Foundation, the Leenaards Foundation, and the Santos-Suarez Foundation for Medical Research.

References

1. Holliday R. (1987) The inheritance of epigenetic defects. *Science* **238**(4824): 163–170.
2. Bonasio R, Tu S, Reinberg D. (2010) Molecular signals of epigenetic states. *Science* **330**(6004): 612–616.
3. Feng S, Jacobsen SE, Reik W. (2010) Epigenetic reprogramming in plant and animal development. *Science* **330**(6004): 622–627.
4. Meissner A. (2010) Epigenetic modifications in pluripotent and differentiated cells. *Nat Biotechnol* **28**(10): 1079–1088.
5. Margueron R, Reinberg D. (2010) Chromatin structure and the inheritance of epigenetic information. *Nat Rev Genet* **11**(4): 285–296.
6. Gibney ER, Nolan CM. (2010) Epigenetics and gene expression. *Heredity* **105**(1): 4–13.
7. Portela A, Esteller M. (2010) Epigenetic modifications and human disease. *Nat Biotechnol* **28**(10): 1057–1068.
8. Bernstein BE, Stamatoyannopoulos JA, Costello JF *et al.* (2010) The NIH roadmap epigenomics mapping consortium. *Nat Biotechnol* **28**(10): 1045–1048.
9. Satterlee JS, Schubeler D, Ng HH. (2010) Tackling the epigenome: Challenges and opportunities for collaboration. *Nat Biotechnol* **28**(10): 1039–1044.
10. Wu SC, Zhang Y. (2010) Active DNA demethylation: Many roads lead to Rome. *Nat Rev Mol Cell Biol* **11**(9): 607–620.

11. Fazzari MJ, Greally JM. (2004) Epigenomics: Beyond CpG islands. *Nat Rev Genet* **5**(6): 446–455.
12. Feinberg AP, Vogelstein B. (1983) Hypomethylation distinguishes genes of some human cancers from their normal counterparts. *Nature* **301**(5895): 89–92.
13. Esteller M. (2007) Cancer epigenomics: DNA methylomes and histone-modification maps. *Nat Rev Genet* **8**(4): 286–298.
14. Esteller M. (2008) Epigenetics in cancer. *N Engl J Med* **358**(11): 1148–1159.
15. Fenaux P, Mufti GJ, Hellstrom-Lindberg E *et al.* (2009) Efficacy of azacitidine compared with that of conventional care regimens in the treatment of higher-risk myelodysplastic syndromes: A randomised, open-label, phase III study. *Lancet Oncol* **10**(3): 223–232.
16. Issa JP, Kantarjian HM. (2009) Targeting DNA methylation. *Clin Cancer Res* **15**(12): 3938–3946.
17. Kelly TK, De Carvalho DD, Jones PA. (2010) Epigenetic modifications as therapeutic targets. *Nat Biotechnol* **28**(10): 1069–1078.
18. Inoue A, Fujimoto D. (1969) Enzymatic deacetylation of histone. *Biochem Biophys Res Commun* **36**(1): 146–150.
19. Glozak MA, Seto E. (2007) Histone deacetylases and cancer. *Oncogene* **26**(37): 5420–5432.
20. Xu WS, Parmigiani RB, Marks PA. (2007) Histone deacetylase inhibitors: Molecular mechanisms of action. *Oncogene* **26**(37): 5541–5552.
21. Yang XJ, Seto E. (2008) The Rpd3/Hda1 family of lysine deacetylases: From bacteria and yeast to mice and men. *Nat Rev Mol Cell Biol* **9**(3): 206–218.
22. Bolden JE, Peart MJ, Johnstone RW. (2006) Anticancer activities of histone deacetylase inhibitors. *Nat Rev Drug Discov* **5**(9): 769–784.
23. Minucci S, Pelicci PG. (2006) Histone deacetylase inhibitors and the promise of epigenetic (and more) treatments for cancer. *Nat Rev Cancer* **6**(1): 38–51.
24. Glozak MA, Seto E. (2007) Histone deacetylases and cancer. *Oncogene* **26**(37): 5420–5432.
25. Choi JH, Oh SW, Kang MS *et al.* (2005) Trichostatin A attenuates airway inflammation in mouse asthma model. *Clin Exp Allergy* **35**(1): 89–96.
26. Glauben R, Batra A, Fedke I *et al.* (2006) Histone hyperacetylation is associated with amelioration of experimental colitis in mice. *J Immunol* **176**(8): 5015–5022.
27. Roger T, Lugrin J, Le RD *et al.* (2010) Histone deacetylase inhibitors impair innate immune responses to Toll-like receptor agonists and to infection. *Blood* **117**(4): 1205–1217.
28. Reilly CM, Mishra N, Miller JM *et al.* (2004) Modulation of renal disease in MRL/lpr mice by suberoylanilide hydroxamic acid. *J Immunol* **173**(6): 4171–4178.

29. Nishida K, Komiyama T, Miyazawa S *et al.* (2004) Histone deacetylase inhibitor suppression of autoantibody-mediated arthritis in mice via regulation of p16INK4a and p21(WAF1/Cip1) expression. *Arthritis Rheum* **50**(10): 3365–3376.
30. Mishra N, Reilly CM, Brown DR *et al.* (2003) Histone deacetylase inhibitors modulate renal disease in the MRL-lpr/lpr mouse. *J Clin Invest* **111**(4): 539–552.
31. Camelo S, Iglesias AH, Hwang D *et al.* (2005) Transcriptional therapy with the histone deacetylase inhibitor trichostatin A ameliorates experimental autoimmune encephalomyelitis. *J Neuroimmunol* **164**(1–2): 10–21.
32. Haberland M, Montgomery RL, Olson EN. (2009) The many roles of histone deacetylases in development and physiology: Implications for disease and therapy. *Nat Rev Genet* **10**(1): 32–42.
33. Bucala R, Donnelly SC. (2007) Macrophage migration inhibitory factor: A probable link between inflammation and cancer. *Immunity* **26**(3): 281–285.
34. Calandra T, Roger T. (2003) Macrophage migration inhibitory factor: A regulator of innate immunity. *Nat Rev Immunol* **3**(10): 791–800.
35. Rendon BE, Willer SS, Zundel W, Mitchell RA. (2009) Mechanisms of macrophage migration inhibitory factor (MIF)-dependent tumor microenvironmental adaptation. *Exp Mol Pathol* **86**(3): 180–185.
36. Weiser WY, Temple PA, Witek-Giannotti JS *et al.* (1989) Molecular cloning of a cDNA encoding a human macrophage migration inhibitory factor. *Proc Natl Acad Sci USA* **86**(19): 7522–7526.
37. Paralkar V, Wistow G. (1994) Cloning the human gene for macrophage migration inhibitory factor (MIF). *Genomics* **19**(1): 48–51.
38. Kozak CA, Adamson MC, Buckler CE *et al.* (1995) Genomic cloning of mouse MIF (macrophage inhibitory factor) and genetic mapping of the human and mouse expressed gene and nine mouse pseudogenes. *Genomics* **27**(3): 405–411.
39. Esumi N, Budarf M, Ciccarelli L *et al.* (1998) Conserved gene structure and genomic linkage for D-dopachrome tautomerase (DDT) and MIF. *Mamm Genome* **9**(9): 753–757.
40. Bacher M, Metz CN, Calandra T *et al.* (1996) An essential regulatory role for macrophage migration inhibitory factor in T-cell activation. *Proc Natl Acad Sci USA* **93**(15): 7849–7854.
41. Calandra T, Bernhagen J, Mitchell RA, Bucala R. (1994) The macrophage is an important and previously unrecognized source of macrophage migration inhibitory factor. *J Exp Med* **179**(6): 1895–1902.
42. Cao WG, Morin M, Metz C *et al.* (2005) Stimulation of macrophage migration inhibitory factor expression in endometrial stromal cells by interleukin 1, beta involving the nuclear transcription factor NFkappaB. *Biol Reprod* **73**(3): 565–570.
43. Cao WG, Morin M, Sengers V *et al.* (2006) Tumour necrosis factor-alpha up-regulates macrophage migration inhibitory factor expression in endometrial

stromal cells via the nuclear transcription factor NF-kappaB. *Hum Reprod* **21**(2): 421–428.
44. Hirokawa J, Sakaue S, Furuya Y *et al.* (1998) Tumor necrosis factor-alpha regulates the gene expression of macrophage migration inhibitory factor through tyrosine kinase-dependent pathway in 3T3-L1 adipocytes. *J Biochem* **123**(4): 733–739.
45. Bernhagen J, Calandra T, Mitchell RA *et al.* (1993) MIF is a pituitary-derived cytokine that potentiates lethal endotoxaemia. *Nature* **365**(6448): 756–759.
46. Bacher M, Meinhardt A, Lan HY *et al.* (1997) Migration inhibitory factor expression in experimentally induced endotoxemia. *Am J Pathol* **150**(1): 235–246.
47. Calandra T, Spiegel LA, Metz CN, Bucala R. (1998) Macrophage migration inhibitory factor is a critical mediator of the activation of immune cells by exotoxins of Gram-positive bacteria. *Proc Natl Acad Sci USA* **95**(19): 11383–11388.
48. Waeber G, Calandra T, Roduit R *et al.* (1997) Insulin secretion is regulated by the glucose-dependent production of islet beta cell macrophage migration inhibitory factor. *Proc Natl Acad Sci USA* **94**(9): 4782–4787.
49. Santini E, Lupi R, Baldi S *et al.* (2008) Effects of different LDL particles on inflammatory molecules in human mesangial cells. *Diabetologia* **51**(11): 2117–2125.
50. Burger-Kentischer A, Goebel H, Seiler R *et al.* (2002) Expression of macrophage migration inhibitory factor in different stages of human atherosclerosis. *Circulation* **105**(13): 1561–1566.
51. Shimizu T, Abe R, Ohkawara A, Nishihira J. (1999) Ultraviolet B radiation upregulates the production of macrophage migration inhibitory factor (MIF) in human epidermal keratinocytes. *J Invest Dermatol* **112**(2): 210–215.
52. Baugh JA, Gantier M, Li L *et al.* (2006) Dual regulation of macrophage migration inhibitory factor (MIF) expression in hypoxia by CREB and HIF-1. *Biochem Biophys Res Commun* **347**(4): 895–903.
53. Copple BL, Bai S, Burgoon LD, Moon JO. (2010) Hypoxia-inducible factor-1alpha regulates the expression of genes in hypoxic hepatic stellate cells important for collagen deposition and angiogenesis. *Liver Int* **31**(2): 230–244.
54. Leng L, Wang W, Roger T *et al.* (2009) Glucocorticoid-induced MIF expression by human CEM T cells. *Cytokine* **48**(3): 177–185.
55. Calandra T, Bernhagen J, Metz CN *et al.* (1995) MIF as a glucocorticoid-induced modulator of cytokine production. *Nature* **377**(6544): 68–71.
56. Bacher M, Schrader J, Thompson N *et al.* (2003) Up-regulation of macrophage migration inhibitory factor gene and protein expression in glial tumor cells during hypoxic and hypoglycemic stress indicates a critical role for angiogenesis in glioblastoma multiforme. *Am J Pathol* **162**(1): 11–17.
57. Tomiyasu M, Yoshino I, Suemitsu R *et al.* (2002) Quantification of macrophage migration inhibitory factor mRNA expression in non-small cell lung cancer tissues and its clinical significance. *Clin Cancer Res* **8**(12): 3755–3760.

58. Munaut C, Boniver J, Foidart JM, Deprez M. (2002) Macrophage migration inhibitory factor (MIF) expression in human glioblastomas correlates with vascular endothelial growth factor (VEGF) expression. *Neuropathol Appl Neurobiol* **28**(6): 452–460.
59. Meyer-Siegler KL, Bellino MA, Tannenbaum M. (2002) Macrophage migration inhibitory factor evaluation compared with prostate specific antigen as a biomarker in patients with prostate carcinoma. *Cancer* **94**(5): 1449–1456.
60. White ES, Strom SR, Wys NL, Arenberg DA. (2001) Non-small cell lung cancer cells induce monocytes to increase expression of angiogenic activity. *J Immunol* **166**(12): 7549–7555.
61. Ogawa H, Nishihira J, Sato Y *et al.* (2000) An antibody for macrophage migration inhibitory factor suppresses tumour growth and inhibits tumour-associated angiogenesis. *Cytokine* **12**(4): 309–314.
62. Shimizu T, Abe R, Nakamura H *et al.* (1999) High expression of macrophage migration inhibitory factor in human melanoma cells and its role in tumor cell growth and angiogenesis. *Biochem Biophys Res Commun* **264**(3): 751–758.
63. Chesney J, Metz C, Bacher M *et al.* (1999) An essential role for macrophage migration inhibitory factor (MIF) in angiogenesis and the growth of a murine lymphoma. *Mol Med* **5**(3): 181–191.
64. Waeber G, Thompson N, Chautard T *et al.* (1998) Transcriptional activation of the macrophage migration-inhibitory factor gene by the corticotropin-releasing factor is mediated by the cyclic adenosine 3′,5′-monophosphate responsive element-binding protein CREB in pituitary cells. *Mol Endocrinol* **12**(5): 698–705.
65. Elsby LM, Donn R, Alourfi Z *et al.* (2009) Hypoxia and glucocorticoid signaling converge to regulate macrophage migration inhibitory factor gene expression. *Arthritis Rheum* **60**(8): 2220–2231.
66. Roger T, Ding X, Chanson AL *et al.* (2007) Regulation of constitutive and microbial pathogen-induced human macrophage migration inhibitory factor (MIF) gene expression. *Eur J Immunol* **37**(12): 3509–3521.
67. Baugh JA, Chitnis S, Donnelly SC *et al.* (2002) A functional promoter polymorphism in the macrophage migration inhibitory factor (MIF) gene associated with disease severity in rheumatoid arthritis. *Genes Immun* **3**(3): 170–176.
68. Donn RP, Shelley E, Ollier WE, Thomson W. (2001) A novel 5′-flanking region polymorphism of macrophage migration inhibitory factor is associated with systemic-onset juvenile idiopathic arthritis. *Arthritis Rheum* **44**(8): 1782–1785.
69. Chen YC, Zhang XW, Niu XH *et al.* (2010) Macrophage migration inhibitory factor is a direct target of HBP1-mediated transcriptional repression that is over-expressed in prostate cancer. *Oncogene* **29**(21): 3067–3078.
70. Lugrin J, Ding XC, Le RD *et al.* (2009) Histone deacetylase inhibitors repress macrophage migration inhibitory factor (MIF) expression by targeting MIF gene

transcription through a local chromatin deacetylation. *Biochim Biophys Acta* **1793**(11): 1749–1758.

71. Irizarry RA, Ladd-Acosta C, Wen B *et al.* (2009) The human colon cancer methylome shows similar hypo- and hypermethylation at conserved tissue-specific CpG island shores. *Nat Genet* **41**(2): 178–186.
72. De Larco JE, Wuertz BR, Yee D *et al.* (2003) Atypical methylation of the interleukin-8 gene correlates strongly with the metastatic potential of breast carcinoma cells. *Proc Natl Acad Sci USA* **100**(24): 13988–13993.
73. Strathdee G, Davies BR, Vass JK *et al.* (2004) Cell type-specific methylation of an intronic CpG island controls expression of the MCJ gene. *Carcinogenesis* **25**(5): 693–701.
74. Shun CT, Lin JT, Huang SP *et al.* (2005) Expression of macrophage migration inhibitory factor is associated with enhanced angiogenesis and advanced stage in gastric carcinomas. *World J Gastroenterol* **11**(24): 3767–3771.
75. Krockenberger M, Dombrowski Y, Weidler C *et al.* (2008) Macrophage migration inhibitory factor contributes to the immune escape of ovarian cancer by down-regulating NKG2D. *J Immunol* **180**(11): 7338–7348.
76. Marks PA, Breslow R. (2007) Dimethyl sulfoxide to vorinostat: Development of this histone deacetylase inhibitor as an anticancer drug. *Nat Biotechnol* **25**(1): 84–90.
77. Chen J, Odenike O, Rowley JD. (2010) Leukaemogenesis: More than mutant genes. *Nat Rev Cancer* **10**(1): 23–36.
78. Ma X, Ezzeldin HH, Diasio RB. (2009) Histone deacetylase inhibitors: Current status and overview of recent clinical trials. *Drugs* **69**(14): 1911–1934.
79. Kuendgen A, Lubbert M. (2008) Current status of epigenetic treatment in myelodysplastic syndromes. *Ann Hematol* **87**(8): 601–611.
80. Mitchell RA, Metz CN, Peng T, Bucala R. (1999) Sustained mitogen-activated protein kinase (MAPK) and cytoplasmic phospholipase A2 activation by macrophage migration inhibitory factor (MIF). Regulatory role in cell proliferation and glucocorticoid action. *J Biol Chem* **274**(25): 18100–18106.
81. Swant JD, Rendon BE, Symons M, Mitchell RA. (2005) Rho GTPase-dependent signaling is required for macrophage migration inhibitory factor-mediated expression of cyclin D1. *J Biol Chem* **280**(24): 23066–23072.
82. Piette C, Deprez M, Roger T *et al.* (2009) The dexamethasone-induced inhibition of proliferation, migration, and invasion in glioma cell lines is antagonized by macrophage migration inhibitory factor (MIF) and can be enhanced by specific MIF inhibitors. *J Biol Chem* **284**(47): 32483–32492.
83. Chesney J, Metz C, Bacher M *et al.* (1999) An essential role for macrophage migration inhibitory factor (MIF) in angiogenesis and the growth of a murine lymphoma. *Mol Med* **5**(3): 181–191.

84. Meyer-Siegler KL, Iczkowski KA, Leng L *et al.* (2006) Inhibition of macrophage migration inhibitory factor or its receptor (CD74) attenuates growth and invasion of DU-145 prostate cancer cells. *J Immunol* **177**(12): 8730–8739.
85. Repp AC, Mayhew ES, Apte S, Niederkorn JY. (2000) Human uveal melanoma cells produce macrophage migration-inhibitory factor to prevent lysis by NK cells. *J Immunol* **165**(2): 710–715.
86. Abe R, Peng T, Sailors J *et al.* (2001) Regulation of the CTL response by macrophage migration inhibitory factor. *J Immunol* **166**(2): 747–753.
87. Zhou Q, Yan X, Gershan J *et al.* (2008) Expression of macrophage migration inhibitory factor by neuroblastoma leads to the inhibition of antitumor T cell reactivity *in vivo*. *J Immunol* **181**(3): 1877–1886.
88. Calandra T, Echtenacher B, Roy DL *et al.* (2000) Protection from septic shock by neutralization of macrophage migration inhibitory factor. *Nat Med* **6**(2): 164–170.
89. Mikulowska A, Metz CN, Bucala R, Holmdahl R. (1997) Macrophage migration inhibitory factor is involved in the pathogenesis of collagen type II-induced arthritis in mice. *J Immunol* **158**(11): 5514–5517.
90. de Jong YP, Abadia-Molina AC, Satoskar AR *et al.* (2001) Development of chronic colitis is dependent on the cytokine MIF. *Nat Immunol* **2**(11): 1061–1066.
91. Hoi AY, Hickey MJ, Hall P *et al.* (2006) Macrophage migration inhibitory factor deficiency attenuates macrophage recruitment, glomerulonephritis, and lethality in MRL/lpr mice. *J Immunol* **177**(8): 5687–5696.
92. Ferguson M, Henry PA, Currie RA. (2003) Histone deacetylase inhibition is associated with transcriptional repression of the Hmga2 gene. *Nucleic Acids Res* **31**(12): 3123–3133.
93. Duan H, Heckman CA, Boxer LM. (2005) Histone deacetylase inhibitors downregulate bcl-2 expression and induce apoptosis in t(14;18) lymphomas. *Mol Cell Biol* **25**(5): 1608–1619.

II-2

Regulation of MIF Gene Expression in the Lung

Lili Li and John Baugh*

1. Introduction

Although macrophage migration inhibitory factor (MIF) was originally described in 1966,[1,2] it was not until the cloning of the MIF cDNA in 1989, the subsequent expression of recombinant protein, and the generation of specific anti-MIF antibodies that the full extent of MIF actions began to become unraveled. Since this time, MIF has been well described as a potent proinflammatory mediator and has emerged as a key integrator of the immunoneuroendocrine interface and a potential regulator of tumorigenesis.[3–5]

Early in the history of MIF research, it was demonstrated that MIF expression increased in inflammatory tissues and in the serum of patients with inflammatory disease. Interestingly, it also became apparent that within both normal and diseased patient populations there was a large range in serum MIF levels. These data raised the postulate that MIF is regulated both locally and systemically and that genetic variations may render individuals prone to exaggerated MIF expression. The cloning of the human and mouse MIF genes in 1994 and 1995, respectively, allowed the initiation of more detailed studies into the transcriptional regulation of MIF expression and the starting place for analysis of genetic variation. This chapter will highlight the advances that have been made in the understanding of the regulation of MIF gene expression including the potential roles played by transcription factors and promoter polymorphisms. As few studies specifically address pulmonary expression of MIF, this chapter will provide a general overview of the mechanisms that regulate MIF gene expression and highlight those most pertinent to lung biology.

*Corresponding author: UCD Conway Institute of Biomolecular and Biomedical Research, School of Medicine and Medical Science, University College Dublin, Belfield, Dublin 4, Ireland. Email: john.baugh@ucd.ie

2. MIF Gene Structure

2.1 *Human*

A cDNA encoding MIF was isolated in 1989 from a lectin-stimulated T cell hybridoma library using functional expression cloning in Cos-1 cells.[6] The MIF cDNA was isolated from a library representing approximately 60,000 cloned mRNAs that were screened by transfection of pooled plasmid DNA into Cos-1 cells followed by functional analysis of Cos-1 supernatants in an *in vitro* macrophage migration assay. Plasmids that induced the secretion of a candidate factor from transfected Cos-1 were identified, enriched, and subdivided until a single active clone was identified. Sequencing of the identified plasmid revealed a 345-base pair (bp) open reading frame encoding a 115 amino acid polypeptide with a calculated molecular weight of 12,650 Da, which was in agreement with the apparent molecular weight of the MIF-specific protein band observed in metabolic labeling experiments. Analysis of this sequence also revealed that MIF did not contain the usual hydrophobic leader sequence normally associated with secreted proteins and implied a nonclassical method of protein secretion. Although this study identified a cDNA sequence for MIF, the actual gene product remained unpurified, and studies with supernatants from an "MIF" expressing transfected cell line were questioned because of the presence of phytohemagglutinin, the mitogen that induces MIF activity. It was not until the cloning and expression of recombinant mouse MIF in 1993[7] and human MIF in 1994[8] that advances into the biological activity of MIF were made.

Following the cloning of MIF, it became apparent that MIF was expressed in various nonlymphoid tissues, suggesting a role for MIF beyond the immune system. However, in order to gain a greater understanding of the tissue- and stimulus-specific expression of MIF, a more thorough analysis of the MIF gene was required. In 1994 Paralkar and Wistow were the first to clone the human *MIF* gene (GeneBank Accession number: L19686).[9] Restriction digested human genomic DNA revealed a single band when probed with human MIF cDNA, implying a single *MIF* gene. Screening of a human placental genomic library with the same MIF cDNA then identified a single phage, which was isolated and subcloned for sequencing. Sequence analysis revealed the full MIF coding sequence, and comparison with the human MIF cDNA sequence identified three exons of 204, 172, and 182 bp separated by introns of 189 and 95 bp. In addition, the isolated genomic DNA also revealed 1 kb of 5′ flanking sequence and about 250 bp of 3′ flanking sequence. Primer extension analysis from a primer that was complementary to a region in exon 1 produced a single faint product consistent with a start site 97 bp upstream of the initiator methionine. This start site was confirmed independently using 5′RACE and indicated that transcription of MIF started from a region that was flanked by a cAMP response element (CRE) and multiple sites for transcription factor specificity protein (Sp) 1 but not a classical TATA box. Indeed, analysis of the 5′ flanking region revealed putative CCAAT and TATA boxes that were 1,000 bp upstream of the transcriptional start site, but their role in MIF expression remains unclear.

Initial genetic mapping using PCR screening of a human chromosome panel suggested that the human MIF gene lay on chromosome 19.[10] Subsequent detailed analysis by the same group, however, showed the localization to actually be on chromosome 22.[11] Further PCR screening of a chromosome 22 somatic cell hybrid panel, which divides the chromosome into 24 separate regions, localized the MIF gene to region 11 in 22q11.2. This location was confirmed using fluorescence *in situ* hybridization (FISH).[11]

2.2 *Mouse*

A 543-bp cDNA encoding a murine homologue to human MIF was cloned in 1992 as part of a study analyzing growth factor-induced delayed early response genes in mouse fibroblasts.[12] The cloned sequence revealed a single open reading frame for a 115 amino acid polypeptide that shared an 88% sequence identity to human MIF and yielded a product of approximately 12 kDa upon *in vitro* transcription and translation. At the time, however, as the cDNA probe used to identify MIF hybridized to multiple rat, mouse, and human genomic restriction fragments, it was concluded that the identified cDNA belonged to a member of an MIF-related family. Independent studies cloned a cDNA for MIF from mouse pituitary in 1993, but it was not until 1995 that the full mouse MIF gene was cloned.[10,13] Southern blotting of a mouse liver genomic DNA with a probe derived from amplification of mouse MIF cDNA revealed multiple bands, indicative of either a closely homologous gene family or multiple pseudogenes. In contrast, high stringency Southern analysis with an apparent intronic probe that was generated by PCR amplification of mouse genomic DNA with MIF-specific exon primers yielded a more simple hybridization pattern indicative of a single gene copy.[13]

Screening of a genomic DNA phage library ultimately identified the full mouse MIF gene and two MIF pseudogenes. Independent studies using a PCR based approach also identified the functional full-length mouse MIF gene as well as nine processed pseudogenes.[10] The structure of the mouse MIF gene is found to be highly similar to that of humans with three exons separated by two short introns (201 and 145 nucleotides) and maps to chromosome 10.[10,13] 5′RACE and primer extension indicated that the transcriptional start site for mouse MIF was 85 bases upstream of the start methionine, and similar to the human MIF promoter, there is no classical TATA box in the immediate 5′ flanking region.[13]

3. Regulation of MIF Expression

MIF is ubiquitously expressed in response to various hormones, growth factors, and inflammatory stimuli such as lipopolysaccharide. Furthermore, MIF is expressed at high levels in many resting inflammatory cells and is stored preformed, ready for immediate secretion. Specific release of MIF is seen from macrophages in response to LPS, T cells in response to anti-CD3/CD28 stimulation, and fibroblasts in response to adhesion, integrin ligation, and growth factor stimulation.

Elevated expression of MIF has been described in many inflammatory pathologies, including acute respiratory distress syndrome, asthma, pulmonary fibrosis, septic shock, rheumatoid arthritis, and glomerulonephritis.[7,14–20] In contrast, decreased expression of MIF is associated with bronchopulmonary dysplasia,[21] suggesting that deficiencies in MIF are also pathologically relevant. Interestingly, elevated MIF expression is also found in a large variety of human malignancies. Prostate-, breast-, colon-, brain-, skin-, and lung-derived tumors have all been shown to contain significantly higher levels of MIF mRNA and protein than their noncancerous cell counterparts.[22–24] Despite the compelling evidence that MIF is over expressed in, and contributes to, the pathology of so many inflammatory and malignant diseases, the mechanisms that contribute to exaggerated expression of MIF have not been fully elucidated. Indeed, although more than 20 years have passed since the cloning of a human MIF cDNA, there is much to be learned about the transcriptional regulation of MIF gene during development, in normal health, and in disease.

3.1 *Transcription factors involved in MIF gene expression*

While species-specific putative DNA-binding sites have been identified in the human and in the mouse MIF promoters, promoter scanning software suggests several putative transcription factor binding sites in the immediate 5′ flanking region that are conserved between the murine and human MIF promoters. To date, however, a very limited number of response elements have been functionally tested. These include two cAMP response elements, a binding site for the transcription factor Sp1, a hypoxia inducible factor

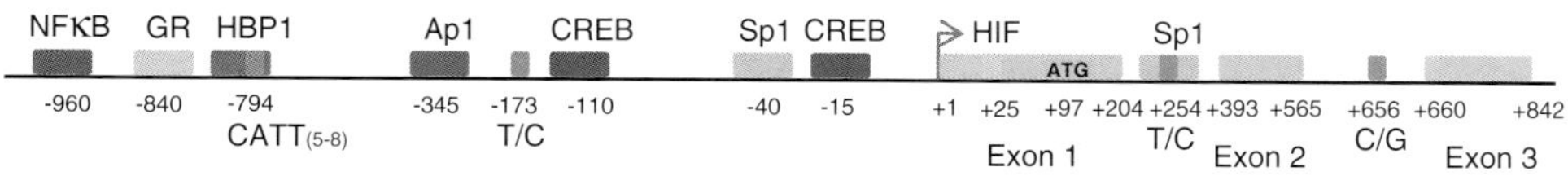

Fig. 1. Schematic representation of human MIF gene structure including promoter region, identified polymorphisms and transcription factor binding sites. Nuclear Factor Kappa B (NFκB), Glucocorticoid Receptor (GR), High Mobility Group Box-Containing Protein 1 (HBP1), Activator protein 1 (Ap1), Cyclic AMP Response Element Binding Protein (CREB), Specific Protein 1 (Sp1), Hypoxia-inducible Factor (HIF)

(HIF-1α) response element (HRE), a DNA-binding element for HMG box-containing transcription factor (HBP) 1, a glucocorticoid receptor element (GRE), and a DNA-binding element for transcription factor AP-1 in the human MIF promoter region.[25–30] The proximal CRE and the HRE within the 5′UTR have also been tested in the murine promoter.[26,31]

3.1.1 *CREB and Sp1*

Studies analyzing the regulation of MIF expression in the pituitary identified a proximal CRE (–41 to –48 nucleotides upstream of the transcriptional start site) that is essential for corticotrophin releasing factor (CRF)-induced transcription from the murine MIF promoter in rat pituitary cells.[26] The –1033 to +63 bp of the murine MIF promoter was cloned 5′ to a luciferase reporter gene and transiently transfected into freshly isolated rat anterior pituitary cells. This construct drove high basal transcriptional activity that was further enhanced by the activation of the cAMP protein kinase A pathway. These effects were paralleled by a similar increase in MIF mRNA expression.[26] Furthermore, a putative CRE was identified within the proximal MIF promoter region which, once mutated, abolished the cAMP responsiveness of the MIF promoter. Electromobility shift assay (EMSA) demonstrated that DNA-binding complexes were detected in pituitary nuclear extracts using the putative MIF CRE. It was also shown that this binding activity was specific and competed by an excess of unlabeled probe, but not with the mutated probe. CREB antibody disrupts the DNA-binding complex, implying that this factor is part of the complex.[26] Furthermore, recombinant CREB was shown to be able to bind to the CRE.[26] Taken together, these data demonstrate that CREB is the mediator of CRF-induced MIF gene transcription in anterior pituitary cells through an identified CRE in the proximal region of the MIF promoter.

The human MIF promoter also contains a proposed CRE at around the same location. A recent study carried out by Roger *et al.* reported that the proximal CRE within the human MIF promoter is also functional.[29] Truncation analysis using 5′ and 3′ end deletions suggested a possible *cis*-acting regulatory region located between –81 and +47 bp controlling the transcription of the MIF gene. Computer-assisted sequence analysis revealed a potential Sp1 binding site (–42 to –34) and a potential CRE (–20 to –13). Disruption of either the Sp1 or CRE sites reduced promoter activity significantly in human monocytic THP-1 cells. However, reduction in the MIF promoter activity by CRE deletion in human cervical epithelial HeLa cells was not statistically significant, suggesting that the CRE site might be less critical in HeLa cells that in THP-1 cells. In contrast, dual disruption of the Sp1 and CRE sites nearly fully abolished promoter activity in all the cell lines tested (THP-1, HeLa, A549, and HaCat). The binding of Sp1 and CREB to their corresponding sites was tested *in vitro* by EMSA and their identities were confirmed by supershift experiments. The binding of Sp1 and CREB to the native MIF promoter was further demonstrated *in vivo* by ChIP assays. The authors also investigated the possible contribution of methylation to MIF gene expression because, interestingly, the MIF gene was found to be located within a CpG island, and both Sp1 and CRE sites contained CpG sequences.[29,32,33] However, MIF mRNA levels in primary BM-derived macrophages, HeLa, A549, HaCat or THP-1 cells did not change upon treatment of cells with the demethylating agent 5-aza-2′deoxycitidine, suggesting that DNA methylation does not influence MIF gene expression despite its location within a CpG island. These findings indicate that Sp1 and CREB are critical regulators of constitutive expression of the MIF gene in a fairly broad range of cells, including myeloid (THP-1) cells, epithelial (HeLa and A549) cells, and keratinocytes (HaCat),[29] resulting in the accumulation of preformed intracellular pools of the MIF cytokine that is rapidly released upon exposure to microbial products.[7,34,35] Given that MIF promotes proinflammatory responses of innate immune cells and plays a critical role in host innate immune defenses against a broad range of microbial pathogens,[7,18,34,36–40] Roger and colleagues also examined whether MIF gene expression was affected by microbial products and which molecular mechanisms were involved.[29] It was shown that stimulation of THP-1 and PBMCs with *Escherichia coli* upregulated phosphorylated Sp1 nuclear content, Sp1 DNA-binding activity, MIF promoter activity, and MIF mRNA levels in a MEK1/2- and Sp1-dependent manner. Interestingly, while mutation of the proximal CRE site also markedly reduced MIF promoter activity in response to *E. coli*, DNA-binding activity to the proximal CRE site did not increase in cells stimulated with microbial products. It was previously shown that exposure of macrophages to LPS increased CREB-phosphorylation and CREB-dependent gene transcription without concomitant increase of CREB levels.[41] It was therefore proposed by the authors that microbial stimulation may increase MIF promoter activity and promote MIF expression in a CREB-dependent fashion without increasing CREB DNA-binding activity to the proximal CRE site. Taken together, these analyses have revealed that the binding of Sp1 and CREB to *cis*-acting regulatory sequences located in the proximal MIF promoter is essential for constitutive transcription of the MIF gene in several cell types and MIF gene upregulation in cells of the monocyte/macrophage lineage and PBMC after stimulation with microbial pathogens.

Sp-1 has also been implicated in the regulation of a cell type-specific enhancer region located in intron 1 of the MIF gene. Beaulieu *et al.* identified a DNAse I hypersensitivity site (HS) centered on intron 1 that contained two Sp-1 GC box sites.[42] The HS acted as an enhancer in human T lymphoblasts (CEMC7A), human embryonic kidney cells (HEK293T), and human monocytic cells (THP-1), but not in a fibroblast-like synoviocyte (FLS) cell line (SW982) or cultured FLS derived from rheumatoid arthritis (RA) patients. Sp-1 binding to one of the GC boxes was confirmed with EMSA, and mutation of either GC box inhibited enhancer activity. Furthermore, treatment of cells with mithramycin, which binds GC motifs and blocks SP-1 binding, inhibited MIF expression in CEMC7A cells in a manner that was dependent upon the intronic enhancer region.[42]

From the studies described, it is clear that both Sp-1 and CREB play a role in regulating both basal and stimulus-specific expression of MIF but that their mechanism of action may vary in different cell types. In addition to driving basal MIF expression or upregulating its expression in response to stimulation, CREB may also exert negative regulatory effects in HIF-1-mediated hypoxic induction of MIF,[25] which will be discussed in the following section.

3.1.2 *HIF-1*

Several recent studies have implicated MIF as a hypoxia-responsive gene in cancer, cardiac ischemic injury, and inflammatory responses. Gene array studies have shown MIF to be upregulated by hypoxia in squamous cell carcinoma and in the breast cancer cell line MCF-7.[43,44] In colon cancer, hypoxic induction of MIF has been shown to correlate with sensitivity to apoptosis.[45] Furthermore, a role for MIF in ischemic responses in cardiac tissues was proposed by Takahashi *et al.*, who described elevated MIF expression in patients with acute myocardial infarction and demonstrated that hypoxia induced the secretion of MIF from rat cardiac myocytes.[46] More recently, Bacher *et al.* have shown that MIF is highly expressed in glioblastoma multiforme and that hypoxic stimulation of a glioblastoma cell line resulted in increased expression of MIF mRNA and protein.[47] The mechanism by which hypoxia induces MIF expression has recently been shown to be dependent upon a functional HRE in the 5′UTR of the human MIF gene but also influenced by CREB activity.[25] It was shown that hypoxia is a potent inducer of MIF expression in both transformed and immortalized carcinoma cell line HeLa cells and nontransformed primary human lung fibroblast CCD-19Lu cells, which was accompanied by hypoxia-dependent induction of MIF promoter activity. Further studies by promoter truncation indicated that the hypoxia-responsive region is retained in the minimal promoter fragment (–25 to +103), within which a putative HRE was predicted at +25 in the 5′UTR by promoter analysis, in addition to a putative CRE at –20.[25] The putative HRE was naturally the primary candidate to explain the mechanism by which hypoxia drives MIF expression. However, several recent studies have highlighted the potential role for CREB. It was previously reported that CREB in approximately the same region is essential for CRF-induced MIF promoter activity in rat pituitary cells.[26] Furthermore, it is noteworthy that in a model of hypoxia-induced TNF-α production it was found that in the absence of PKA activity, CREB acted as a transcriptional repressor in normoxia, while during exposure to hypoxia this repressor activity was lost due to hyperphosphorylation, ubiquitination, and degradation of CREB.[48,49] Indeed, similar mechanisms were reported in hypoxia-driven MIF gene transcription.[25] Under conditions of hypoxia HIF-1 upregulated MIF promoter activity through activation of the putative HRE and mutating the HRE inhibited hypoxic response by 90%. Hypoxia-induced promoter activity could also be blocked by overexpression of CREB, while mutation of the CRE alone had no effect on hypoxia-induced transcriptional activity. As mutating the CRE would block CREB binding, these data support the view that in hypoxia CREB-mediated repression is lost. Furthermore, under normoxia, overexpression of HIF-1 stimulated both MIF promoter activity and endogenous MIF gene expression. This robust induction of promoter activity was completely blocked by mutation of the HRE in the 5′UTR, and mutation of the CRE significantly augmented HIF-1-induced activation of the MIF promoter, also supporting the hypothesis that binding of CREB in normoxia suppresses the activity of transcriptional complexes. Overall, it was proposed that hypoxia-induced MIF expression is driven by HIF-1 but amplified by hypoxia-induced degradation of CREB.[25]

The role of the HRE has also been tested in a murine model studying premature senescence where similar responses showed that the HRE in the 5′UTR is essential for hypoxia-induced murine MIF promoter activity.[31] A very recent study also demonstrated that hypoxia-induced MIF gene expression is HIF-1α dependent in human vascular smooth muscle cells (VSMCs).[50]

Given the importance of MIF in inflammatory and malignant disease, these data reveal a HIF-1-mediated pathway that is potentially responsible for the exaggerated MIF expression in hypoxic tissues.

3.1.3 *NF-κB*

The role of NF-κB in MIF transcription was proposed by Cao *et al.*, who showed an association between NF-κB and the expression of MIF in response to IL-1β and TNF-α in human endometrial cells.[51,52] IL-1β and TNF-α were convincingly shown to induce MIF mRNA expression and protein secretion in a dose-dependent manner in parallel with an increase in NF-κB translocation and IκB phosphorylation, but a direct interaction between NF-κB and the MIF promoter was not shown.

NF-κB has also been implicated in the regulation of MIF expression by estrogen. Studies addressing the delay in wound healing in the elderly associated decreased estrogen levels with increased serum and wound MIF level.[53] Extending from these data it was shown that estrogen inhibits basal and LPS-induced MIF promoter activity and that the effect of estrogen was abrogated by the NF-κB inhibitor pyrrolidine dithiocarbamate. Therefore, it was proposed that in collaboration with an active estrogen receptor (ER), NF-κB may also be essential in downregulating MIF expression, in contrast to the proposed role of NF-κB in driving IL-1β and TNF-α-induced upregulation of MIF.[51,52] The precise interaction between ER and NF-κB on the MIF promoter was not investigated and is yet to be fully elucidated.

The role of NF-κB in MIF gene expression was further explored in an attempt to understand MIF upregulation in atherosclerotic lesions. It was shown that oxidized low-density lipoprotein (oxLDL), a proatherogenic factor, induces MIF mRNA and protein expression accompanied by activated NF-κB pathway in rabbit vascular smooth muscle cells (VSMCs).[54] OxLDL-dependent induction of MIF promoter activity was mapped to two consensus NF-κB binding sites (–964 to –955 and –973 to –965) and mutation or deletion of these two sites in murine MIF promoter abolished oxLDL-enhanced MIF transcription. It was further shown that the induction of MIF by oxLDL can be blocked by IkB overexpression, implying a role of NF-κB in oxLDL-mediated MIF expression.

3.1.4 *Hormone receptors*

Over recent years there has been great interest in hormonal regulation of MIF expression, particularly the relationship between MIF and glucocorticoid biology. Studies in the early 1990s rediscovered MIF as a pituitary-derived mediator of systemic stress responses.[55] In addition to being secreted from the pituitary, MIF was found to be released from immune/inflammatory cells as a consequence of glucocorticoid stimulation.[36,56] Further analysis in rodents suggested that MIF protein content was induced in several organs by parenteral glucocorticoid administration, suggesting a direct effect on MIF production, as well as on MIF secretion. However, this induction of MIF was not accompanied by changes in MIF gene expression, and post-translational mechanisms were proposed.[57]

More recent studies, however, demonstrated that glucocorticoids regulate MIF gene transcription in a cell-type-dependent and concentration-sensitive manner. Alourfi *et al.* reported that dexamethasone (Dex) represses MIF promoter activity by 50% at 10^{-6} M in CEM C7A cells, a diploid clone from glucocorticoid-sensitive human T lymphoblast cell line CEM C7.[58] Endogenous MIF mRNA was also repressed by glucocorticoids at the same concentration in CEM C7A cells.[58] However, there was no such regulation in human lung epithelial A549 cells, suggesting that this suppression functions in a cell-type-specific manner.[58] Extending from these findings, it was later identified by the same group that two *cis* elements, within the proximal MIF 5′ region, are required for basal MIF gene expression and glucocorticoid-dependent MIF inhibition in CEM C7A cells.[59] When both sites were deleted, a significant reduction was observed in the repressive effect of glucocorticoids on MIF promoter activity, while deletion of one of the two sites alone had no significant effect. While neither of these *cis* elements presents a consensus GRE, a tethering mechanism was proposed for glucocorticoid receptor (GR) association with the MIF promoter. Analysis by EMSA confirmed that the 5′ element, the known proximal CRE, can bind CREB/activating transcription factor (ATF) 1, and the 3′ element, the HRE with 5′UTR, can bind HIF-1α in normoxic CEM C7A cells. Furthermore, it was shown by ChIP assays that GRα was recruited to the MIF gene proximal promoter after stimulation with 10^{-6} M dexamethasone, in parallel with a reduced expression of the MIF gene as demonstrated by a progressive decline in histone H3 acetylation of the proximal MIF promoter. Therefore, it was hypothesized that GRα can be recruited to the CRE *via* CREB/ATF-1 and to the HRE *via* HIF-1α and that recruitment of GRα to the MIF promoter subsequently assembles transcription co-modulators and histone deacetylases and inhibits the recruitment of HATs, resulting in inhibition of MIF gene transcription. Interestingly, deletion studies suggested that the proximal CRE also serves to potentiate basal gene expression of MIF in A549 cells, but no binding of GRα to the MIF promoter was seen. The authors concluded that this repression is a cell-type specific mechanism, likely due to altered transcription factor loading, supported by significantly lower abundance of HIF-1α in normoxic A549 cells while functional GRα expression was observed.[59]

MIF regulation by glucocorticoids is complex. Low, physiologic levels of glucocorticoids promote MIF release by macrophages, synoviocytes, and neurons.[36,60,61] This release response follows a bell-shaped dose response curve with high anti-inflammatory concentrations of glucocorticoids ($>10^{-6}$ M) suppressing MIFsecretion.[36] A recent study by Leng *et al.* suggested that at the transcriptional level, glucocorticoid regulation of MIF may also be concentration-sensitive, in addition to being cell-type-specific, at least in lymphoid tissue.[28] In contrast to the proposed role of GR in mediating glucocorticoid-induced MIF repression (10^{-6} M Dex) in CEM C7A cells, GR also may be essential in activating MIF gene expression in CEM C7 cells stimulated with low concentrations of dexamethasone.[28] It was shown that dexamethasone in low concentrations (10^{-8} to 10^{-14}) induced MIF secretion from the glucocorticoid-sensitive CEM T cells (C7 clone), but not the glucocorticoid-resistant C1 clone. Glucocorticoid stimulation of CEM C7 T cells was accompanied by an MIF transcriptional response. The highest levels of MIF mRNA were observed at a dexamethasone concentration (10^{-12} M) that corresponded with the highest levels of MIF protein production. Promoter analysis of the 1177-bp upstream segment of the MIF promoter identified two putative transcription factor binding sites (a GRE at –847 to –837 and a ATF/CRE at –116 to –109), which are both involved in glucocorticoid-dependent MIF induction.[28] Removal of either the GRE or the ATF/CRE site resulted in a 10–15% reduction in baseline MIF transcriptional activity and a loss of glucocorticoid inducibility, with the ATF/CRE site possibly contributing more to this responsiveness that the GRE site. Furthermore, DNA binding at the ARF/CRE site increased significantly upon dexamethasone treatment of T cells. A role of GR in this transcriptional response was confirmed by co-transfection of a GR expression plasmid, which enhanced MIF promoter activity in a dose-dependent fashion.[28] Glucocorticoids also influence gene expression by regulating the degradation of mRNA.[62] However, in the study by Leng *et al.* it was shown that dexamethasone did not influence the turnover of MIF mRNA in CEM C7 T cells under the experimental conditions; therefore, a direct transcriptional regulatory mechanism was proposed.[28]

Taken together, while prior studies have attributed changes in the MIF content of different organs to post-translational regulatory mechanisms, these observations that MIF release is accompanied by transcriptional regulation in a cell-specific and concentration-dependent manner is noteworthy and point to a role for glucocorticoids in both constitutive and inducible MIF expression, particularly in lymphoid tissue.

3.1.5 *HBP1*

Much work has focused on transcription factors that promote MIF expression but more recently an important association has been made between HBP1, a transcriptional repressor and a member of the sequence-specific HMG box family of transcription factors, and repression of MIF expression. Promoter analysis by Chen *et al.* identified a long HBP1 high-affinity binding site in the MIF promoter at positions –811 to –792.[27] Previous studies have shown that HBP1 and MIF exert opposite effects in regard to tumor cell growth.[63–67] It was therefore proposed by Chen *et al.* that HBP1 may act as a transcriptional repressor of MIF expression. It was demonstrated that RWPE2-W99 cells (prostate epithelial cell lines) express higher levels of HBP1 and significantly lower levels of MIF than do DU-145 cells (prostate carcinoma cell lines), suggesting a link between HBP1 and MIF expression. It was further shown that the activity of a promoter construct containing a 1069-bp human MIF promoter region was significantly reduced when transfected into RWPE2-W99 cells compared to DU-145 cells. Reporter constructs lacking the predicted HBP1-binding site or with a mutated HBP-1 binding site did not exhibit HBP1-mediated repression. ChIP assays confirmed that HBP1 binds to the predicted HBP1-binding site in the MIF promoter and inhibits endogenous MIF expression.[27] Mutations within the HBP1-binding site prevented its ability to inhibit MIF expression. Both DNA binding and repression domains are required for HBP1-mediated repression of MIF expression, demonstrated by mutation and deletion studies, respectively. Furthermore, HBP1 expression correlated negatively with MIF expression in prostate tumor samples, and the repressive effect of HBP1 on cancer cell growth and invasion could be partially rescued by the addition of rMIF to the culture medium.[27] Compelling evidence shows that MIF overexpression is associated with, and contributes to, the pathology of malignant disorders.[68–73] This study is one of the first functional investigations into defective or aberrant repression mechanisms that contribute to MIF overexpression in tumors.

3.1.6 *AP1*

Compelling evidence has collectively established the remarkable activity of MIF in the development of atherosclerosis. MIF is highly expressed in advanced atherosclerotic lesions and exhibits a tight correlation with the recruitment of inflammatory cells and neointimal microvessel formation.[74] Neutralizing MIF with a blocking antibody was shown to induce a regression of established atherosclerotic lesions and stabilize atherosclerotic plaques,[75,76] suggesting an important role for MIF in atherosclerosis and a need for better understanding of the mechanism leading to elevated MIF expression in the atherosclerotic lesions. As previously discussed in this chapter, the proatherogenic factor oxLDL can upregulate MIF gene expression *via* NF-κB in SMCs; it has also been reported that angiotensin (Ang) II, the main effector of the renin-angiotensin system, strongly co-expressed with MIF in human atherosclerotic plaques and induced MIF expression in neurons in rats and endothelial cells in humans.[74,77–79] A very recent study has revealed that Ang II-dependent MIF expression is mediated by the transcription factor AP-1 in human endothelial cells.[30] It was shown that Ang II induces MIF expression in human umbilical vascular endothelial cells (HUVECs) in a time- and dose-dependent manner. By using promoter analysis and gene reporters, it was demonstrated that the putative AP-1, located at –349 to –339, is essential for MIF promoter activity. In HUVECs with inhibited expression of *c-Jun*, a component of AP-1, both MIF promoter activity and endogenous MIF expression were significantly inhibited. The AP-1 inhibitor CHX also efficiently inhibited MIF promoter activity. Given that Ang II specifically increased AP1 activity in HUVECs, these data suggest that Ang II mediated the upregulation of MIF gene expression in HUVECs cells *via* AP-1.[30]

3.2 *MIF gene polymorphism*

With the advent of sensitive assays for quantifying serum MIF levels, it became apparent that significantly elevated MIF levels were associated with many inflammatory and malignant pathologies. Interestingly, it also became apparent that within both normal and diseased populations there was a broad range in serum MIF levels. Such observed variation in expression could be partly contributed to by the circadian regulation of MIF[80] but raised the question of whether genetic variation determined susceptibility to exaggerated MIF expression.

The concept that polymorphisms in immune response genes contribute to the pathogenesis of certain human autoimmune/inflammatory diseases has received increasing interest over the last decade. At present, few gene polymorphisms have been shown to be functionally significant and to be of prognostic value in specific disease states. Previously defined examples include polymorphisms in TNF-α and IL-1 receptor antagonist (IL-1ra), which have been shown to have certain prognostic significance in malaria and ischemic heart disease, respectively.[81,82] Similarly, a number of polymorphisms in TNF-α and IL-1β have been reported to be associated with rheumatoid arthritis severity.[83–85]

Several polymorphic regions of the human MIF gene have been identified, including −794 $CATT_{5-8}$, −173 G/C, +254 T/C, and +656 C/G. Sequencing of the MIF gene in six individuals highlighted a tetranucleotide CATT-repeat at position −794, which was originally described by Paralkar and Wistow,[9] and revealed a variation in the number of repeats.[86] Further screening of a larger population identified individuals that were homozygous or heterozygous for 5-, 6-, 7- or 8-CATT repeats. Independent studies using high performance denaturing liquid chromatography also revealed the existence of a G-to-C single nucleotide polymorphism (SNP) at position −173 of the human MIF promoter[87] as well as a T-to-C SNP at position +254 (intron 1), and a C-to-G SNP at position +656 (intron 2).[86] It is currently unclear whether the intronic SNPs play any role in regulating MIF expression, but it has become well established that both the −794 CATT repeat polymorphism and the −173 SNP affect transcriptional activity of the human MIF promoter.

In vitro analysis has shown that MIF promoter activity is proportional to the number of CATT repeats at position −794 with the 5-CATT allele conferring significantly lower basal as well as serum- and forskolin-induced transcriptional activity.[88] Similarly, it has been shown that the −173 C variant also increases MIF promoter activity.[86] Multiple studies have correlated carriage of non-5 CATT alleles and −173 C alleles with exaggerated inflammatory disease, supporting the hypothesis that these variants are responsible for elevated MIF expression, but the precise mechanism by which these polymorphisms affect MIF promoter activity is currently unknown. Promoter scanning software shows that the pituitary-specific transcription factor Pit-1 could potentially bind to the CATT-repeat region. Such an interaction has been confirmed *in vitro* using electromobility shift assays (*unpublished*), but the functional significance of this interaction is unclear. Given the proposed role of pituitary-derived MIF in determining circulating MIF levels, it is attractive to propose that Pit-1 may play a role in determining an individual's genetic susceptibility to enhanced MIF expression and exaggerated inflammatory responses, but further studies are required to address this.

Analysis of the −173 position of the MIF promoter reveals that the mutation to a cytosine residue generates a putative activator protein (AP) 4 binding site.[86] Despite clear evidence that this site is involved in exaggerated MIF expression, the transcriptional complexes involved have not been identified and the role of AP-4 is unproven.

4. Summary

Since the cloning of MIF cDNA, significant advances have been made in the understanding of MIF biology. Identification of the specific transcription factors and their associated binding sites within the MIF promoter region has further added to our knowledge in terms of the tissue- and stimulus-specific regulation of MIF expression. Within these data, it has been proposed that Sp1 and CREB are responsible for the constitutive expression of the MIF gene in a broad range of cell types, including human lung epithelial cells.[29] The proximal Sp1 and CRE sites have also been detected in mouse MIF promoter, and the CRE site is essential for CRF-induced murine MIF gene transcription.[26] Furthermore, hypoxic upregulation of MIF expression is mediated by the interaction of HIF-1α with the HRE in the 5′UTR in both human and mouse.[25,31] One therefore can postulate that common regulatory elements are implicated in constitutive and induced expression of the human and mouse MIF gene. In particular, the observation that both Sp1 and CREB are required for optimal transcription of MIF is consistent with the fact that the MIF gene lacks a TATA box. In the absence of a TATA box, Sp1 facilitates the binding of the transcription factor TFIID and assembly of the preinitiation transcriptional complex needed for RNA polymerase II binding to the promoter.[89,90] Likewise, CREB, CREB-binding protein (CBP) and its paralogue p300 may interact with TFIID and trigger the recruitment of RNA polymerase II complexes.[32,91] The roles of NF-κB, GR, HBP1, and AP-1 have only been investigated in a limited range of tissues and species, and to date it is considered to function in a tissue-specific manner.

This chapter has also highlighted the association between MIF promoter polymorphisms and MIF gene expression. The CATT and −173 SNP variants have been individually linked with regulation of MIF expression and disease severity in many inflammatory diseases. Specifically, it has been shown that the 5-CATT is a low expressing allele that protects against disease progression in cystic fibrosis and asthma.[15,92] In addition, the −173 C allele has been linked with the increased MIF expression.[86] Nevertheless, it is important to note that several studies have also made an association between the 7-CATT allele and the −173 C allele.[93–98] Taken together, these data highlight the possibility that genetic susceptibility to enhanced MIF expression may be a critical determinant in the outcome of inflammatory diseases, including asthma and cystic fibrosis, and strongly influence the efficacy of glucocorticoid treatment. Furthermore, identification of patients susceptible to exaggerated MIF expression may play an important role in targeting focused anti-MIF therapeutic strategies to those patients most likely to benefit.

References

1. Bloom BR, Bennett B. (1966) Mechanism of a reaction *in vitro* associated with delayed-type hypersensitivity. *Science* **153**: 80–82.

2. David JR. (1966) Delayed hypersensitivity *in vitro*: Its mediation by cell-free substances formed by lymphoid cell-antigen interaction. *Proc Natl Acad Sci USA* **56**: 72–77.
3. Baugh JA, Bucala R. (2002) Macrophage migration inhibitory factor. *Crit Care Med* **30**: S27–S35.
4. Baugh JA, Donnelly SC. (2003) Macrophage migration inhibitory factor: A neuroendocrine modulator of chronic inflammation. *J Endocrinol* **179**: 15–23.
5. Mitchell RA. (2004) Mechanisms and effectors of MIF-dependent promotion of tumourigenesis. *Cell Signal* **16**: 13–19.
6. Weiser WY *et al.* (1989) Molecular cloning of a cDNA encoding a human macrophage migration inhibitory factor. *Proc Natl Acad Sci USA* **86**: 7522–7526.
7. Bernhagen J *et al.* (1993) MIF is a pituitary-derived cytokine that potentiates lethal endotoxaemia. *Nature* **365**: 756–759.
8. Bernhagen J *et al.* (1994) Purification, bioactivity, and secondary structure analysis of mouse and human macrophage migration inhibitory factor (MIF). *Biochemistry* **33**: 14144–14155.
9. Paralkar V, Wistow G. (1994) Cloning the human gene for macrophage migration inhibitory factor (MIF). *Genomics* **19**: 48–51.
10. Kozak CA *et al.* (1995) Genomic cloning of mouse MIF (macrophage inhibitory factor) and genetic mapping of the human and mouse expressed gene and nine mouse pseudogenes. *Genomics* **27**: 405–411.
11. Budarf M *et al.* (1997) Localization of the human gene for macrophage migration inhibitory factor (MIF) to chromosome 22q11.2. *Genomics* **39**: 235–236.
12. Lanahan A, Williams JB, Sanders LK, Nathans D. (1992) Growth factor-induced delayed early response genes. *Mol Cell Biol* **12**: 3919–3929.
13. Mitchell R *et al.* (1995) Cloning and characterization of the gene for mouse macrophage migration inhibitory factor (MIF). *J Immunol* **154**: 3863–3870.
14. Donnelly SC *et al.* (1997) Regulatory role for macrophage migration inhibitory factor in acute respiratory distress syndrome. *Nat Med* **3**: 320–323.
15. Mizue Y *et al.* (2005) Role for macrophage migration inhibitory factor in asthma. *Proc Natl Acad Sci USA* **102**: 14410–14415.
16. Rossi AG *et al.* (1998) Human circulating eosinophils secrete macrophage migration inhibitory factor (MIF). Potential role in asthma. *J Clin Invest* **101**: 2869–2874.
17. Magi B *et al.* (2002) Bronchoalveolar lavage fluid protein composition in patients with sarcoidosis and idiopathic pulmonary fibrosis: A two-dimensional electrophoretic study. *Electrophoresis* **23**: 3434–3444.
18. Calandra T *et al.* (2000) Protection from septic shock by neutralization of macrophage migration inhibitory factor. *Nat Med* **6**: 164–170.
19. Morand EF *et al.* (2002) Macrophage migration inhibitory factor in rheumatoid arthritis: Clinical correlations. *Rheumatology (Oxford)* **41**: 558–562.

20. Lan HY *et al.* (2000) Expression of macrophage migration inhibitory factor in human glomerulonephritis. *Kidney Int* **57**: 499–509.
21. Kevill KA *et al.* (2008) A role for macrophage migration inhibitory factor in the neonatal respiratory distress syndrome. *J Immunol* **180**: 601–608.
22. Takahashi N *et al.* (1998) Involvement of macrophage migration inhibitory factor (MIF) in the mechanism of tumor cell growth. *Mol Med* **4**: 707–714.
23. Meyer-Siegler K, Fattor RA, Hudson PB. (1998) Expression of macrophage migration inhibitory factor in the human prostate. *Diagn Mol Pathol* **7**: 44–50.
24. Tomiyasu M, Yoshino I, Suemitsu R *et al.* (2002) Quantification of macrophage migration inhibitory factor mRNA expression in non-small cell lung cancer tissues and its clinical significance. *Clin Cancer Res* **8**: 3755–3760.
25. Baugh JA *et al.* (2006) Dual regulation of macrophage migration inhibitory factor (MIF) expression in hypoxia by CREB and HIF-1. *Biochem Biophys Res Commun* **347**: 895–903.
26. Waeber G *et al.* (1998) Transcriptional activation of the macrophage migration-inhibitory factor gene by the corticotropin-releasing factor is mediated by the cyclic adenosine 3′,5′-monophosphate responsive element-binding protein CREB in pituitary cells. *Mol Endocrinol* **12**: 698–705.
27. Chen YC *et al.* (2010) Macrophage migration inhibitory factor is a direct target of HBP1-mediated transcriptional repression that is overexpressed in prostate cancer. *Oncogene* **29**: 3067–3078.
28. Leng L *et al.* (2009) Glucocorticoid-induced MIF expression by human CEM T cells. *Cytokine* **48**: 177–185.
29. Roger T, Ding X, Chanson AL *et al.* (2007) Regulation of constitutive and microbial pathogen-induced human macrophage migration inhibitory factor (MIF) gene expression. *Eur J Immunol* **37**: 3509–3521.
30. Shan ZX *et al.* (2010) Transcription factor Ap-1 mediates proangiogenic MIF expression in human endothelial cells exposed to Angiotensin II. *Cytokine* **53**(1): 35–41.
31. Welford SM *et al.* (2006) HIF1alpha delays premature senescence through the activation of MIF. *Genes Dev* **20**: 3366–3371.
32. Mayr B, Montminy M. (2001) Transcriptional regulation by the phosphorylation-dependent factor CREB. *Nat Rev Mol Cell Biol* **2**: 599–609.
33. Suske G. (1999) The Sp-family of transcription factors. *Gene* **238**: 291–300.
34. Bacher M *et al.* (1997) Migration inhibitory factor expression in experimentally induced endotoxemia. *Am J Pathol* **150**: 235–246.
35. Calandra T, Bernhagen J, Mitchell RA, Bucala R. (1994) The macrophage is an important and previously unrecognized source of macrophage migration inhibitory factor. *J Exp Med* **179**: 1895–1902.
36. Calandra T *et al.* (1995) MIF as a glucocorticoid-induced modulator of cytokine production. *Nature* **377**: 68–71.

37. Roger T, David J, Glauser MP, Calandra T. (2001) MIF regulates innate immune responses through modulation of Toll-like receptor 4. *Nature* **414**: 920–924.
38. Roger T, Froidevaux C, Martin C, Calandra T. (2003) Macrophage migration inhibitory factor (MIF) regulates host responses to endotoxin through modulation of Toll-like receptor 4 (TLR4). *J Endotoxin Res* **9**: 119–123.
39. Mitchell RA, Metz CN, Peng T, Bucala R. (1999) Sustained mitogen-activated protein kinase (MAPK) and cytoplasmic phospholipase A2 activation by macrophage migration inhibitory factor (MIF). Regulatory role in cell proliferation and glucocorticoid action. *J Biol Chem* **274**: 18100–18106.
40. Calandra T, Froidevaux C, Martin C, Roger T. (2003) Macrophage migration inhibitory factor and host innate immune defenses against bacterial sepsis. *J Infect Dis* **187**(Suppl 2): S385–S390.
41. Park JM *et al.* (2005) Signaling pathways and genes that inhibit pathogen-induced macrophage apoptosis — CREB and NF-kappaB as key regulators. *Immunity* **23**: 319–329.
42. Beaulieu E *et al.* Identification of a novel cell type-specific intronic enhancer of macrophage migration inhibitory factor (MIF) and its regulation by mithramycin. *Clin Exp Immunol* **163**: 178–188.
43. Bando H, Toi M, Kitada K, Koike M. (2003) Genes commonly upregulated by hypoxia in human breast cancer cells MCF-7 and MDA-MB-231. *Biomed Pharmacother* **57**: 333–340.
44. Koong AC *et al.* (2000) Candidate genes for the hypoxic tumor phenotype. *Cancer Res* **60**: 883–887.
45. Yao K *et al.* (2005) Macrophage migration inhibitory factor is a determinant of hypoxia-induced apoptosis in colon cancer cell lines. *Clin Cancer Res* **11**: 7264–7272.
46. Takahashi M *et al.* (2001) Macrophage migration inhibitory factor as a redox-sensitive cytokine in cardiac myocytes. *Cardiovasc Res* **52**: 438–445.
47. Bacher M *et al.* (2003) Up-regulation of macrophage migration inhibitory factor gene and protein expression in glial tumor cells during hypoxic and hypoglycemic stress indicates a critical role for angiogenesis in glioblastoma multiforme. *Am J Pathol* **162**: 11–17.
48. Taylor CT, Fueki N, Agah A *et al.* (1999) Critical role of cAMP response element binding protein expression in hypoxia-elicited induction of epithelial tumor necrosis factor-alpha. *J Biol Chem* **274**: 19447–19454.
49. Taylor CT, Furuta GT, Synnestvedt K, Colgan SP. (2000) Phosphorylation-dependent targeting of cAMP response element binding protein to the ubiquitin/proteasome pathway in hypoxia. *Proc Natl Acad Sci USA* **97**: 12091–12096.

50. Fu H, Luo F, Yang L *et al.* (2010) Hypoxia stimulates the expression of macrophage migration inhibitory factor in human vascular smooth muscle cells via HIF-1alpha dependent pathway. *BMC Cell Biol* **11**: 66.
51. Cao WG, Morin M, Metz C *et al.* (2005) Stimulation of macrophage migration inhibitory factor expression in endometrial stromal cells by interleukin 1, beta involving the nuclear transcription factor NFkappaB. *Biol Reprod* **73**: 565–570.
52. Cao WG *et al.* (2006) Tumour necrosis factor-alpha up-regulates macrophage migration inhibitory factor expression in endometrial stromal cells via the nuclear transcription factor NF-kappaB. *Hum Reprod* **21**: 421–428.
53. Hardman MJ *et al.* (2005) Macrophage migration inhibitory factor: A central regulator of wound healing. *Am J Pathol* **167**: 1561–1574.
54. Chen L *et al.* (2009) Induction of MIF expression by oxidized LDL via activation of NF-kappaB in vascular smooth muscle cells. *Atherosclerosis* **207**: 428–433.
55. Bucala R. (1996) MIF rediscovered: Cytokine, pituitary hormone, and glucocorticoid-induced regulator of the immune response. *FASEB J* **10**: 1607–1613.
56. Leech M, Metz C, Bucala R, Morand EF. (2000) Regulation of macrophage migration inhibitory factor by endogenous glucocorticoids in rat adjuvant-induced arthritis. *Arthritis Rheum* **43**: 827–833.
57. Fingerle-Rowson G *et al.* (2003) Regulation of macrophage migration inhibitory factor expression by glucocorticoids *in vivo*. *Am J Pathol* **162**: 47–56.
58. Alourfi Z *et al.* (2005) Glucocorticoids suppress macrophage migration inhibitory factor (MIF) expression in a cell-type-specific manner. *J Mol Endocrinol* **34**: 583–595.
59. Elsby LM *et al.* (2009) Hypoxia and glucocorticoid signaling converge to regulate macrophage migration inhibitory factor gene expression. *Arthritis Rheum* **60**: 2220–2231.
60. Santos L, Hall P, Metz C *et al.* (2001) Role of macrophage migration inhibitory factor (MIF) in murine antigen-induced arthritis: Interaction with glucocorticoids. *Clin Exp Immunol* **123**: 309–314.
61. Vedder H, Krieg JC, Gerlach B *et al.* (2000) Expression and glucocorticoid regulation of macrophage migration inhibitory factor (MIF) in hippocampal and neocortical rat brain cells in culture. *Brain Res* **869**: 25–30.
62. Rhen T, Cidlowski JA. (2005) Antiinflammatory action of glucocorticoids — New mechanisms for old drugs. *N Engl J Med* **353**: 1711–1723.
63. Yee AS *et al.* (2004) The HBP1 transcriptional repressor and the p38 MAP kinase: Unlikely partners in G1 regulation and tumor suppression. *Gene* **336**: 1–13.
64. Zhang X *et al.* (2006) The HBP1 transcriptional repressor participates in RAS-induced premature senescence. *Mol Cell Biol* **26**: 8252–8266.

65. Hudson JD *et al.* (1999) A proinflammatory cytokine inhibits p53 tumor suppressor activity. *J Exp Med* **190**: 1375–1382.
66. Mitchell RA, Bucala R. (2000) Tumor growth-promoting properties of macrophage migration inhibitory factor (MIF). *Semin Cancer Biol* **10**: 359–366.
67. Petrenko O, Fingerle-Rowson G, Peng T *et al.* (2003) Macrophage migration inhibitory factor deficiency is associated with altered cell growth and reduced susceptibility to Ras-mediated transformation. *J Biol Chem* **278**: 11078–11085.
68. Bini L *et al.* (1997) Protein expression profiles in human breast ductal carcinoma and histologically normal tissue. *Electrophoresis* **18**: 2832–2841.
69. Kamimura A *et al.* (2000) Intracellular distribution of macrophage migration inhibitory factor predicts the prognosis of patients with adenocarcinoma of the lung. *Cancer* **89**: 334–341.
70. Meyer-Siegler K, Hudson PB. (1996) Enhanced expression of macrophage migration inhibitory factor in prostatic adenocarcinoma metastases. *Urology* **48**: 448–452.
71. Meyer-Siegler KL, Leifheit EC, Vera PL. (2004) Inhibition of macrophage migration inhibitory factor decreases proliferation and cytokine expression in bladder cancer cells. *BMC Cancer* **4**: 34.
72. Ren Y *et al.* (2003) Macrophage migration inhibitory factor: Roles in regulating tumor cell migration and expression of angiogenic factors in hepatocellular carcinoma. *Int J Cancer* **107**: 22–29.
73. Shimizu T *et al.* (1999) High expression of macrophage migration inhibitory factor in human melanoma cells and its role in tumor cell growth and angiogenesis. *Biochem Biophys Res Commun* **264**: 751–758.
74. Schmeisser A *et al.* (2005) The expression of macrophage migration inhibitory factor 1alpha (MIF 1alpha) in human atherosclerotic plaques is induced by different proatherogenic stimuli and associated with plaque instability. *Atherosclerosis* **178**: 83–94.
75. Bernhagen J *et al.* (2007) MIF is a noncognate ligand of CXC chemokine receptors in inflammatory and atherogenic cell recruitment. *Nat Med* **13**: 587–596.
76. Schober A *et al.* (2004) Stabilization of atherosclerotic plaques by blockade of macrophage migration inhibitory factor after vascular injury in apolipoprotein E-deficient mice. *Circulation* **109**: 380–385.
77. Busche S, Gallinat S, Fleegal MA *et al.* (2001) Novel role of macrophage migration inhibitory factor in angiotensin II regulation of neuromodulation in rat brain. *Endocrinology* **142**: 4623–4630.
78. Gallinat S, Busche S, Yang H *et al.* (2001) Gene expression profiling of rat brain neurons reveals angiotensin II-induced regulation of calmodulin and synapsin I: Possible role in neuromodulation. *Endocrinology* **142**: 1009–1016.

79. Zhong JC *et al.* (2008) Enhanced angiotensin converting enzyme 2 regulates the insulin/Akt signalling pathway by blockade of macrophage migration inhibitory factor expression. *Br J Pharmacol* **153**: 66–74.
80. Petrovsky N *et al.* (2003) Macrophage migration inhibitory factor exhibits a pronounced circadian rhythm relevant to its role as a glucocorticoid counter-regulator. *Immunol Cell Biol* **81**: 137–143.
81. McGuire W, Hill AV, Allsopp CE *et al.* (1994) Variation in the TNF-alpha promoter region associated with susceptibility to cerebral malaria. *Nature* **371**: 508–510.
82. Francis SE *et al.* (1999) Interleukin-1 receptor antagonist gene polymorphism and coronary artery disease. *Circulation* **99**: 861–866.
83. Kaijzel EL *et al.* (1998) Functional analysis of a human tumor necrosis factor alpha (TNF-alpha) promoter polymorphism related to joint damage in rheumatoid arthritis. *Mol Med* **4**: 724–733.
84. Fabris M *et al.* (2002) Tumor necrosis factor-alpha gene polymorphism in severe and mild-moderate rheumatoid arthritis. *J Rheumatol* **29**: 29–33.
85. Cox A *et al.* (1999) Combined sib-TDT and TDT provide evidence for linkage of the interleukin-1 gene cluster to erosive rheumatoid arthritis. *Hum Mol Genet* **8**: 1707–1713.
86. Donn R *et al.* (2002) Mutation screening of the macrophage migration inhibitory factor gene: Positive association of a functional polymorphism of macrophage migration inhibitory factor with juvenile idiopathic arthritis. *Arthritis Rheum* **46**: 2402–2409.
87. Donn RP, Shelley E, Ollier WE, Thomson W. (2001) A novel 5′-flanking region polymorphism of macrophage migration inhibitory factor is associated with systemic-onset juvenile idiopathic arthritis. *Arthritis Rheum* **44**: 1782–1785.
88. Baugh JA *et al.* (2002) A functional promoter polymorphism in the macrophage migration inhibitory factor (MIF) gene associated with disease severity in rheumatoid arthritis. *Genes Immun* **3**: 170–176.
89. Black AR, Black JD, Azizkhan-Clifford J. (2001) Sp1 and kruppel-like factor family of transcription factors in cell growth regulation and cancer. *J Cell Physiol* **188**: 143–160.
90. Bouwman P, Philipsen S. (2002) Regulation of the activity of Sp1-related transcription factors. *Mol Cell Endocrinol* **195**: 27–38.
91. Kalkhoven E. (2004) CBP and p300: HATs for different occasions. *Biochem Pharmacol* **68**: 1145–1155.
92. Plant BJ *et al.* (2005) Cystic fibrosis, disease severity, and a macrophage migration inhibitory factor polymorphism. *Am J Respir Crit Care Med* **172**: 1412–1415.
93. Nunez C *et al.* (2007) Involvement of macrophage migration inhibitory factor gene in celiac disease susceptibility. *Genes Immun* **8**: 168–170.

94. Wu SP *et al.* (2006) Macrophage migration inhibitory factor promoter polymorphisms and the clinical expression of scleroderma. *Arthritis Rheum* **54**: 3661–3669.
95. Sanchez E *et al.* (2006) Evidence of association of macrophage migration inhibitory factor gene polymorphisms with systemic lupus erythematosus. *Genes Immun* **7**: 433–436.
96. Donn RP *et al.* (2004) Macrophage migration inhibitory factor gene polymorphism is associated with psoriasis. *J Invest Dermatol* **123**: 484–487.
97. Donn R *et al.* (2004) A functional promoter haplotype of macrophage migration inhibitory factor is linked and associated with juvenile idiopathic arthritis. *Arthritis Rheum* **50**: 1604–1610.
98. Barton A *et al.* (2003) Macrophage migration inhibitory factor (MIF) gene polymorphism is associated with susceptibility to but not severity of inflammatory polyarthritis. *Genes Immun* **4**: 487–491.

II-3

Hypoxic Adaptation Facilitated by MIF

Robert A. Mitchell*

1. Introduction

The availability of oxygen within a tumor's microenvironment dictates both the aggressiveness and therapeutic responsiveness of human tumors. Physiologic and pathophysiologic hypoxic adaptation responses are characterized by hypoxia-inducible factor (HIF)-directed transcription of gene products associated with erythropoiesis, anaerobic metabolism, vascular remodeling and cell survival. Macrophage migration inhibitory factor (MIF) has recently been identified as a direct transcriptional target of hypoxia-induced, HIF-dependent transcription. Increased MIF expression in low oxygen microenvironments contributes to both genotypic and phenotypic events that are necessary for appropriate adaptation to hypoxic environments and associated stresses. Among these MIF-mediated effects on hypoxic adaptation is the facilitated transcription of vascular remodeling growth factors — such as vascular endothelial growth factor (VEGF) and CXCL-8 — and the coordinated antagonism of hypoxia-induced p53 expression and activity. MIF also functionally promotes hypoxia-induced HIF-1α stabilization and subsequent transcription creating a feed-forward amplification loop. Finally, the MIF homologue, D-dopachrome tautomerase (D-DT = MIF-2), is similarly regulated by hypoxia and cooperates with MIF in facilitating VEGF and CXCL-8 transcription and secretion. Recent studies additionally indicate that D-DT/MIF-2 may also cooperate with MIF in antagonizing the p53 tumor suppressor. Thus, it is becoming evident that the MIF family of cytokines/growth factors represents a critical functional intermediate in physiologic and pathophysiologic adaptation to low oxygen tensions.

*Corresponding author: University of Louisville, Clinical and Translational Research Building, 505 S. Hancock St., Suite 404, Louisville, KY 40202, USA. Email: robert.mitchell@louisville.edu

2. HIF-α Regulation

In order to fully understand MIF family member contributions to hypoxic adaptation, it is first important to understand how HIF-α transcription factor stability is physiologically regulated by oxygen. HIF is a heterodimeric transcription factor in which both the α- and β-subunits are basic helix-loop-helix Per-ARNT-Sim (PAS) proteins.[1] Isozymes of HIF-1α and HIF-1β subunits have been identified with HIF-1α and HIF-2α being regulated by dioxygen levels.[1] The HIF-β subunit, identical to the aryl hydrocarbon receptor nuclear translocator (ARNT), is a constitutively expressed nuclear protein involved in other transcriptional responses under normoxic conditions. While HIF-β protein is readily detectable in most tissues under physiologic oxygen tensions, protein levels of the HIF-α subunit are virtually undetectable under normoxic conditions, thus indicating that oxygen-dependent protein stability is a key mechanism regulating HIF function. As such, under hypoxic conditions, protein levels of HIF-α rise, allowing HIF-α nuclear translocation, heterodimerization with HIF-β, and subsequent transcriptional activation.

Several molecular events involved in oxygen-dependent HIF-α degradation have been identified, leading to the identification of a family of Fe(II) and 2-oxoglutarate (2-OG) dependent dioxygenase enzymes that act as dioxygen sensors and regulators.[1] Under normoxic conditions, HIF-1α undergoes trans-4-hydroxylation at Pro-564 in the C-terminal oxygen-dependent degradation (CODD) domain and Pro-402 in the N-terminal ODD (NODD) domain.[1,2] Hydroxylation allows recognition of HIF-1α by the von Hippel–Lindau tumor suppressor protein (pVHL), which serves as the recognition component of the ubiquitin E3 ligase complex consisting of VHL/Elongin C/Elongin B (VCB), Cullin 2, and the RING-H2 finger protein Rbx-1.[3] This HIF-degradation complex is distinct from SCF complexes, which are made up of Cullin 1 (Cull), Skp1 and F-box proteins. Structural analysis of the HIF-CODD and pVHL reveals that all five pVHL residues lining the 4-hydroxyproline-binding pocket are affected by missense mutations in VHL disease,[4] suggesting that failure to capture HIF-1α, and/or other hydroxylated targets, is important to the tumor-promoting mechanism associated with VHL disease. Subsequent ubiquitylation of HIF-α by the Cdc34/Ubc5 E2 ubiquitin conjugating complex targets HIF-α for transport to the proteasome and degradation.

3. MIF, HIF-1α and Tumor-Associated Angiogenesis

MIF is abundantly expressed in a number of human solid cancers. Breast, lung, prostate, pancreatic, colon, skin, and brain-derived tumors have all been shown to contain significantly higher levels of MIF message and protein than their noncancerous cell counterparts.[5–11] Many studies also report that MIF expression closely correlates with tumor aggressiveness and metastatic potential, suggesting an important contribution to disease severity and survival by MIF.[11–15] By way of example, immunohistochemical staining for MIF in pre-chemotherapy osteosarcoma biopsy specimens shows a strong and significant correlation between both disease and metastases-free survival.[15] In a separate study, increased MIF expression in glioblastoma multiformae (GBM) tumor biopsies localizes predominantly to necrotic areas of GBM lesions and within tumor cells surrounding blood vessels.[16] Because necrotic regions within GBM and other neoplasms are commonly associated with very low oxygen tensions,[17] it is not unreasonable to speculate that MIF expression in GBM is positively regulated by hypoxia.[16]

MIF was first linked to tumor hypoxic responses when it was reported that MIFs mRNA levels are induced by low oxygen tensions in human squamous carcinoma cell lines.[18] Although subsequent studies confirmed these findings,[16,19,20] it was not until a report by Baugh and colleagues that MIF was confirmed as being a direct target of HIF-mediated transcription.[21] This study described a hypoxia-induced transcriptional mechanism where a hypoxia response element (HRE) in the 5′UTR of the MIF gene is necessary for hypoxia-induced transcription of MIF. Interestingly, the cAMP response element-binding (CREB) transcription factor was identified as having a repressive function in hypoxia-induced MIF transcription, and hypoxia-induced degradation of CREB was suggested to result in an amplification of hypoxia-induced, HIF-dependent MIF transcription. A subsequent study validated a role for HIF-1α-dependent transcription of MIF under low oxygen tensions.[22] More importantly, this study revealed that loss of MIF phenocopies loss of HIF-1α in inducing hypoxia-induced, p53-dependent premature senescence,[22] providing the first description for a functional contribution by MIF as an effector of hypoxic adaptation.

The second description of hypoxia-induced MIF acting as a functional effector of HIF-dependent hypoxic adaptation came when Winner and colleagues reported that hypoxia-induced vascular endothelial growth factor (VEGF) expression is dramatically reduced in MIF-deficient cells.[5] This defect in HIF-1α-dependent VEGF transcription was found to be due to an inherent requirement for MIF in hypoxia-induced HIF-1α protein stabilization, creating, in essence, an amplification loop between hypoxia, MIF, and HIF-1α.[5] Consistent with these findings, MIF overexpression in human breast cancer cell lines was found to promote hypoxia-induced HIF-1α stabilization.[23] Interestingly, this study revealed that the MIF receptor, CD74, is necessary for MIF-dependent HIF-1α stabilization.[23]

As mentioned above, hypoxia-induced VEGF expression is significantly reduced in MIF-deficient cells and increased in MIF overexpressing cells consistent with its contribution to HIF-1α stabilization.[5,23] In accord with these findings, numerous studies report that MIF intratumoral expression correlates with VEGF expression, tumor vessel density, and risk of recurrence after resection.[15,24–29] In mouse models, *Mif*-deficient mice crossed to adenomatous polyposis coli (*Apc*$^{Min/+}$) "oncomice" are characterized by significant reductions in both the number and size of adenomas that correspond to diminished tumor microvessel density.[30] Additionally, *Mif*-deficient mice show a 45% reduction in chronic ultraviolet B (UVB) irradiation-induced epidermal tumorigenesis.[31] Decreased tumor incidence and delayed tumor outgrowth in *Mif*-deficient mice exposed to UVB correlated with significantly less VEGF expression and intratumoral microvessel density. Thus, one of the most consistent phenotypes associated with loss or inhibition of MIF in tumorigenesis is decreased angiogenic growth factor expression and microvascular density reminiscent of an impaired ability to adapt to hypoxia. While no studies to date have evaluated hypoxia either directly or indirectly with respect to intratumoral MIF, the invariability of this angiogenic phenotype suggests that MIF strongly influences tumoral hypoxic adaptation and associated neovascularization.

4. MIF/D-DT-Dependent Angiogenic Growth Factor Expression

Despite the aforementioned studies linking MIF to intratumoral angiogenesis, none provided a mechanistic link between MIF, VEGF, and tumor vascularization until Coleman and colleagues reported that MIF is also an important regulator of *normoxic* VEGF expression.[32] Interestingly, MIF and MIF's only known homologue, D-dopachrome tautomerase, were found to coordinately support CXCL8 and VEGF expression and secretion from lung adenocarcinoma cells. Although several prior studies had demonstrated an important contribution by MIF to CXCL8/VEGF expression and maintenance of angiogenic phenotypes in malignant cells and tissue,[25,26,29,33,34] this was the first demonstration of a functional overlap between MIF and its only known homologue, D-DT/MIF-2.

MIF and D-DT/MIF-2 were shown to be necessary for maximal c-Jun-N-terminal kinase (JNK)-dependent AP-1 transactivation and subsequent CXCL8 and VEGF transcription in lung adenocarcinoma cells. Perhaps more importantly, the cognate MIF receptor, CD74, was found to be necessary for CXCL8 expression and maximal JNK and c-Jun phosphorylation induced by MIF and

D-DT/MIF-2. These findings are consistent with a study demonstrating that CD74 is necessary for MIF-dependent contributions to prostatic adenocarcinoma cell invasion, anchorage-independence, and tumor-associated neovascularization.[35]

What is less clear is how JNK is activated by CD74. An earlier study revealed that MIF functionally regulates Rac1 effector binding by stabilizing cholesterol-enriched membrane microdomains.[36] Although JNK is a well-known effector of Rac1, there is no evidence, as yet, that the defective JNK observed with loss of MIF is linked to Rac1. However, a more recent study demonstrates that MIF:CD74-mediated JNK activation requires the presence of the alternate MIF receptor, CXCR4, in conjunction with cell surface-associated CD74.[37] Importantly, this study also revealed that JNK activation through the CXCR4/CD74 receptor complex requires both c-Src and phosphatidylinositol-3-kinase (PI 3-kinase) activities. Because PI 3-kinase is a well-known activator of Rac1,[38,39] it is not unreasonable to speculate that MIF — and likely D-DT/MIF-2 — induce JNK activation in a PI 3-kinase $\rightarrow$ Rac1-dependent fashion.

Although results indicate that AP-1 activity is important for MIF and D-DT/MIF-2 contributions to CXCL8 and VEGF expression, it is possible that other signaling pathways may be involved. Of note, MIF:CD74 signaling has recently been suggested to modulate CXCL8 expression in an NF-κB-dependent manner.[33]

Studies with human umbilical vein endothelial cells (HUVECs) reveal that supernatants from lung adenocarcinoma cells deficient in MIF and/or D-DT/MIF-2 result in significant reductions in both endothelial cell migration and tube formation.[32] While these findings indicate that the depleted CXCL8 and VEGF levels present in NSCLC supernatants are responsible for the observed defective angiogenic phenotype, the possibility that endothelial cell-derived MIF and/or D-DT/MIF-2 may influence endothelial cell activation in an autocrine manner cannot be ruled out.[40] This possibility, coupled with the fact that tumor-derived MIF stimulates CXCL8 and VEGF from tumor stromal macrophages,[41] suggests that MIF family members modulate intratumoral neoangiogenesis on a number of different levels.

5. CSN5/JAB-1 and the COP9 Signalosome

Both extracellular and intracellular signaling effects of MIF have been shown to be regulated via a direct interaction with the intracellular protein, Jab1/CSN5.[5,42–44] There is substantial evidence that CD74 represents a necessary intermediate step in this functional interaction between MIF and Jab1/CSN5.[42,45]

Jab1/CSN5 is a 38-kDa protein and an essential component of the COP9 signalosome (CSN), which is composed of eight subunits designated CSN1–CSN8 (for review see Ref. 46). The function of the CSN was obscure until it was discovered that it appeared to control proteins that had high turnover rates. Mutational analysis in *Schizosaccharomyces pombe*, for example, revealed that disruption of CSN1 resulted in the accumulation of neddylated cullins.[46] The conjugation of the small, ubiquitin-like protein Nedd8 to cullins is required for E2 ligase-recruitment and targeted substrate ubiquitylation. CSN5 contains a JAB-1/MPN domain Metalloenzyme motif (JAMM) that forms the catalytic region of CSN5 which confers an isopeptidase activity. In CSN5, the JAMM domain is responsible for the cleavage of Nedd8 from cullins. Alternate cycles of cullin neddylation (required for RING finger, ubiquitin E2 ligase recruitment and ubiquitin transfer) and deneddylation are required for cullin-dependent ubiquitin E3 ligase function.[46] Thus, altering CSN function directly or indirectly has significant effects on the protein stability of cullin-dependent E3 ubiquitin ligase targets.

Prior studies revealed that CSN5 binds both the CODD of HIF-1α and the pVHL tumor suppressor.[47] High CSN5 expression generates a pVHL-independent form of CSN5 that stabilizes HIF-1α aerobically by inhibiting HIF-1α prolyl-564 hydroxylation. Aerobic CSN5 association with HIF-1α occurs independently of the CSN holocomplex, leading to HIF-1α stabilization independent of Cullin 2 deneddylation. CSN5 also associates with HIF-1α under hypoxia and is required for optimal hypoxia-mediated HIF-1α stabilization.[47] Less clear from this study is whether the anaerobic binding of CSN5 to HIF-1α occurs independently of the signalosome and/or pVHL.

Several studies have shown that CSN5 exists in small subunits or in monomeric form outside of the CSN in various species.[48–52] Both monomeric and CSN-associated CSN5 have been found to functionally interact with a number of intracellular proteins and, in almost all cases, regulates their turnover.[53] A notable exception to this is MIF.[42] Jab1/CSN5 was initially identified by yeast two-hybrid screening to interact with MIF.[42] MIF modulates CSN5 function and subsequent CSN5-dependent effects on p27 degradation, JNK activation, and AP-1-mediated transcription.[42] As alluded to above, extracellular MIF functionally regulates the activities of intracellular CSN5.[5,42,54–56] Although the precise mechanism remains unclear, it is likely that there is a CD74-dependent internalization step resulting in an MIF/CSN5 interaction.[5,42,54,57]

Consistent with the finding that MIF functionally regulates Jab1/CSN5-dependent p27 ubiquitylation and proteasomal degradation,[42] MIF was found to be necessary for DNA damage checkpoint responses in developing lymphomas.[43] Specifically, MIF controls Jab1/CSN5-dependent deneddylation and subsequent Cullin 1-containing ubiquitin E3 (SCF) complex stability. Aberrant neddylation of the SCF complex in MIF-deficient B cell lymphomas results in accumulation of defective checkpoint response proteins (Chk1, Chk2, Cdc25A), resulting in defective DNA repair. Moreover, MIF-dependent HIF-1α stabilization is suggested to require a physical and functional interaction with Jab1/CSN5.[5] In contrast to studies involving MIF's regulation of the SCF complex resulting in enhanced degradation/decreased stability of p27 and Cdc25A proteins, MIF/CSN5 is suggested to regulate HIF-1α turnover, resulting in enhanced stability or decreased degradation of HIF-1α.[5] These differences in p27 (the target of a Cullin-1 E3 ubiquitin ligase) and HIF-1α (the target of a Cullin-2 E3 ubiquitin ligase), underscore the fact that because CSN5 influences its targets differently, perturbation of its activity — whether directly or via MIF — can produce different effects.

While the study described above suggested a necessary role for Jab1/CSN5 in MIF-dependent modulation of HIF-α protein stability, a separate study investigating MIF-dependent promotion of hypoxia-induced HIF-1α stabilization suggested that MIF-associated HIF stabilization was independent of Jab1/CSN5.[23] MIF's potentiation of HIF-1α stability was instead proposed to be due to an MIF-dependent inhibition of the tumor suppressor, p53.[23] A relationship between p53 and HIF-1α co-regulation is well documented, but the complexities of the pathways involved have made it difficult to delineate a consistent regulatory theme. For example, prior studies reveal that hypoxia-induced HIF-α acts to stabilize p53[58] while others show that p53 regulates HIF-α expression and activity.[59,60] Regardless of this complexity, it is interesting to note that both p53 and HIF-1α are regulated by the proinflammatory peptide, MIF.

6. MIF and p53

Of the many protumorigenic activities attributed to the cytokine MIF, one of the most compelling is its ability to inhibit p53 expression and activity. Several years ago, a phenotypic screen for p53 protein antagonists yielded, among other hits, five independent cDNA clones that contained the MIF open reading frame.[61] Subsequent studies validated the earlier finding that exogenously added recombinant MIF is capable of inhibiting the stabilization, transcriptional activity, and subsequent apoptotic effects of p53.[62] The importance of this pathway *in vivo* was confirmed when fibrosarcomas induced in $Mif^{-/-}$ mice were found to be significantly smaller and have reduced mitotic indices compared to wildtype littermates.[63] Crossing $Mif^{-/-}$

onto a $p53^{-/-}$ background rescues the MIF phenotype, indicating an important negative regulatory role for MIF in p53 activity *in situ*.[63]

A study from the Giaccia laboratory demonstrated how MIF functionally links HIF-1α and p53. In this study, conditionally-deficient HIF-1α mouse embryonic fibroblasts were found to be substantially more sensitive to hypoxia and gamma irradiation-induced cell senescence compared to the relatively resistant HIF-1α competent cells.[22] Intriguingly, a hypoxia-induced, HIF-1α-dependent increase in MIF transcription was found to be necessary for the ability of HIF-1α to inhibit hypoxia or gamma irradiation-induction of a p53-dependent senescent phenotype. This was demonstrated when MIF-deficient cells were found to phenocopy HIF-1α-deficiency in sensitizing primary rodent fibroblasts to hypoxia and radiation-induced cell senescence.[22] It is interesting to note that the MIF homologue, D-dopachrome tautomerase (D-DT/MIF-2), is also induced by hypoxia, and D-DT/MIF-2 — like MIF — actively suppresses p53 stabilization and transcriptional activity in p53 wildtype tumor cell lines (E. Brock, D. Xin, and R.A. Mitchell, *unpublished observations*) suggesting a potential compensatory role for D-DT/MIF-2 in MIF-mediated p53 antagonism. These observations are consistent with the cooperative functions discovered for MIF and D-DT/MIF-2 in promoting proangiogenic growth factor expression/potential in non-small cell lung carcinoma (NSCLC) cell lines.[32]

D-DT/MIF-2's potential ability to compensate for MIF — and *vice versa* — in p53 antagonism is consistent with a recent study demonstrating a functional requirement for MIF and D-DT/MIF-2 in maintaining cyclooxygenase-2 expression in tumor cell lines from colorectal cancer patients.[64] Like MIF, D-DT/MIF-2 was found to be necessary for maximal expression of steady-state Cox-2 expression in human colon cancer lines.[64] Similar to the signaling pathway involved in MIF and D-DT/MIF-2-dependent VEGF and CXCL8 expression in NSCLC cell lines,[32] c-Jun-N-terminal kinase (JNK) and subsequent c-Jun phosphorylation were found to be necessary for D-DT/MIF-2-dependent Cox-2 transcription. Interestingly, a functional role for β-catenin-dependent transcription was also identified to be necessary for maximal Cox-2 expression in CRC cell lines,[64] indicating the convergence of two signaling pathways in MIF/D-DT-dependent Cox-2 transcription — the JNK/c-Jun pathway and the β-catenin/TCF pathway. Intriguingly, the mechanism governing D-DT/MIF-2-dependent β-catenin transcription was found to be through stabilization/inhibiting degradation of β-catenin protein thus increasing cellular β-catenin levels and subsequent transcription.[64]

Given the important regulatory functions of Jab1/CSN5 in protein degradatory pathways in mammalian cells, it is not unreasonable to speculate that the mechanism governing MIF-dependent p53 stability,[43,62,63,65] HIF-1α stability,[5,23] and β-catenin stability[64] are either directly or indirectly linked to MIF's — and likely D-DT/MIF-2's — functional interactions with Jab1/CSN5 (Fig. 1).

Although the supposition that MIF is a central effector of Jab1/CSN5-dependent protein turnover is both attractive and generally supported by the literature, divergent phenotypes between Jab1/CSN5 gene targeted mice[48] and MIF gene targeted mice[66] argue that this may not be the case. Jab1/CSN5 *null* embryos die shortly after implantation[48] while MIF *null* embryos develop normally and produce developmentally sound offspring.[66] Similarly, if MIF is necessary for HIF-1α stabilization, one would expect a phenocopy of MIF-*null* mice with HIF-1α *null* mice. This also is not the case: HIF-1α homozygous *null* mice developmentally arrest and die by embryonic day 11.[67,68] Although MIF-deficiency does render these mice more resistant to several

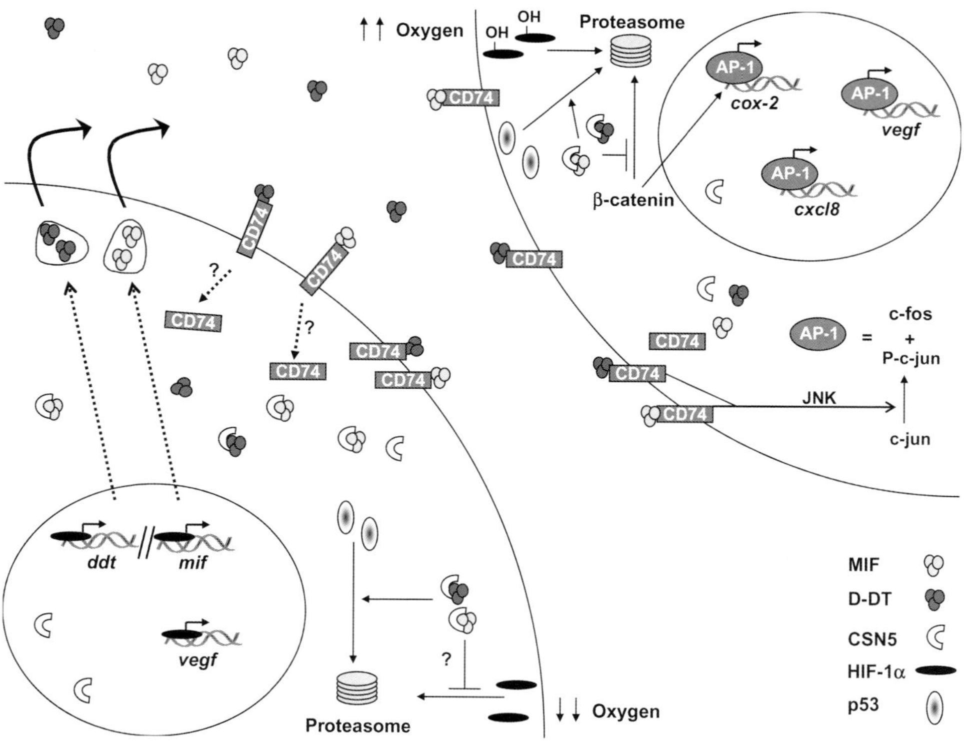

Fig. 1. Proposed scheme for MIF and D-DT/MIF-2 cooperativity in promoting CD74-dependent, Jab1/CSN5-mediated p53 degradation, HIF and β-catenin stability, and AP-1-dependent transcription in normoxic (upper right) and hypoxic (lower left) cells.

disease pathologies, including autoimmune, bacteria and parasitic infections, atherosclerosis and tumorigenesis,[30,63,66,69–74] they do not phenocopy HIF-1α nullizygous mice. Because most data suggest that MIF contributes to, but is not required for, HIF-1α stabilization, it is much more likely that MIF-deficiency more closely resembles HIF-1a heterozygous *null*, rather than HIF-1α homozygous *null*, mice. HIF-1α heterozygous *null* mice are viable and develop normally but do display specific phenotypes, some of which are similar to that of *Mif*$^{-/-}$ mice. For example, a recent study revealed that MIF-deficiency results in defective lung maturation and lethality in prematurely born pups.[75] Moreover, MIF is induced by heart ischemia and contributes to AMPK activation leading to glucose uptake and cardiac repair following ischemia-reperfusion. As these phenotypes are consistent with a role for MIF in hypoxic adaptation associated with HIF-1α activity,[76–79] it is more likely that MIF acts to potentiate HIF-1α stabilization and function but is not absolutely essential for it. As such, when MIF is highly expressed in primary and/or metastatic malignant lesions, there is a correspondingly higher level of HIF-1α stabilization, angiogenic growth factor expression and intratumoral angiogenesis. As discussed earlier, MIF expression within tumors is a predictor of disease survival and prognosis corresponds closely with VEGF expression and microvessel density.[15,24–29] All of that being said, there is also the very real possibility that MIF, in collaboration with its compensating family member, D-dopachrome tautomerase, may in fact be necessary for Jab1/CSN5-dependent HIF-1α — and likely many other substrates' — protein turnover (Fig. 1).

7. MIF/D-DT Compensation and Therapeutic Implications

Multiple studies investigating gene deletion, immunoneutralization, and small molecule antagonism of MIF have found encouraging, albeit modest, reductions in tumor burden and disease survival (reviewed in Refs. 80 and 81). The two most prevalent causes of reduced tumor burden associated with genetic deletion or inhibition of MIF are: (i) reduced tumor-associated angiogenesis[24,25,28,30] and (ii) increased p53-dependent apoptosis and cell cycle inhibition.[22,31,63] While MIF has been studied in relatively great detail, its only known family member, D-DT/MIF-2, has been largely ignored. Recent studies investigating the biology of D-DT/MIF-2 reveal that there is a significant degree of overlap between activities, expression patterns, and phenotypic influences between these two proteins.[32,64] This raises the question as to whether D-DT/MIF-2 may functionally compensate for MIF and *vice versa*. Given accumulating evidence, compensation does, in fact, exist between MIF and D-DT/MIF-2 for at least two of MIF's most important functional contributions to malignant disease maintenance and progression: (i) promotion of tumor-associated neovascularization and (ii) p53 inhibition. Studies involving gene-targeted deletions of both MIF and D-DT/MIF-2 will allow for greater clarity and understanding as to the extent of overlap and compensation between these two family members in both physiologic and pathophysiologic conditions. If extensive MIF/D-DT compensation is confirmed, it would follow that therapeutic targeting of MIF and D-DT/MIF-2 would simultaneously have far better translational potential than targeting either individually. Simultaneous inhibition of MIF and D-DT/MIF-2 as a novel anticancer therapeutic approach would be expected to have the added benefit of inducing p53 (limited to lesions that have functional p53) while compromising tumor vascularization.

Three-dimensional X-ray crystallographic studies reveal that both MIF and D-DT/MIF-2 exist as homotrimers and are structurally related to the bacterial isomerases, 4-OT and CHMI.[82–84] The only known substrate with a reasonably low K_m for MIF and D-DT/MIF-2s' catalytic activities is the D stereoisomer of dopachrome, a product of tyrosine metabolism. Of note, D-DT/MIF-2 also decarboxylates the D-dopachrome substrate giving rise to a decarboxylated dihydroxyindole product.[85] Because mammals only utilize L-tyrosine, human MIF and D-DT/MIF-2 catalytic activities are presumed to be non-physiologic. The general consensus behind this catalytic active site is that evolutionary conservation has resulted in maintaining conserved amino acid residues that are necessary for CD74 receptor binding leading to receptor-mediated endocytosis[42,54] and/or chemokine receptor/CD44 complex formation and subsequent activation.[86–88]

Several groups have identified small molecule inhibitors of MIF's enzymatic activity using virtual screening approaches,[89–91] although only a few have validated them as being inhibitory to MIFs biologic activities.[91,92] Indeed, we have identified a potent and selective suicide substrate for MIF (4-iodo-6-phenylpyrimidine, 4-IPP) that covalently modifies the N-terminal proline (Pro2), thereby inhibiting its enzymatic *and* — by virtue of blocking other conserved residues within the active site resulting in reduced CD74 receptor binding — inhibition of its tumor-promoting bioactivities.[92] Follow-up studies revealed that 4-IPP is nontoxic, can be delivered orally, is bioavailable, and readily crosses the blood brain barrier (R. Mitchell, *unpublished observations*). Nonetheless, despite having activity against NSCLC *outgrowth*, consistent with its ability to inhibit NSCLC anchorage-independence,[92] its activity against established tumors is relatively modest (R. Mitchell, *unpublished observations*).

While it is abundantly clear that small molecule targeting of MIF is of great interest to both academicians[89,91–93] and those in industry,[90,94–96] there is no evidence that D-DT/MIF-2 — a protein that cooperates with, and whose functions overlap with, MIF — is being targeted similarly. There is now sufficient rationale to suggest that a novel regulatory paradigm exists in which two cytokine members of the same family cooperate to promote malignant disease progression. The identification of small molecule compounds or dual targeting monoclonal antibodies that specifically target *both* MIF and D-DT/MIF-2 could represent unique and clinically efficacious therapeutic modalities.

8. Dopachrome Tautomerases as p53 Antagonists

MIF was one of the first cytokine/chemokine activities ever described.[97,98] In addition to its unusual, presumably nonphysiologic enzymatic activity, MIF also has the equally unusual ability — as a soluble peptide — to elicit hypoxia-induced prosurvival pathways by virtue of its ability to negatively regulate p53 tumor suppressor expression and activity.

A recent study reveals that MIF, and likely D-dopachrome tautomerase, represent only two of at least three dopachrome tautomerases that act as soluble proteins induced by hypoxia/HIF-1α to negatively regulate p53. Sendoel *et al.* reported that constitutive expression of HIF-1α protects germ cells against ionizing or UVC irradiation-induced apoptosis in *C. elegans.*[99] This HIF-1α-dependent, antiapoptotic pathway was found to require the inhibition of the worm orthologue of p53 in a cell nonautonomous manner. The HIF-1α-dependent transcription and subsequent secretion of a tyrosinase-related protein-2 (TYR-2) was determined to be responsible for HIF-1α-dependent p53 antagonism. Intriguingly, TYR-2 is the orthologue of the human TRP-2 (aka DCT) and is an L-dopachrome tautomerase.[99] The human orthologue of TYR-2, DCT, was additionally found to actively suppress cisplatin-induced p53 expression and apoptosis in human melanoma cells, suggesting an evolutionary conservation of this p53 antagonist activity among L-dopachrome tautomerases.[99] This study is compelling to MIF researchers for two important reasons. First, MIF,[22] D-dopachrome tautomerase (R. Mitchell, *unpublished observations*), and this newly discovered L-dopachrome tautomerase (DCT)[99] are *all induced by hypoxia and all inhibit p53 as secreted proteins.* Second, because there is negligible primary sequence homology between MIF/D-DT and DCT (and no structures of DCT exist to compare with MIF/D-DT), the only apparent connection between these three hypoxia-induced, p53-inhibiting enzymes is the identical catalytic activities all three possess, albeit acting on substrates of opposite chirality. Also of interest is the fact that the ability of DCT to inhibit cisplatin-induced p53 expression and apoptosis in melanoma cells was found to require its L-dopachrome tautomerase catalytic activity.[99] These findings reinforce the supposition that therapeutic targeting of the catalytic activities of MIF and D-DT/MIF-2 (and perhaps, L-dopachrome tautomerase) may represent a unique strategy to amplify p53-dependent cell cycle inhibition and/or apoptosis in developing malignancies.

Despite this intriguing connection to L-dopachrome tautomerase, there remains the fact that MIF and D-DT/MIF-2 catalysis does not act on the physiologic substrate, L-dopachrome,[100,101] and it is not immediately clear whether D-dopachrome exists in mammalian cells at all. So the question remains: How do these extracellular dopachrome tautomerase inhibit intracellular p53 stabilization/transcriptional activity? We envision one of two mechanistic scenarios by which this evidently well-conserved dopachrome tautomerase-mediated inhibition of p53 occurs. The first scenario assumes that a catalytic activity *is* required for dopachrome tautomerase-mediated p53 inhibition. Many aspects of plant cell division, expansion, and differentiation are controlled by small chemical phytohormones called auxins. The mechanism of action of the best-studied auxin, indole-3-acetic acid (IAA), has been elegantly delineated over the past decade. IAA is now known to control the stabilities of several F-box containing proteins by tethering them to SCF ubiquitin ligase.[102–104] Because the core structure of IAA is moderately similar to that of dopachrome, it is not complete folly to envision a similar "hormone"-like system existing in mammals that is regulated by enzymatic control of these hormone levels (e.g., dopachrome or its tautomerized product, dihydroxyindole carboxylic acid) thus dictating a subset of protein stabilities. In this context, it is worth noting that MIF regulates F-box protein stabilities through SCF modulation by an MIF/CSN5 functional interaction.[42,43,105,106] To investigate if this scenario might have some validity, experiments could be designed to determine whether p53 expression/activity is affected by exposing cells to exogenously added L-dopachrome or D-dopachrome to recapitulate the substrate buildup expected when the respective enzymes are lost or inhibited. In parallel, one could also evaluate the potential suppressive effects on p53 by these respective enzymes' products: dihydroxyindole carboxylic acid (for MIF) or dihydroxyindole (for D-DT/MIF-2 and DCT).

The second scenario assumes that the catalytic active sites of MIF, D-dopachrome tautomerase and L-dopachrome tautomerase, have evolved to allow them to elicit signals through cell surface receptor-mediated signaling and/or endocytosis. The fact that DCT was found to only be able to inhibit p53 after endocytosis and that active site mutations render it inactive[99] strongly argues for this scenario. There is also considerable evidence that this is also the case for MIF, as mutation of residues in and around the active site partially blocks CD74 receptor binding, activation and/or internalization, and inhibitors of the catalytic active site are, to varying degrees, inhibitory to MIF:CD74 interaction.[91,107–109] Also in support of this latter scenario, Fingerle-Rowson and colleagues elegantly demonstrated that tautomerase *null* MIF knock-in mice have a phenotype that is intermediate between that of $Mif^{+/+}$ and $Mif^{-/-}$ mice. This phenotype, it was concluded, was due not to MIF's complete loss of catalytic activity, but rather to the reduced — though not entirely absent — ability of catalytically inactive MIF to bind to CD74.[110]

While further studies are clearly needed to elucidate the distinction between catalytic activities, receptor binding activities and p53 inhibiting activities for both MIF and D-DT/MIF-2, the similarities with L-dopachrome tautomerases suggest an important and previously unrecognized biological paradigm involving MIF family members. Because MIF and D-DT/MIF-2 act in an additive fashion in regulating neoangiogenic processes,[32] a similar level of cooperative activity in antagonizing steady state, hypoxia- and/or cisplatin-induced p53 is likely. Therapeutic targeting of both of these family members simultaneously may have significant translational potential as a novel cancer treatment modality.

References

1. Kim W, Kaelin WG Jr. (2003) The von Hippel–Lindau tumor suppressor protein: New insights into oxygen sensing and cancer. *Curr Opin Genet Dev* **13**: 55–60.
2. Chan DA, Sutphin PD, Yen SE, Giaccia AJ. (2005) Coordinate regulation of the oxygen-dependent degradation domains of hypoxia-inducible factor 1 alpha. *Mol Cell Biol* **25**: 6415–6426.
3. Hon WC *et al.* (2002) Structural basis for the recognition of hydroxyproline in HIF-1alpha by pVHL. *Nature* **417**: 975–978.
4. Kim WY, Kaelin WG. (2004) Role of VHL gene mutation in human cancer. *J Clin Oncol* **22**, 4991–5004.
5. Winner M, Koong AC, Rendon BE *et al.* (2007) Amplification of tumor hypoxic responses by macrophage migration inhibitory factor-dependent hypoxia-inducible factor stabilization. *Cancer Res* **67**: 186–193.

6. Meyer-Siegler K, Fattor RA, Hudson PB. (1998) Expression of macrophage migration inhibitory factor in the human prostate. *Diagn Mol Pathol* **7**: 44–50.
7. Bando H *et al.* (2002) Expression of macrophage migration inhibitory factor in human breast cancer: Association with nodal spread. *Jpn J Cancer Res* **93**: 389–396.
8. Takahashi N *et al.* (1998) Involvement of macrophage migration inhibitory factor (MIF) in the mechanism of tumor cell growth. *Mol Med* **4**: 707–714.
9. Markert JM *et al.* (2001) Differential gene expression profiling in human brain tumors. *Physiol Genomics* **5**: 21–33.
10. Shimizu T *et al.* (1999) High expression of macrophage migration inhibitory factor in human melanoma cells and its role in tumor cell growth and angiogenesis. *Biochem Biophys Res Commun* **264**: 751–758.
11. Kamimura A *et al.* (2000) Intracellular distribution of macrophage migration inhibitory factor predicts the prognosis of patients with adenocarcinoma of the lung. *Cancer* **89**: 334–341.
12. del Vecchio MT *et al.* (2000) Macrophage migration inhibitory factor in prostatic adenocarcinoma: Correlation with tumor grading and combination endocrine treatment-related changes. *Prostate* **45**: 51–57.
13. Meyer-Siegler KL, Bellino MA, Tannenbaum M. (2002) Macrophage migration inhibitory factor evaluation compared with prostate specific antigen as a biomarker in patients with prostate carcinoma. *Cancer* **94**: 1449–1456.
14. Tomiyasu M, Yoshino I, Suemitsu R *et al.* (2002) Quantification of macrophage migration inhibitory factor mRNA expression in non-small cell lung cancer tissues and its clinical significance. *Clin Cancer Res.* **8**: 3755–3760.
15. Han I *et al.* (2008) Expression of macrophage migration inhibitory factor relates to survival in high-grade osteosarcoma. *Clin Orthop Relat Res* **466**: 2107–2113.
16. Bacher M *et al.* (2003) Up-regulation of macrophage migration inhibitory factor gene and protein expression in glial tumor cells during hypoxic and hypoglycemic stress indicates a critical role for angiogenesis in glioblastoma multiforme. *Am J Pathol* **162**: 11–17.
17. Louis DN. (2006) Molecular pathology of malignant gliomas. *Annu Rev Pathol* **1**: 97–117.
18. Koong AC *et al.* (2000) Candidate genes for the hypoxic tumor phenotype. *Cancer Res* **60**: 883–887.
19. Takahashi M *et al.* (2001) Macrophage migration inhibitory factor as a redox-sensitive cytokine in cardiac myocytes. *Cardiovasc Res* **52**: 438–445.
20. Schmeisser A *et al.* (2005) The expression of macrophage migration inhibitory factor 1alpha (MIF 1alpha) in human atherosclerotic plaques is induced by different proatherogenic stimuli and associated with plaque instability. *Atherosclerosis* **178**: 83–94.

21. Baugh JA *et al.* (2006) Dual regulation of macrophage migration inhibitory factor (MIF) expression in hypoxia by CREB and HIF-1. *Biochem Biophys Res Commun* **347**: 895–903.
22. Welford SM *et al.* (2006) HIF1alpha delays premature senescence through the activation of MIF. *Genes Dev* **20**: 3366–3371.
23. Oda S *et al.* (2008) Macrophage migration inhibitory factor activates hypoxia-inducible factor in a p53-dependent manner. *PLoS One* **3**: e2215.
24. Hagemann T *et al.* (2007) Ovarian cancer cell-derived migration inhibitory factor enhances tumor growth, progression, and angiogenesis. *Mol Cancer Ther* **6**: 1993–2002.
25. Xu X *et al.* (2007) Overexpression of macrophage migration inhibitory factor induces angiogenesis in human breast cancer. *Cancer Lett* **261**(2):147–157.
26. Ren Y *et al.* (2005) Macrophage migration inhibitory factor stimulates angiogenic factor expression and correlates with differentiation and lymph node status in patients with esophageal squamous cell carcinoma. *Ann Surg* **242**: 55–63.
27. Shun CT, Lin JT, Huang SP *et al.* (2005) Expression of macrophage migration inhibitory factor is associated with enhanced angiogenesis and advanced stage in gastric carcinomas. *World J Gastroenterol* **11**: 3767–3771.
28. Hira E *et al.* (2005) Overexpression of macrophage migration inhibitory factor induces angiogenesis and deteriorates prognosis after radical resection for hepatocellular carcinoma. *Cancer* **103**: 588–598.
29. White ES *et al.* (2003) Macrophage migration inhibitory factor and CXC chemokine expression in non-small cell lung cancer: Role in angiogenesis and prognosis. *Clin Cancer Res* **9**: 853–860.
30. Wilson JM *et al.* (2005) Macrophage migration inhibitory factor promotes intestinal tumorigenesis. *Gastroenterology* **129**: 1485–1503.
31. Martin J *et al.* (2009) Macrophage migration inhibitory factor (MIF) plays a critical role in pathogenesis of ultraviolet-B (UVB)-induced nonmelanoma skin cancer (NMSC). *FASEB J* **23**(3): 720–730.
32. Coleman AM *et al.* (2008) Cooperative regulation of non-small cell lung carcinoma angiogenic potential by macrophage migration inhibitory factor and its homolog, D-dopachrome tautomerase. *J Immunol* **181**: 2330–2337.
33. Binsky I *et al.* (2007) IL-8 secreted in a macrophage migration-inhibitory factor- and CD74-dependent manner regulates B cell chronic lymphocytic leukemia survival. *Proc Natl Acad Sci USA* **104**: 13408–13413.
34. Ren Y *et al.* (2004) Upregulation of macrophage migration inhibitory factor contributes to induced N-Myc expression by the activation of ERK signaling pathway and increased expression of interleukin-8 and VEGF in neuroblastoma. *Oncogene* **23**: 4146–4154.

35. Meyer-Siegler KL, Iczkowski KA, Leng L *et al.* (2006) Inhibition of macrophage migration inhibitory factor or its receptor (CD74) attenuates growth and invasion of DU-145 prostate cancer cells. *J Immunol* **177**: 8730–8739.
36. Rendon BE *et al.* (2007) Regulation of human lung adenocarcinoma cell migration and invasion by macrophage migration inhibitory factor. *J Biol Chem* **282**: 29910–29918.
37. Lue H, Dewor M, Leng L *et al.* (2011) Activation of the JNK signalling pathway by macrophage migration inhibitory factor (MIF) and dependence on CXCR4 and CD74. *Cell Signal* **23**: 135–144.
38. Hawkins PT *et al.* (1995) PDGF stimulates an increase in GTP-Rac via activation of phosphoinositide 3-kinase. *Curr Biol* **5**: 393–403.
39. Clark EA, King WG, Brugge JS *et al.* (1998) Integrin-mediated signals regulated by members of the rho family of GTPases. *J Cell Biol* **142**: 573–586.
40. Chesney J *et al.* (1999) An essential role for macrophage migration inhibitory factor (MIF) in angiogenesis and the growth of a murine lymphoma. *Mol Med* **5**: 181–191.
41. White ES, Strom SR, Wys NL, Arenberg DA. (2001) Non-small cell lung cancer cells induce monocytes to increase expression of angiogenic activity. *J Immunol* **166**: 7549–7555.
42. Kleemann R *et al.* (2000) Intracellular action of the cytokine MIF to modulate AP-1 activity and the cell cycle through Jab1. *Nature* **408**: 211–216.
43. Nemajerova A, Mena P, Fingerle-Rowson G *et al.* (2007) Impaired DNA damage checkpoint response in MIF-deficient mice. *EMBO J* **26**: 987–997.
44. Lue H *et al.* (2006) Rapid and transient activation of the ERK MAPK signalling pathway by macrophage migration inhibitory factor (MIF) and dependence on JAB1/CSN5 and Src kinase activity. *Cell Signal* **18**, 688–703.
45. Lue H *et al.* (2007) Macrophage migration inhibitory factor (MIF) promotes cell survival by activation of the Akt pathway and role for CSN5/JAB1 in the control of autocrine MIF activity. *Oncogene* **26**: 5046–5059.
46. Wolf DA, Zhou C, Wee S. (2003) The COP9 signalosome: An assembly and maintenance platform for cullin ubiquitin ligases? *Nat Cell Biol* **5**: 1029–1033.
47. Bemis L *et al.* (2004) Distinct aerobic and hypoxic mechanisms of HIF-alpha regulation by CSN5. *Genes Dev* **18**: 739–744.
48. Tomoda K, Yoneda-Kato N, Fukumoto A *et al.* (2004) Multiple functions of Jab1 are required for early embryonic development and growth potential in mice. *J Biol Chem* **279**: 43013–43018.
49. Tomoda K *et al.* (2002) The cytoplasmic shuttling and subsequent degradation of p27Kip1 mediated by Jab1/CSN5 and the COP9 signalosome complex. *J Biol Chem* **277**: 2302–2310.

50. Oron E *et al.* (2002) COP9 signalosome subunits 4 and 5 regulate multiple pleiotropic pathways in *Drosophila melanogaster. Development* **129**: 4399–4409.
51. Kwok SF *et al.* (1998) *Arabidopsis* homologs of a c-Jun coactivator are present both in monomeric form and in the COP9 complex, and their abundance is differentially affected by the pleiotropic cop/det/fus mutations. *Plant Cell* **10**: 1779–1790.
52. Freilich S *et al.* (1999) The COP9 signalosome is essential for development of *Drosophila melanogaster. Curr Biol* **9**: 1187–1190.
53. Richardson KS, Zundel W. (2005) The emerging role of the COP9 signalosome in cancer. *Mol Cancer Res* **3**: 645–653.
54. Kleemann R, Grell M, Mischke R *et al.* (2002) Receptor binding and cellular uptake studies of macrophage migration inhibitory factor (MIF): Use of biologically active labeled MIF derivatives. *J Interferon Cytokine Res* **22**: 351–363.
55. Berndt K, Kim M, Meinhardt A, Klug J. (2008) Macrophage migration inhibitory factor does not modulate co-activation of androgen receptor by Jab1/CSN5. *Mol Cell Biochem* **307**: 265–271.
56. Lue H *et al.* (2007) Macrophage migration inhibitory factor (MIF) promotes cell survival by activation of the Akt pathway and role for CSN5/JAB1 in the control of autocrine MIF activity. *Oncogene* **26**: 5046–5059.
57. Leng L *et al.* (2003) MIF signal transduction initiated by binding to CD74. *J Exp Med* **197**: 1467–1476.
58. An WG *et al.* (1998) Stabilization of wild-type p53 by hypoxia-inducible factor 1alpha. *Nature* **392**: 405–408.
59. Ravi R *et al.* (2000) Regulation of tumor angiogenesis by p53-induced degradation of hypoxia-inducible factor 1alpha. *Genes Dev* **14**: 34–44.
60. Blagosklonny MV *et al.* (1998) p53 inhibits hypoxia-inducible factor-stimulated transcription. *J Biol Chem* **273**: 11995–11998.
61. Hudson JD *et al.* (1999) A proinflammatory cytokine inhibits p53 tumor suppressor activity. *J Exp Med* **190**: 1375–1382.
62. Mitchell RA *et al.* (2002) Macrophage migration inhibitory factor (MIF) sustains macrophage proinflammatory function by inhibiting p53: Regulatory role in the innate immune response. *Proc Natl Acad Sci USA* **99**: 345–350.
63. Fingerle-Rowson G. *et al.* (2003) The p53-dependent effects of macrophage migration inhibitory factor revealed by gene targeting. *Proc Natl Acad Sci USA* **100**: 9354–9359.
64. Xin D *et al.* (2010) The MIF homolog, D-dopachrome tautomerase (D-DT), promotes COX-2 expression through beta-catenin-dependent and independent mechanisms. *Mol Cancer Res* **8**(12): 1601–1609.

65. Nemajerova A, Moll UM, Petrenko O, Fingerle-Rowson G. (2007) Macrophage migration inhibitory factor coordinates DNA damage response with the proteasomal control of the cell cycle. *Cell Cycle* **6**: 1030–1034.
66. Bozza M *et al.* (1999) Targeted disruption of migration inhibitory factor gene reveals its critical role in sepsis. *J Exp Med* **189**: 341–346.
67. Iyer NV *et al.* (1998) Cellular and developmental control of O2 homeostasis by hypoxia-inducible factor 1 alpha. *Genes Dev* **12**: 149–162.
68. Ryan HE, Lo J, Johnson RS. (1998) HIF-1alpha is required for solid tumor formation and embryonic vascularization. *EMBO J* **17**: 3005–3015.
69. Santos LL, Dacumos A, Yamana J *et al.* (2008) Reduced arthritis in MIF deficient mice is associated with reduced T cell activation: Down-regulation of ERK MAP kinase phosphorylation. *Clin Exp Immunol* **152**: 372–380.
70. de Jong YP *et al.* (2001) Development of chronic colitis is dependent on the cytokine MIF. *Nat Immunol* **2**: 1061–1066.
71. Koebernick H *et al.* (2002) Macrophage migration inhibitory factor (MIF) plays a pivotal role in immunity against *Salmonella typhimurium*. *Proc Natl Acad Sci USA* **99**: 13681–13686.
72. McDevitt MA *et al.* (2006) A critical role for the host mediator macrophage migration inhibitory factor in the pathogenesis of malarial anemia. *J Exp Med* **203**: 1185–1196.
73. Pan JH *et al.* (2004) Macrophage migration inhibitory factor deficiency impairs atherosclerosis in low-density lipoprotein receptor-deficient mice. *Circulation* **109**: 3149–3153.
74. Taylor JA III *et al.* (2007) Null mutation for macrophage migration inhibitory factor (MIF) is associated with less aggressive bladder cancer in mice. *BMC Cancer* **7**: 135.
75. Kevill KA *et al.* (2008) A role for macrophage migration inhibitory factor in the neonatal respiratory distress syndrome. *J Immunol* **180**: 601–608.
76. Compernolle V *et al.* (2002) Loss of HIF-2alpha and inhibition of VEGF impair fetal lung maturation, whereas treatment with VEGF prevents fatal respiratory distress in premature mice. *Nat Med* **8**: 702–710.
77. Land SC, Wilson SM. (2005) Redox regulation of lung development and perinatal lung epithelial function. *Antioxid Redox Signal* **7**: 92–107.
78. Cai Z *et al.* (2008) Complete loss of ischaemic preconditioning-induced cardioprotection in mice with partial deficiency of HIF-1alpha. *Cardiovasc Res* **77**: 463–470.
79. Loor G, Schumacker PT (2008) Role of hypoxia-inducible factor in cell survival during myocardial ischemia-reperfusion. *Cell Death Differ* **15**: 686–690.
80. Bifulco C, McDaniel K, Leng L, Bucala R. (2008) Tumor growth-promoting properties of macrophage migration inhibitory factor. *Curr Pharm Des* **14**: 3790–3801.

81. Rendon BE, Willer SS, Zundel W, Mitchell RA. (2009) Mechanisms of macrophage migration inhibitory factor (MIF)-dependent tumor microenvironmental adaptation. *Exp Mol Pathol* **86**(3): 180–185.
82. Sugimoto H *et al.* (1999) Crystal structure of human D-dopachrome tautomerase, a homologue of macrophage migration inhibitory factor, at 1.54 Å resolution. *Biochemistry* **38**: 3268–3279.
83. Sugimoto H, Suzuki M, Nakagawa A *et al.* (1996) Crystal structure of macrophage migration inhibitory factor from human lymphocyte at 2.1 Å resolution. *FEBS Lett* **389**: 145–148.
84. Sun HW, Bernhagen J, Bucala R, Lolis E. (1996) Crystal structure at 2.6-Å resolution of human macrophage migration inhibitory factor. *Proc Natl Acad Sci USA* **93**: 5191–5196.
85. Nishihira J *et al.* (1998) Molecular cloning of human D-dopachrome tautomerase cDNA: N-terminal proline is essential for enzyme activation. *Biochem Biophys Res Commun* **243**: 538–544.
86. Bernhagen J *et al.* (2007) MIF is a noncognate ligand of CXC chemokine receptors in inflammatory and atherogenic cell recruitment. *Nat Med* **13**: 587–596.
87. Schwartz V *et al.* (2009) A functional heteromeric MIF receptor formed by CD74 and CXCR4. *FEBS Lett* **583**: 2749–2757.
88. Shi X *et al.* (2006) CD44 is the signaling component of the macrophage migration inhibitory factor-CD74 receptor complex. *Immunity* **25**: 595–606.
89. Orita M *et al.* (2001) Coumarin and chromen-4-one analogues as tautomerase inhibitors of macrophage migration inhibitory factor: Discovery and X-ray crystallography. *J Med Chem* **44**: 540–547.
90. McLean LR *et al.* (2009) Discovery of covalent inhibitors for MIF tautomerase via cocrystal structures with phantom hits from virtual screening. *Bioorg Med Chem Lett* **19**: 6717–6720.
91. Cournia Z *et al.* (2009) Discovery of human macrophage migration inhibitory factor (MIF)-CD74 antagonists via virtual screening. *J Med Chem* **52**: 416–424.
92. Winner M *et al.* (2008) A novel, macrophage migration inhibitory factor suicide substrate inhibits motility and growth of lung cancer cells. *Cancer Res* **68**: 7253–7257.
93. Ouertatani-Sakouhi H *et al.* (2010) Kinetic-based high-throughput screening assay to discover novel classes of macrophage migration inhibitory factor inhibitors. *J Biomol Screen* **15**: 347–358.
94. Balachandran S *et al.* (2009) Novel derivatives of ISO-1 as potent inhibitors of MIF biological function. *Bioorg Med Chem Lett* **19**: 4773–4776.
95. Dagia NM *et al.* (2009) A fluorinated analog of ISO-1 blocks the recognition and biological function of MIF and is orally efficacious in a murine model of colitis. *Eur J Pharmacol* **607**: 201–212.

96. Ivanenkov YA, Balakin KV, Tkachenko SE. (2008) New approaches to the treatment of inflammatory disease: Focus on small-molecule inhibitors of signal transduction pathways. *Drugs R D* **9**: 397–434.
97. Bloom BR, Bennett B. (1966) Mechanism of a reaction *in vitro* associated with delayed-type hypersensitivity. *Science* **153**: 80–82.
98. David JR. (1966) Delayed hypersensitivity *in vitro*: Its mediation by cell-free substances formed by lymphoid cell-antigen interaction. *Proc Natl Acad Sci USA* **56**: 72–77.
99. Sendoel A, Kohler I, Fellmann C. (2010) HIF-1 antagonizes p53-mediated apoptosis through a secreted neuronal tyrosinase. *Nature* **465**: 577–583.
100. Zhang M *et al.* (1995) Cloning and sequencing of a cDNA encoding rat D-dopachrome tautomerase. *FEBS Lett* **373**: 203–206.
101. Rosengren E *et al.* (1996) The immunoregulatory mediator macrophage migration inhibitory factor (MIF) catalyzes a tautomerization reaction. *Mol Med* **2**: 143–149.
102. Tan X *et al.* (2007) Mechanism of auxin perception by the TIR1 ubiquitin ligase. *Nature* **446**: 640–645.
103. Kepinski S, Leyser O. (2005) The *Arabidopsis* F-box protein TIR1 is an auxin receptor. *Nature* **435**: 446–451.
104. Dharmasiri N, Dharmasiri S, Estelle, M. (2005) The F-box protein TIR1 is an auxin receptor. *Nature* **435**: 441–445.
105. Winner M, Leng L, Zundel W, Mitchell RA. (2007) Macrophage migration inhibitory factor manipulation and evaluation in tumoral hypoxic adaptation. *Method Enzymol* **435**: 355–369.
106. Petrenko O, Moll UM. (2005) Macrophage migration inhibitory factor (MIF) interferes with the Rb-E2F pathway. *Mol Cell* **17**: 225–236.
107. Swope M, Sun HW, Blake PR, Lolis E. (1998) Direct link between cytokine activity and a catalytic site for macrophage migration inhibitory factor. *EMBO J* **17**: 3534–3541.
108. Senter PD *et al.* (2002) Inhibition of macrophage migration inhibitory factor (MIF) tautomerase and biological activities by acetaminophen metabolites. *Proc Natl Acad Sci USA* **99**: 144–149.
109. Dios A *et al.* (2002) Inhibition of MIF bioactivity by rational design of pharmacological inhibitors of MIF tautomerase activity. *J Med Chem* **45**: 2410–2416.
110. Fingerle-Rowson G *et al.* (2009) A tautomerase-null macrophage migration-inhibitory factor (MIF) gene knock-in mouse model reveals that protein interactions and not enzymatic activity mediate MIF-dependent growth regulation. *Mol Cell Biol* **29**: 1922–1932.

PART III

Infectious and Inflammatory Diseases

III-1

MIF in Infectious Diseases

Marcelo Torres Bozza* and Claudia Neto Paiva

1. Introduction

Recognition of molecules from infectious agents by the innate immune system triggers inflammatory responses considered essential to the early control of infection. This initial immune response is also instrumental in promoting the activation of antigen specific lymphocytes, the hallmark of the adaptive immune response. Each infectious agent has a peculiar set of molecules and a *modus vivendi*, its characteristic way to interact with the host. The innate immune system recognizes molecules from infectious agents but also molecules from damaged host cells and tissues, promoting unique immune/inflammatory responses.

The physiological purpose of inflammation during infection is to eliminate the infectious agent and promote tissue repair. It is accepted that the ability to control infectious agents represented a major evolutionary force molding the immune system. In several immune responses to infection, however, the tissue damage is caused by the immune/inflammatory response itself. Mammalian pattern recognition receptors recognize conserved microbial molecules from all classes of microorganisms.[1,2] The activation of these receptors elicits selective intracellular signaling cascades that result in the production of cytokines, chemokines, lipid mediators, and reactive oxygen/nitrogen species. The increased secretion of cytokines is also achieved by combinations of microbial molecules with molecules from host origin such as cytokines, ATP, and ROS.[3,4] This activation of the immune system is considered essential for pathogen killing but is also critically involved in tissue damage and sepsis.[3] Thus, the pathology of infectious diseases can be attributed to either a direct effect of the infectious agents or the result of the immune/inflammatory response, both of which can cause metabolic

* Corresponding author: Laboratório de Inflamação e Imunidade, Departamento de Imunologia, Instituto de Microbiologia, Universidade Federal do Rio de Janeiro, Brazil. E-mail: mbozza@micro.ufrj.br

changes, cellular malfunctioning, and cell death. In fact, the pathology of most infectious diseases is the intricate result of these two forces promoting tissue damage and malfunctioning.

Macrophage migration inhibitory factor (MIF), the very first cytokine activity described almost 50 years ago,[5,6] was cloned several years later in a functional assay based on its ability to inhibit macrophage migration.[7] In 1993, a remarkable study identified MIF as one of the characteristic products of the pituitary gland triggered by LPS.[8] In this study, it was also observed that blockade of MIF protected mice from LPS-induced lethality, the first indication of its critical role in endotoxemia. These studies renewed scientific interest in MIF and opened research avenues in several areas of study. Twenty years of research on MIF have demonstrated a complex scenario of MIF biology, as discussed in this book. Among the biological functions of MIF, it became clear that MIF is an important inflammatory mediator that participates in innate and adaptive immune responses. In this chapter, we will focus on the role of MIF in infectious diseases.

2. MIF is a Critical Mediator in Sepsis

Sepsis is the systemic inflammatory response due to infection, and the resulting dysfunction syndrome is one of the leading causes of death in intensive care units. The physiopathology of sepsis is still poorly understood despite major investment in basic and clinical research. The treatment for sepsis is in the majority of cases supportive, utilizing fluid reposition, vasopressors, and antibiotics. Cytokines act as effectors or modulators of inflammatory responses. Cytokines are fundamental mediators in the development of septic shock, as well as of infection resolution. A number of studies firmly established a critical role for MIF in the physiopathology of sepsis. MIF, initially described as a product of T lymphocytes, was found to be produced *in vitro* and *in vivo* by cells of the pituitary gland stimulated with LPS.[8] MIF localizes to granules present in ACTH and TSH secreting anterior pituitary cells, and upon systemic challenge with LPS, it is massively released.[9] Systemic LPS administration causes an increase in plasma MIF concentrations that peak at 20 hrs.[10] The use T cell-deficient (nude) and hypophysectomized mice challenged with LPS indicated that cells other than T lymphocytes and pituitary cells should be involved in MIF production. In fact, macrophages are major sources of MIF, containing pre-formed protein and mRNA.[10] Stimulation with LPS causes a dose-dependent secretion of MIF by macrophages, while a modest increase in MIF mRNA was observed at late time points. *In vitro,* MIF exhibits several proinflammatory functions, including the induction of TNF, IL-1β and NO release from macrophages.[11–13] The discovery that low concentrations of glucocorticoids induce MIF production from macrophages was unexpected, since most inflammatory cytokines are inhibited by glucocorticoids.[13] MIF, besides being induced by glucocorticoids, can counter-regulate the inhibitory effects of glucocorticoid on cytokine secretion induced by LPS-stimulated monocytes.

MIF protein is also present pre-formed in several organs, and systemic administration of LPS causes an early release of pre-formed MIF from parenchymal cell types and nonlyphoid tissues.[14] In most tissues, the loss of MIF protein early after LPS challenge is accompanied by the induction of MIF mRNA and, at 24 hrs, by the restoration of intracellular MIF stocks.[14] Importantly, blockade of MIF with anti-MIF antibodies protects mice from a lethal dose of LPS, while recombinant MIF along with LPS greatly increases lethality.[8] Recombinant MIF is also capable of overcoming glucocorticoid protection against lethal endotoxemia in mice.[13]

Gram-positive exotoxins from *Staphylococcus aureus* or from *Streptococcus pyogenes* are potent inducers of MIF secretion from mouse macrophages and from a mouse corticotrophic cell line.[15] Challenge with these superantigens causes the activation of macrophages and T lymphocytes, with secretion of high quantities of inflammatory cytokines such as TNF and IFN-γ.[16] Neutralizing anti-MIF antibodies inhibit superantigen-induced lymphocyte proliferation and protect mice from lethal toxic shock.[15]

The generation of $Mif^{-/-}$ mice confirmed the critical role of MIF in endotoxic and superantigen-induced lethality.[17] The genetic deficiency of *Mif* also affects the optimal production of LPS-induced TNF *in vivo* and by macrophages *in vitro*. An elegant study demonstrated that this lower production of TNF by LPS-stimulated macrophages in the absence of MIF is at least in part due to reduced Toll-like receptor 4 (TLR4) expression.[18] Complementation of *Mif* deficient cells with a plasmid encoding MIF reconstitutes the expression of TLR4 and the production of TNF. Similarly, the reduced response of $Mif^{-/-}$ cells to IL-1 and TNF is also attributed to downregulation of cytokine receptor expression, as it can be restored upon MIF reconstitution.[19]

An important role for MIF in controlling glucose metabolic response during endotoxemia has been demonstrated.[20] In contrast to wildtype mice, $Mif^{-/-}$ mice exhibit normal blood glucose and lactate responses following LPS challenge. Upon endotoxemia, $Mif^{-/-}$ mice present increased glucose uptake into white adipose tissue compared to wildtype animals. These effects of MIF are probably due to a modulation of insulin-mediated glucose transport and insulin receptor signal transduction.

Altogether these studies demonstrated that MIF acts as a proinflammatory mediator that controls the cellular response to microbial molecules, cytokines, glucocorticoids, and insulin. MIF was found to be deleterious in several models of systemic inflammatory responses due to the administration of high doses of endotoxins or exotoxins. However, these models do not reflect the complexity of infection or sepsis, as they do not use live microbes. A number of studies addressed this point using different models of bacterial infection. In the pulmonary infection of wildtype and *Mif*$^{-/-}$ mice with *Pseudomonas aeruginosa,* MIF impairs the immune response and bacterial clearance in the lungs. The critical role of MIF in models of sepsis was further demonstrated using the polymicrobial septic model induced by cecal ligation and puncture (CLP) or peritoneal challenge with *E. coli*.[21,22] Peritoneal and plasma concentrations of MIF increased due to CLP or challenge with *E. coli*. Moreover, in models of polymicrobial sepsis, a significant increase in MIF amounts is observed in the lungs late after the induction of sepsis.[23,24] Thus, upon an acute bacterial infection, large quantities of MIF are released, initially at the primary site of infection and later in the systemic circulation and distant tissues, if the infection becomes disseminated. MIF peaks several hours after bacterial challenge, indicating that it might be inhibited in a clinically relevant time frame. Importantly, a significant protection is achieved when anti-MIF antibody is administrated 8 hrs after CLP.[21] Conversely, peritoneal injection of recombinant MIF together with a sublethal dose of *E. coli* dramatically increases lethality. These results indicate that MIF controls the systemic inflammatory response in experimental sepsis and suggest that endogenous MIF hampers the bacterial clearance.

The experimental findings indicating the involvement of MIF in the pathogenesis of sepsis were extended to clinical studies. Several clinical studies have shown an increase in MIF concentrations in the sera of patients with systemic inflammatory response syndrome (SIRS), sepsis, and septic shock.[21,25–29] Intracellular MIF is significantly increased in lymphocytes, B cells, macrophages, and granulocytes of patients with severe sepsis when compared to healthy control individuals.[30] High concentrations of MIF correlate with severity and poor outcome in patients with SIRS[26] and in patients with severe sepsis and septic shock.[21,28,29,31–33] MIF concentrations are an indicative of infection in patients after cardiac surgery, suggesting a possible use as a valuable marker of microbiologically documented sepsis.[34]

Interestingly, it has been suggested that LPS is not a strong inducer of MIF production by human monocytes *in vitro* or in a model of endotoxemia in healthy volunteers, although high concentrations of TNF were observed in these samples.[35] LPS and *S. aureus* cause a low level induction of MIF mRNA and protein in human leukocytes.[36] The induction of MIF by PBMC exposed to *E. coli* peaks at 4 hrs, although it is modest compared to TNF.[37] Gram-positive *Streptococcus pneumoniae* does not induce MIF mRNA expression or protein secretion by human PBMC, and induction by *E. coli* is marginal.[38] Another study showed that MIF production is minimal in healthy controls but marked in septic patients when their blood leukocytes are stimulated with LPS or *E. coli*.[39] These results might suggest that cells able to produce MIF enter the bloodstream during sepsis.

In patients with sepsis or septic shock, including those with meningococcal disease, higher MIF concentrations also correlate with disseminated intravascular coagulation, sepsis severity scores, shock, and a rapidly fatal outcome.[32,33,35] MIF concentrations in the cerebrospinal fluid are significantly higher in patients with purulent meningitis and encephalitis than in patients with lymphocytic meningitis and in patients suspected of having meningitis but without evidence of CNS infection.[40] Moreover, MIF concentrations in cerebrospinal fluid are to some degree related to disease severity but do not correlate with mortality or with the presence of septic shock. In meningococcal sepsis, plasma concentrations of MIF positively correlate with adrenocorticotropic hormone and negatively correlate with cortisol concentrations.[33] An association between MIF and hypothalamo-pituitary-adrenal axis dysfunction was suggested by a study analyzing septic patients who had an ACTH stimulation test. Serum MIF was higher in the adrenal insufficiency group compared to the normal adrenal response group.[41] MIF concentration has an inverse correlation with delta maximum cortisol, which is indicative of free cortisol and an insufficient adrenal response.

There are multiple, consistent lines of evidence from studies with anti-MIF antibodies, *Mif*$^{-/-}$ mice, and also from clinical studies, indicating that MIF is involved in the pathogenesis of sepsis and is a faithful marker of its severity. These observations led to the study of MIF blockade as a therapeutic strategy. The discovery that a small molecule inhibitor, ISO-1, was capable of inhibiting MIF proinflammatory effects by targeting MIF tautomerase activity prompted its use in several models of inflammatory diseases, including sepsis.[22,42] Treatment with ISO-1 initiated 24 hrs after CLP and continued for 3 days resulted in increased survival compared to an untreated group.[22] The improved survival conferred by ISO-1 was comparable to anti-MIF antibodies given in a similar time frame after CLP. Interestingly, treatment with ISO-1 suggested that MIF produced in the lungs probably participates in cardiac depression observed during septic episodes.[24] The use of ISO-1 mimics the phenotype of the *Mif*$^{-/-}$ macrophages, with decreased TNF production in response to LPS.[22] ISO-1 administered to endotoxemic *Mif*$^{-/-}$ mice did not attenuate the macrophage TNF release, indicating that the effect of ISO-1 in reducing macrophage responses to LPS requires endogenous MIF.

The use of a recombinant adenovirus bearing the antisense MIF gene causes a significant reduction in MIF production in response to a combined treatment with Bacille Calmette-Guérin (BCG) and LPS.[43] This treatment is effective in reducing the mortality rate in BCG–LPS-induced liver failure in mice. In another study using gene therapy to reduce MIF expression, mice were immunized with a MIF/tetanus toxin (TTX) DNA vaccine, and sepsis was then induced by LPS or CLP. In both models this treatment protects mice from the lethal effect of sepsis compared with control-vaccinated mice.[44] Targeting MIF with this DNA vaccine causes a significant reduction in systemic TNF concentrations and expression of inflammatory mediators in the lungs.

Melioidosis, caused by the Gram-negative bacteria *Burkholderia pseudomallei,* is an important cause of community-acquired sepsis in Southeast Asia. Patients show a marked increase in MIF plasma levels and a significant, albeit modest, increase in MIF mRNA in leukocytes,[45] and elevated MIF levels are associated with mortality. Mice intranasally infected with *B. pseudomallei* develop pneumonia-derived melioidosis and, similar to human patients, present increased MIF plasma concentrations. Infected mice also present high amounts of MIF in the lungs, epithelial submucosa, bronchial epithelial cells, and alveolar macrophages, and with the use of anti-MIF antibody and recombinant MIF, it was demonstrated that MIF impairs bacterial clearance and is involved in the pathogenesis of sepsis induced by *B. pseudomallei.* The infection with the Gram-negative bacterium *Vibrio vulnificus* causes an inflammatory process involving soft-tissue infections and sepsis. Infection of mice with *V. vulnificus* leads to an increase in MIF that peaks earlier than TNF and IL-6, and *in vivo* treatment with ISO-1 inhibits the production of these cytokines.[46] Infection of human PBMC with *V. vulnificus* also induces MIF, and its blockade with ISO-1 or anti-MIF reduces the production of inflammatory cytokines.

The late phase in a septic response is characterized by reduced production of proinflammatory mediators and increased susceptibility to a secondary infection.[47] Neutralization of MIF during this hyporesponsive immune-suppressed phase makes animals more susceptible to a secondary bacterial infection, while recombinant MIF reconstitutes the ability of the CLP-challenged mice to control the secondary bacterial infection.[48] At the onset of sepsis, blockade of MIF is beneficial due to a reduction of the inflammatory response and tissue damage. Later after CLP, however, the host enters in a hypoinflammatory phase, becoming more susceptible to secondary infections. In this phase, endogenous MIF seems to participate in the containment of bacterial spreading. Interestingly, a recent study indicates that polymorphisms associated with higher MIF expression may have a beneficial effect in community-acquired pneumonia.[49] These results indicate that in the post-sepsis period, endogenous MIF is important to control bacterial growth, such as that of *P. aeruginosa* and *Listeria monocytogenes*.

3. MIF in Extracellular Bacterial Infections

Helicobacter pylori infects approximately 50% of the world's population and is the major cause of gastritis and gastric and duodenal ulcers. The chronic inflammatory response due to persistent infection has been associated with the development of gastric carcinoma. A number of studies demonstrated that infection with *H. pylori* causes the production of MIF by gastric epithelial cells as well as cells of the immune system.[50–52] MIF plays an important role in the pathogenesis of *H. pylori* infection without affecting bacterial loads and clearance. MIF blockade or its genetic deficiency reduces the inflammatory response, the generation of gastritis and the proliferation of gastric epithelial cells.[51,52] These effects are likely to be multifactorial, including the activation and upregulation of the expression of epidermal growth factor receptor (EGFR), and the upregulation of CD74 expression, which is both a receptor for MIF and an adherence receptor for *H. pylori*.[52,53] The reduced inflammatory response and tissue damage observed in $Mif^{-/-}$ mice compared to infected controls is also associated with a reduced Th1 response, with lower T-bet and IFN-γ expression.[52]

The inflammatory response triggered by *P. aeruginosa* in the cornea is responsible for tissue damage and loss of visual acuity. The involvement of MIF in corneal keratitis induced by *P. aeruginosa* infection was recently shown.[54] $Mif^{-/-}$ mice show a more efficient bacterial clearance and have reduced inflammation. Reconstitution of MIF in $Mif^{-/-}$ mice by topical application of rMIF during infection reverses the beneficial effect of the MIF absence. Moreover, treatment with a MIF inhibitor efficiently reduces bacterial burdens and reverses pathology when initiated up to 5 days after infection. The inhibition of MIF expression by siRNA or the blockade of MIF activity on corneal epithelial cells reduces the production of inflammatory mediators and enhances bacterial clearance. Together, these results elegantly demonstrate a pathogenic role for MIF in *P. aeruginosa*-induced keratitis.

MIF is constitutively expressed in epithelial cells of the urinary tract, peripheral ganglia, and lumbosacral spinal cord.[55] Intravesical injection of LPS induces bladder inflammation and increases MIF protein and mRNA in the bladder and lumbosacral spinal cord.[56] Both children and adult patients with urinary tract infection present a significant increase in urinary MIF compared to control subjects, and high concentrations of urinary MIF can allow one to differentiate between upper and lower urinary tract infection in children during the acute phase of the disease.[57–59] Future experimental studies are required to define the role of MIF in the immune response to urinary tract infection and its putative pathogenic role promoting inflammation and tissue damage.

4. MIF in Intracellular Bacterial Infections

MIF is a pivotal proinflammatory cytokine that is capable of increasing macrophage activation and upregulating the production of TNF and IL-12. Thus, it is not surprising that the effect of MIF on pathogen burden in many intracellular infections is mediated primarily by macrophage activation, and in some cases, this effect is associated with Th1 T cell polarization.

Mycobacterium tuberculosis is a highly evolved intracellular bacterial species that infects phagocytic cells. It avoids lysosomal fusion, which is a major reason for its success as a pathogen.[60,61] Effector mechanisms directed against *M. tuberculosis* consist of macrophage activation, which can result from TLR-dependent or IFN-γ-mediated pathways, leading to NO production (in mice,[62] but without a corresponding consensual microbicidal mechanism in humans), expression of vitamin D receptor (and associated microbicidal cathelicidin secretion), and expression of LRG47 (a GTPase of unknown function). The establishment of a Th1 lymphocytic response with secretion of IFN-γ is an effective mechanism for containing *M. tuberculosis* infection.[63] Human macrophages infected with *M. tuberculosis* produce large amounts of MIF and its neutralization with monoclonal antibodies greatly enhances infection, while addition of rMIF reduces infection in a dose-dependent fashion, demonstrating that MIF also contributes to the containment of *M. tuberculosis* infection in macrophages.[64] These results position MIF as a macrophage activator participating in intracellular pathogen clearance, comparable to TNF and IFN-γ. In fact, the levels of MIF are increased in serum of pulmonary tuberculosis patients,[65] suggesting that it might have a role in protection against *M. tuberculosis.* The effector mechanisms activated by MIF in macrophages remain to be elucidated.

Listeria monocytogenes is a foodborne Gram-positive bacterium that invades the gastrointestinal tract, lives inside phagocytic and nonphagocytic cells, and causes a disease that is responsible for many human deaths each year. Activation of macrophages to produce NO and reactive oxygen species

(ROS) is a highly effective defense mechanism against *L. monocytogenes*, and in the course of infection, IFN-γ produced first by NK and γδ T cells, and later by antigen-specific CD4 Th1 and CD8 CTL, helps to activate macrophages.[66] The establishment of a CD8 CTL response is required for complete bacterial clearance from infected hosts, and perforin is the most important effector mechanism against *L. monocytogenes*. Still, production of type I IFN causes susceptibility to *L. monocytogenes* by producing down-regulation of IFN-γR and allowing suppression of macrophage activation by IFN-γ.[67] In murine infection with *L. monocytogenes*, treatment of infected mice with anti-MIF does not alter the course of a sublethal infection.[68] However, in lethal infections, anti-MIF reduces bacterial burden in spleens and livers (measured by CFU assay), which is an effect that can be abrogated by anti-IL-10 and is thus attributed to the inverse correlation observed between IL-10 and MIF levels in listeriosis. Therefore, in lethal listeriosis, MIF production is a factor of susceptibility. How an increase in IL-10 mediates the decrease in bacterial burden produced by treatment with anti-MIF is still to be unraveled. In fact, treatment with anti-IL-10 by itself does not alter spleen and liver CFU, but treatment with IL-10 is known to decrease bacterial burden in spleens during listeriosis through unknown mechanisms.[69] As IL-10 is known to change its effector profile in the presence of IFN-γ, from inhibitor of antigen-presentation and proinflammatory cytokine production to macrophage activator,[70] it remains to be investigated whether treatment with anti-MIF contributes to *L. monocytogenes* clearance by unleashing the macrophage activating properties of IL-10 in the presence of IFN-γ.

Treatment of mice lethally infected with *L. monocytogenes* with anti-MIF produces both a decrease in bacteremia and an increase in glucocorticoids.[68] These mice can be spared from death either by treatment with anti-MIF[68] or glucocorticoids,[71] and treatment with anti-MIF contributed to increase glucocorticoid levels.[68] These results are compatible with a septic-like syndrome caused by listeriosis and mediated by MIF, but the precise mechanisms underlying these findings remain to be elucidated.

Salmonella enterica serovar Typhimurium invades gastrointestinal epithelium causing diarrhea. It then spreads to become systemic infection, which may cause fatal bacteremia or lead to microabscesses in spleen and liver. In these microabscesses, neutrophils and macrophages act to restrict infection. At least in intravenous models of infection, both ROS and NO are involved in bacterial clearance.[72,73] However, the acute intestinal inflammation can enhance the growth of *S. typhimurium*,[74] and it is possible that ROS produced during intestinal inflammation represents a survival advantage to these bacteria, as it can generate a respiratory electron acceptor used by the bacteria.[75]

The establishment of Th1 CD4 and CD8 CTL specific responses is critical to the ultimate clearance of the bacteria. *Mif*$^{-/-}$ mice infected with *S. enterica* serovar Typhimurium fail to control bacterial growth, presenting escalating burdens (measured by CFU assay) in both livers and spleens, and dying more frequently from infection and before wildtype controls.[76] Infected *Mif*$^{-/-}$ mice present decreased IFN-γ serum levels, and spleen cells from *Mif*$^{-/-}$mice fail to produce IFN-γ when stimulated with heat-killed *Salmonella*, suggesting a role for MIF in Th1 polarization. When injected with rIFN-γ, *Mif*$^{-/-}$ mice survive *Salmonella* infection, demonstrating that their susceptibility is related to reduced IFN-γ production. Surprisingly, infected *Mif*$^{-/-}$ mice exhibit increased NO production despite their reduced IFN-γ production, and iNOS inhibitors prolong their survival. As NO can act either as a microbicidal effector mechanism or as an immunosuppressor,[77] these results suggest that MIF protects against salmonelosis by downregulating NO levels and therefore preventing its suppressor effects.

5. MIF in Protozoan Infections

Toxoplasma gondii is an intracellular pathogen that enters its host via the gastrointestinal tract and is highly adapted to infecting many different cell types. In the intestine, microbicidal mechanisms are set into motion by enterocytes, macrophages, and dendritic cells, and the establishment of an antigen-specific Th1 response is both fundamental to protective immunity and potentially lethal, as excess intestinal inflammation promoted by IFN-γ can lead to death.[78] The proinflammatory cytokines IL-12, TNF, IFN-γ, and IL-1β promote resistance against *T. gondii*, and NO production by macrophages is an important effector mechanism in parasite elimination, as is a 47-KDa GTPase of unknown function.[79] *Mif*$^{-/-}$ mice are highly susceptible to intraperitoneal infection with *T. gondii*, presenting increased parasite burden in brains and peritoneal macrophages.[80] In the absence of MIF, the serum concentrations of IL-12, TNF, IFN-γ, nitrite, and IL-1β are decreased during infection.[80] This is a highly expected finding, as MIF is an enhancer of IL-12 and TNF production by macrophages. An impairment of proinflammatory cytokine secretion is also found in the response of bone marrow differentiated dendritic cells (DC) from *Mif*$^{-/-}$ mice to *T. gondii* antigens: they secrete decreased amounts of IL-12p40, IL-12p35, IL-12p19, IL-1β, and TNF.[81] Splenic DC isolated from *Mif*$^{-/-}$ mice and treated *in vitro* with antigens from *T. gondii* also show reduced production of TNF and IL-12 compared to wildtype mice, confirming that MIF is required for the DC response to *T. gondii* antigens.[81] Consistently, DC obtained from spleens and mesenteric lymph nodes from *Mif*$^{-/-}$ mice orally infected with *T. gondii* had impaired maturation, presenting decreased expression of maturation markers CD80, CD86, CD40, and MHC class II.[81] Thus, the protective role of MIF in *T. gondii* infection is apparently related to the production of proinflammatory cytokines and to the activation of macrophages and dendritic cells.

Leishmania major is a trypanosomatid protozoan that is harbored inside macrophages and can be eliminated by macrophage activation with ensuing NO production. Infection with *L. major* causes skin lesions, which in general parallel the parasite load. A highly polarized Th1 response is effective against *L. major*, activating macrophages to produce NO and resulting in resolving skin lesions. Addition of MIF to macrophage cell cultures results in increased *L. major* elimination.[82] Though the MIF concentration required to reduce *L. major* burden in macrophages is high (1 μg/mL, 100 times that of other cytokines with leishmanicidal effects, such as IFN-γ), this concentration is within the range reached in inflammatory conditions. MIF also requires longer intervals to reduce parasite burden than the combined effects of TNF and IFN-γ. The leishmanicidal effects of MIF depend on the production of TNF and NO by infected macrophage cultures and can be reversed by the addition of IL-10, TGF-β or IL-13, indicating that it depends on an M1 activation status.[82] MIF expression increases during *L. major* footpad inoculation in ipsilateral (but not contralateral) popliteal lymph nodes, but the kinetics of its expression compared to that of MIF secretion by T cells upon antigen presentation suggests that lymph node MIF comes from another cellular source.[82] Consistent with the observed role of MIF as an enhancer of macrophage leishmanicidal function, oral administration of *S. typhimurium* transfected with MIF reduces the size of skin lesions,[83] while *Mif*$^{-/-}$ mice are highly susceptible to *L. major*, developing severe skin lesions late after infection.[84] MIF does not affect Th polarization in *L. major* infection, as indicated by the similar IFN-γ and IL-4 production among T cells from *Mif*$^{-/-}$ and wildtype mice. However, IFN-γ-activated macrophages from *Mif*$^{-/-}$ mice infected *in vitro* with *L. major* have slightly decreased parasite clearance,[84] indicating that either they are somewhat insensitive to IFN-γ or MIF production is partially required as an intermediary step to IFN-γ-induced leishmanicidal activity.

Trypanosoma cruzi is an intracellular protozoan that can infect many cell types, including macrophages. The effective response to *T. cruzi* comprises innate macrophage activation to NO production and ultimately, the establishment of antigen-specific Th1 CD4 and CTL CD8 responses.[85] Mice genetically deficient in *Mif* are also more susceptible to *T. cruzi* infection.[86] This increase in susceptibility is accompanied by decreased plasma concentrations of IL-12 and IFN-γ along acute infection and also decreased IL-12 and IFN-γ production by splenocytes stimulated with *T. cruzi* antigens early in the acute phase, indicating that in contrast to the trypanosomatid, *L. major*, MIF participates in Th1 polarization in *T. cruzi* infection. This deficient Th1 polarization is reflected by decreased titers of anti-*T. cruzi* IgG2a (but not IgG1). Also, $Mif^{-/-}$ mice have decreased plasma concentrations of TNF, IL-1β, and IL-18, suggesting that decreased production of proinflammatory cytokines underlies their susceptibility to *T. cruzi* infection. The deficient Th1 polarization, specific IgG, and proinflammatory cytokine secretion are all highly compatible with susceptibility to *T. cruzi* infection, but there is currently no functional data to support this hypothesis. In fact, IFN-γ-activated macrophages have a prominent role in *T. cruzi* clearance through NO production, a function that can be enhanced by TNF production. As MIF controls TNF production by macrophages in a number of cases, and along with TNF enhances production of NO by macrophages and the elimination of trypanosomatid *L. major*,[82] it seems likely that MIF enhances macrophage trypanocidal activity. This possibility awaits experimental testing.

Sepsis can result from systemic inflammation caused by an intracellular infection, though this is an issue that has seldom been exploited. A prior intracellular infection can also sensitize the organism to septic shock by priming monocytes to overreact in the presence of very low amounts of TLR ligands, as happens in influenza,[87] VSV,[88] and LCMV infection,[89] among others. *T. cruzi*-infected mice are highly susceptible to septic shock, which can be caused by infection itself in mice lineages that develop severe inflammatory response or by administration of TNF, anti-CD3,[90] SEB[91] or LPS.[92] The lethal synergism between *T. cruzi* infection and LPS inoculation likely results from redundant lethal pathways induced by TNF and MIF: although both $Mif^{-/-}$ and $Tnfr1^{-/-}$ infected mice succumb to LPS administration, treatment with anti-MIF rescues *Tnfr1*-deficient mice from lethal shock.[92] However, there are no clues as to the specific role of MIF, if any, in the septic condition caused by *T. cruzi* infection itself, or if it is a contributor to human mortality in Chagas disease or in other parasitic infections that can cause a sepsis-like condition.

Plasmodium falciparum is an intracellular parasite that causes malaria in humans. Four species of rodent *Plasmodium* can infect mice, but infection with *P. chabaudi* offers a model that more closely resembles infection in humans. In *P. chabaudi* infection, $CD4^+$ and $CD8^+$ αβ T cells, γδ T cells, B cells, and macrophages are activated, but protective immunity against asexual blood stages is mainly conferred by CD4 cells.[93] Distinct from other infections, both Th1 and Th2 seem to contribute to immunity, Th1 by activating

Table 1. Role of MIF in intracellular pathogen burden.

Intracellular Pathogen	Experimental System of MIF Manipulation	Effects of MIF on Parasite Burden	Attributed Immune Mechanism	References
Listeria monocytogenes	Mice, anti-MIF	↑	MIF inhibits secretion of IL-10, a cytokine that reduces burden	68
Salmonella enterica (serovar Typhimurium)	*Mif*$^{-/-}$ mice	↓	MIF promotes a Th1 response and downregulates NO, which has a suppressor role	76
Mycobacterium tuberculosis	Human macrophages, anti-MIF, rMIF	↓	MIF increases macrophage activation	64
Leishmania major	Murine macrophages, anti-MIF, rMIF	↓	MIF increases macrophage activation through enhancement of TNF and NO production	82
	Mif$^{-/-}$ mice		MIF decreases lesion sizes and partially mediates leishmanicidal effects of IFN-γ on macrophages but does not alter Th1 polarization	84
Toxoplasma gondii	*Mif*$^{-/-}$ mice	↓	MIF stimulates production of IL-1β, IL-12, TNF, NO, IFN-γ and improves maturation of DC	80, 81
Trypanosoma cruzi	*Mif*$^{-/-}$ mice	↓	MIF stimulates production of IL-1β, IL-12, IL-18, Th1 polarization and specific IgG2a production	86
Plasmodium chabaudi	*Mif*$^{-/-}$ mice	=	—	95
Plasmodium falciparum	Human volunteers; correlation	=	—	96
	Infected children in endemic zone; correlation	↓	—	97

macrophages to NO production and Th2 by stimulating the production of parasite-specific IgG1. Macrophages secrete large amounts of MIF in response to red blood cells infected with *P. chabaudi*,[94] and also to hemozoin,[95] the heme pigment generated as a byproduct of hemoglobin degradation by the parasite. In Zambia, *P. falciparum*-infected patients showed elevated serum amounts of MIF, similar to those observed in bacterial sepsis patients.[95] However, *Mif*$^{-/-}$ mice infected with *P. chabaudi* develop a parasitemia similar to that of wildtype controls, indicating that MIF is not necessary to the immune response against this parasite.[95] Consistently, in short experiments with humans (10 days), MIF decreases in response to infection and increases in response to antimalarial treatment, but there is no correlation between MIF and parasitemia at the start of the treatment.[96] A study of *P. falciparum*-infected children in Kenya, however, showed an inverse correlation between MIF levels and parasite burden,[97] suggesting a role for MIF in *P. falciparum* elimination.

5.1. *The role of host MIF in the pathogenesis of malaria*

Infection with *P. falciparum* causes severe anemia in human malaria, as does infection with *P. chabaudi* in mice. Although the parasite does cause red blood cell lysis as a result of schizogony, the anemia in malaria apparently results from a much more complex phenomenon that comprises loss of red blood cells by several mechanisms and dysfunctional erythropoiesis.[98] Several proinflammatory cytokines have been implicated in the genesis of malarial anemia by mediating erythroid suppression, such as IFN-γ, IL-1β, and TNF, but experimental evidence has ruled out a role for each of these cytokines, at least in mice.[99] Plasma MIF concentrations are elevated in *P. chabaudi*-infected mice and correlate with disease severity.[94] MIF inhibits erythroid colony formation in bone marrow cell cultures containing erythropoietin, as do TNF and IFN-γ, but together, they synergize to suppress erythroid colony formation at very low concentrations.[95] In *P. chabaudi* infection, anemia greatly contributes to mortality, and infected *Mif*$^{-/-}$ mice are spared from death despite parasitemias similar to wildtype controls, presenting less severe anemia and inhibition of erythroid colony formation.[95] These results strongly support a role for MIF in the mouse experimental model of severe malarial anemia.

Macrophages from naïve mice release MIF after contact with *P. chabaudi*-infected red blood cells[94] or hemozoin. Human normal macrophages also release MIF when stimulated with hemozoin,[95] and hemozoin contributes to the suppression of erythropoiesis in several ways,[100–102] suggesting that

MIF might act as a mediator of hemozoin-induced anemia. In children affected by severe anemia, however, there is an inverse correlation between hemozoin accumulation and plasma MIF concentrations, indicating that hemozoin may decrease MIF production in malaria.[97] The explanation to this paradox may lie in a feedback loop involving long-lasting hemozoin activation of macrophages in these children. In fact, PBMC from malaria-naïve patients can react to hemozoin by either increasing or decreasing MIF production,[97] depending on whether they have a generally well-preserved MIF production response, and *MIF* polymorphisms also give rise to variable magnitudes of response to hemozoin,[95] which may explain the variability found in these studies regarding MIF production in response to hemozoin.

Plasma MIF concentrations were found to be elevated in the endemic region of Zambia, to the levels found in septic patients.[95] In contrast with these results, decreased MIF levels correlates with severity of malarial anemia in sub-Saharan African children.[97] Moreover, in normal adult human volunteers submitted to *P. falciparum* infection, MIF amounts decreases sharply.[96] The differences found in MIF production in response to infection among these populations may lie in polymorphisms of the MIF gene. In fact, there is an association between certain *MIF* haplotypes of the −173 G/C and −794 $CATT_{5-8}$ polymorphisms and susceptibility to severe malarial anemia.[103]

6. MIF in Viral Infections

Recent clinical studies indicated that patients with viral infections, such as those caused by hepatitis B virus, West Nile virus, dengue virus or Ebola virus, have higher MIF plasma concentrations than control subjects.[104–108] A number of studies using *in vitro* and *in vivo* models of viral infection have established the production of MIF. However, the mechanisms involved in the production of MIF seem to be particular to each viral infection, and overall this important subject requires further investigations.

Early studies demonstrated that the infection of lung epithelial cells with influenza A virus does not induce MIF gene transcription but causes the release of pre-formed MIF, probably dependent on necrotic cell death.[109] The addition of recombinant MIF does not influence the release of CXCL8/IL-8 in a virus-infected epithelial cell line, while blocking endogenous MIF with ISO-1 inhibits the virus-dependent secretion of CXCL8/IL-8. In the mouse model of H5N1 influenza virus infection, increased MIF mRNA amounts in the lungs and in MIF protein concentrations in the serum are observed.[110]

Blocking MIF functions with ISO-1 significantly inhibits proinflammatory cytokines in the lung, although the regimen of treatment tested showed no beneficial effects on survival. Together, these results indicate that MIF participates in the control of the inflammatory response evoked by influenza virus infection.

Infection of fibroblasts with human cytomegalovirus triggers an early and sustained induction of MIF mRNA and protein production, with subsequent MIF secretion.[111] In the infection of human macrophages with cytomegalovirus, however, the release of pre-formed MIF is not accompanied by an increase in MIF gene transcription.[112] The endogenous MIF released in this context inhibits macrophage chemotaxis. Interestingly, infection of rats with Borna disease virus increases the amount of MIF protein in astrocytes and is associated with reduced macrophage infiltration in the areas of MIF immunostaining.[113] The accumulation of MIF protein in astrocytes during Borna disease virus infection occurs without the presence of detectable MIF mRNA in these cells, thus suggesting a possible translocation of MIF from other cells such as neurons. However, no direct evidence of MIF transfer in this context has been demonstrated.

Persistent infection by hepatitis B virus and consequent chronic inflammation is a major cause of hepatic cirrhosis and hepatic cellular carcinoma. An increase in serum MIF concentrations has been shown in patients with hepatitis B virus infection and a critical role for MIF in the pathogenesis of viral hepatitis also was demonstrated. In a clinical study, a significant increase in serum MIF concentrations in patients with chronic B virus hepatitis and hepatitis cirrhosis was observed compared to healthy controls.[104] MIF has no effect on hepatitis B viral replication either *in vivo* or *in vitro*, but it contributes to liver injury in a transgenic model of cytotoxic T lymphocyte-induced hepatitis.[114] Under normal conditions, MIF expression is constitutively detected in hepatocytes surrounding the central vein and parenchyma. The number of MIF-positive hepatocytes increases 24 hrs after cytotoxic T lymphocyte injection, and macrophages also show increased MIF expression. In this model, treatment with anti-MIF provides partial protection against liver injury, probably due to a reduction in proinflammatory cytokine production, leukocyte infiltration in the liver, and tissue damage.[114] A physical association between MIF and HBx, a transcriptional transactivating protein of hepatitis B virus, was demonstrated using the yeast two-hybrid system and confirmed by co-immunoprecipitation, GST pull-down, and cellular co-localization. MIF protein and mRNA expression were upregulated by HBV infection but not HBx transfection alone. Recombinant MIF causes HepG2 cell G0/G1 phase arrest, proliferation inhibition, and apoptosis. However, MIF could inhibit the apoptotic effect of HBx. Considering the role of MIF in cell transformation, this study suggests a link between MIF in HBV infection and hepatocellular carcinoma.[115]

In vivo infection with Japanese encephalitis virus, West Nile virus (flaviviruses), or Venezuelan equine encephalitis virus (alphavirus) causes a significant, albeit modest, increase in MIF mRNA in mouse brains.[106,116,117] The peak of MIF gene transcripts in the mouse brains occurs 2 days after Japanese encephalitis virus, 3 days after Venezuelan equine encephalitis virus, and 6 days after West Nile virus infection is initiated. Neurons and glial cells express MIF transcripts upon infection and some of the cells are co-labeled by double staining for Japanese encephalitis virus antigens and MIF mRNA.[116] A critical role for MIF in the pathogenesis of viral encephalitis was demonstrated in the mouse model of West Nile virus infection.[106] Genetic deficiency or MIF blockade with antibodies or with ISO-1 had a beneficial effect upon West Nile infection. Infected *Mif*$^{-/-}$ mice present reduced viral loads in the brain, reduced systemic and local inflammation, and increased survival compared to infected controls. Various evidence supports the notion that the reduced viral loads in the brains of *Mif*$^{-/-}$ mice are due to reduced viral entry and not to a direct effect of MIF on viral replication. These observations, together with an increase in MIF concentrations in the plasma and cerebrospinal fluid of patients suffering from West Nile virus infection, suggest that blocking MIF might in the future constitute a valuable tool for the treatment of viral encephalitis.[106]

More recently, an important role for MIF was demonstrated in the infection caused by another member of flaviviruses, the dengue virus.[105,107] Dengue fever is an emerging viral disease transmitted by arthropods to humans in tropical countries. Dengue hemorrhagic fever is escalating in frequency and mortality rate, and in some physiopathological aspects resembles sepsis. Patients with dengue hemorrhagic fever have higher MIF plasma concentrations than control subjects.[105,107] Plasma concentrations of MIF were higher in all dengue hemorrhagic fever patients who died than in survivors. The secretion of MIF by macrophages and hepatocytes requires a productive infection and occurs without an increase in gene transcription or cell death, thus indicating active secretion from pre-formed pools. Similarly, infection of macrophages with Sindbis virus, a member of the alphavirus genus and the most widely distributed of all known arboviruses, resulted in MIF secretion from intracellular pools, without an increase in MIF gene expression or any effect on cell viability.[118] Blocking MIF with anti-MIF antibody or with ISO-1 *in vitro* inhibits the production of inflammatory mediators but has no effect in viral loads.[107,118] *Mif*$^{-/-}$ mice had a significant delay in lethality and a reduction in all parameters of severity upon dengue virus infection compared with wildtype mice,[107] reinforcing the role of MIF in the pathogenesis of dengue. The mild pathological condition of *Mif*$^{-/-}$ mice might reflect both the reduced viral load observed in the initial days and the lower production of inflammatory mediators. The reduction in viral load could be related to the better hemodynamic status of *Mif*$^{-/-}$ mice, thus facilitating leukocyte circulation. At later time points, however, the viremia became similar to that of the wildtype mice, and eventually the *Mif*$^{-/-}$ mice died.

Zaire ebolavirus causes highly lethal hemorrhagic fever and a cytokine storm that resembles other viral infections as well as bacterial sepsis. An increase in plasma MIF concentrations together with several proinflammatory mediators was observed in a recent clinical study analyzing Ebola patients.[108] Significant increases in MIF concentrations early after diagnosis (1–4 days) correlated with fatal cases compared to survivors. The very high concentrations of inflammatory cytokines and the massive lymphocyte apoptosis observed in this study probably triggers intravascular coagulation, plasma leakage, hypotension, and immunoparalysis, thus contributing to multiple organ failure and death of patients with Zaire ebolavirus infection.

Human immunodeficiency virus (HIV)-infected patients show a significant increase in plasma MIF concentrations, although no correlation with viral load could be observed.[119] *In vitro* infection of human PBMCs, but not human macrophages with HIV, causes secretion of MIF starting 7 days after infection, when elevated amounts of HIV-1 production are also detected. In HIV infection, MIF seems to contribute to viral growth, since blockade of MIF with anti-MIF or with ISO-1 reduces viral load, while treatment with recombinant MIF enhances HIV transcription. Together, the results of this study suggest that HIV infection of cells triggers MIF release, which then augments HIV-1 replication.

An interesting study demonstrated that transfection with various silence RNAs (siRNA), including one against MIF itself, caused a dose-dependent increase in MIF production by mammary adenocarcinoma cells that could be attenuated in the presence of a double-stranded RNA-dependent protein kinase (PKR) inhibitor, 2-aminopurine.[120] Transfection with siRNA induces MIF mRNA transcription, and transfection of poly(I:C), a prototypic double-stranded RNA, also stimulates a PKR-dependent increase in MIF production from adenocarcinoma cells. This study highlights a molecular mechanism of MIF induction that might take place during certain viral infections. Further studies are required to define the mechanisms of MIF production upon viral infections.

7. The Role of MIF in Helminth Infections

Taenia crassiceps is a unique helminth infection in that there is no clear correlation between Th2 response and protective immunity; on the contrary, proinflammatory cytokines appear to have a role in parasite elimination.[121] *Mif*$^{-/-}$ mice are much more susceptible to infection with *T. crassiceps*, developing high parasite burdens.[122] The peritoneal cavity of infected *Mif*$^{-/-}$ mice exhibit increased numbers of eosinophils, which contrasts with the role of MIF in eosinophil recruitment found in *Schistosoma mansoni* infection[123] and after administration of a helminthic protein orthologue of MIF to mice.[124,125] The antigen-specific IgG2a titers are reduced in infected *Mif*$^{-/-}$ mice, but IgE levels are similar to those found in infected wildtypes.[122] Surprisingly, despite decreased levels of Th1-associated IgG2a, splenocytes from infected mice produce increased amounts of IFN-γ when stimulated with specific antigens. No alteration in the serum concentrations of IL-12, IL-4 or IFN-γ were found, but macrophages produce less TNF, IL-12, and NO, which in this singular helminth infection can possibly be detrimental.

In mice infected with *S. japonicum*, two-week treatment with anti-MIF, beginning after female worms began laying eggs, results in increased worm burden, as measured by worm pairs.[126] These data suggest that MIF participates in the anthelminthic response. The number of eggs in livers does not change in response to treatment with anti-MIF, but the number of eggs per worm pair is reduced, suggesting that treatment alters egg deposition. There are no clues to the mechanisms that contribute to increased worm burden in this case. In contrast, in *Mif*$^{-/-}$ mice infected with *S. mansoni,* the worm and egg burden are similar to wildtype controls.[123] In both *S. japonicum* and *S. mansoni* infections, MIF does not seem to influence the serum concentrations of cytokines that may be indicative of T cell Th polarization.

7.1. *Host MIF in the pathogenesis in helminth infections*

In mice infected with *S. japonicum* and treated with anti-MIF, the area of the granuloma remains similar to IgG-treated controls.[126] On the other hand, *Mif*$^{-/-}$ mice infected with *S. mansoni* present greatly decreased egg granuloma sizes.[123] These granulomas contain fewer eosinophils, which is a phenomenon that is paralleled by decreased bone marrow eosinophilopoiesis in *Mif*$^{-/-}$ mice infected with *S. mansoni* mice and increased migration of granuloma eosinophils towards MIF in chemotaxis assays. Together, these results indicate that in *S. mansoni* infection, MIF orchestrates bone marrow eosinophil differentiation and recruitment to granulomas. Though the area of granulomas is decreased in *S. mansoni*-infected *Mif*$^{-/-}$ mice, they present no decrease in fibrosis, the pivot of portal hypertension, and therefore, MIF is not a candidate target to treat liver pathology in this infection.

8. Future Directions

The past decade has witnessed a revolution in the field of innate immunity with emphasis on the mechanisms of microbial molecule recognition by pattern recognition receptors, the dissection of these signaling pathways that culminate in the production of cytokines, and the role of innate immunity in the instruction of adaptive immunity. However, the molecular mechanism of MIF production upon recognition of microorganisms is still largely unknown. Research in the coming years will fulfill these gaps, including also a better characterization of the role of MIF in the transition from innate to adaptive immunity.

In this chapter we described the involvement of MIF in several models of infectious diseases and in each case there are distinctive features. Interestingly, we could not find studies characterizing the role of MIF in fungal infections. Another interesting and unexplored area that MIF is likely involved refers to the physiological interactions of host cells with the commensal microflora. Thus, future studies using other infectious models are still required to determine the participation of MIF in pathogenesis and protection in infection.

The essential role of MIF in the pathogenesis of infectious diseases and the concept that MIF is a therapeutic target will require clinical studies using strategies to block MIF. The great amount of evidence on the role of MIF in sepsis makes this condition one of the first candidates to test this assumption. Thus, several highways of research on MIF and infectious diseases are open for the coming years, promising beautiful and intriguing scientific landscapes.

References

1. Medzhitov R. (2007) Recognition of microorganisms and activation of the immune response. *Nature* **449**: 819–826.
2. Akira S, Uematsu S, Takeuchi O. (2006) Pathogen recognition and innate immunity. *Cell* **124**: 783–801.
3. Nathan C. (2002) Points of control in inflammation. *Nature* **420**: 846–852.
4. Powers KA, Szaszi K, Khadaroo RG *et al.* (2006) Oxidative stress generated by hemorrhagic shock recruits Toll-like receptor 4 to the plasma membrane in macrophages. *J Exp Med* **203**: 1951–1961.

5. David JR. (1966) Delayed hypersensitivity *in vitro*: Its mediation by cell-free substances formed by lymphoid cell-antigen interaction. *Proc Natl Acad Sci USA* **56**: 72–77.
6. Bloom BR, Bennett B. (1966) Mechanism of a reaction *in vitro* associated with delayed-type hypersensitivity. *Science* **153**: 80–82.
7. Weiser WY, Temple PA, Witek-Giannotti JS *et al.* (1989) Molecular cloning of a cDNA encoding a human macrophage migration inhibitory factor. *Proc Natl Acad Sci USA* **86**: 7522–7526.
8. Bernhagen J, Calandra T, Mitchell RA *et al.* (1993) MIF is a pituitary-derived cytokine that potentiates lethal endotoxaemia. *Nature* **365**: 756–759.
9. Nishino T, Bernhagen J, Shiiki H *et al.* (1995) Localization of macrophage migration inhibitory factor (MIF) to secretory granules within the corticotrophic and thyrotrophic cells of the pituitary gland. *Mol Med* **1**: 781–788.
10. Calandra T, Bernhagen J, Mitchell RA, Bucala R. (1994) The macrophage is an important and previously unrecognized source of macrophage migration inhibitory factor. *J Exp Med* **179**: 1895–1902.
11. Herriott MJ, Jiang H, Stewart CA *et al.* (1993) Mechanistic differences between migration inhibitory factor (MIF) and IFN-gamma for macrophage activation. MIF and IFN-gamma synergize with lipid A to mediate migration inhibition but only IFN-gamma induces production of TNF-alpha and nitric oxide. *J Immunol* **150**: 4524–4531.
12. Bernhagen J, Mitchell RA, Calandra T *et al.* (1994) Purification, bioactivity, and secondary structure analysis of mouse and human macrophage migration inhibitory factor (MIF). *Biochemistry* **33**: 14144–14155.
13. Calandra T, Bernhagen J, Metz CN *et al.* (1995) MIF as a glucocorticoid-induced modulator of cytokine production. *Nature* **377**: 68–71.
14. Bacher M, Meinhardt A, Lan HY *et al.* (1997) Migration inhibitory factor expression in experimentally induced endotoxemia. *Am J Pathol* **150**: 235–246.
15. Calandra T, Spiegel LA, Metz CN, Bucala R. (1998) Macrophage migration inhibitory factor is a critical mediator of the activation of immune cells by exotoxins of Gram-positive bacteria. *Proc Natl Acad Sci USA* **95**: 11383–11388.
16. Fleischer B, Gerlach D, Fuhrmann A, Schmidt KH. (1995) Superantigens and pseudosuperantigens of Gram-positive cocci. *Med Microbiol Immunol* **184**: 1–8.
17. Bozza M, Satoskar AR, Lin G *et al.* (1999) Targeted disruption of migration inhibitory factor gene reveals its critical role in sepsis. *J Exp Med* **189**: 341–346.
18. Roger T, David J, Glauser MP, Calandra T. (2001) MIF regulates innate immune responses through modulation of Toll-like receptor 4. *Nature* **414**: 920–924.

19. Toh ML, Aeberli D, Lacey D *et al.* (2006) Regulation of IL-1 and TNF receptor expression and function by endogenous macrophage migration inhibitory factor. *J Immunol* **177**: 4818–4825.
20. Atsumi T, Cho YR, Leng L *et al.* (2007) The proinflammatory cytokine macrophage migration inhibitory factor regulates glucose metabolism during systemic inflammation. *J Immunol* **179**: 5399–5406.
21. Calandra T, Echtenacher B, Roy DL *et al.* (2000) Protection from septic shock by neutralization of macrophage migration inhibitory factor. *Nat Med* **6**: 164–170.
22. Al-Abed Y, Dabideen D, Aljabari B *et al.* (2005) ISO-1 binding to the tautomerase active site of MIF inhibits its pro-inflammatory activity and increases survival in severe sepsis. *J Biol Chem* **280**: 36541–36544.
23. Lin X, Sakuragi T, Metz CN *et al.* (2005) Macrophage migration inhibitory factor within the alveolar spaces induces changes in the heart during late experimental sepsis. *Shock* **24**: 556–563.
24. Sakuragi T, Lin X, Metz CN *et al.* (2007) Lung-derived macrophage migration inhibitory factor in sepsis induces cardio-circulatory depression. *Surg Infect (Larchmt)* **8**: 29–40.
25. Joshi PC, Poole GV, Sachdev V *et al.* (2000) Trauma patients with positive cultures have higher levels of circulating macrophage migration inhibitory factor (MIF). *Res Commun Mol Pathol Pharmacol* **107**: 13–20.
26. Gando S, Nishihira J, Kobayashi S *et al.* (2001) Macrophage migration inhibitory factor is a critical mediator of systemic inflammatory response syndrome. *Intensive Care Med* **27**: 1187–1193.
27. Lehmann LE, Novender U, Schroeder S *et al.* (2001) Plasma levels of macrophage migration inhibitory factor are elevated in patients with severe sepsis. *Intensive Care Med* **27**: 1412–1415.
28. Beishuizen A, Thijs LG, Haanen C, Vermes I. (2001) Macrophage migration inhibitory factor and hypothalamo-pituitary-adrenal function during critical illness. *J Clin Endocrinol Metab* **86**: 2811–2816.
29. Bozza FA, Gomes RN, Japiassu AM *et al.* (2004) Macrophage migration inhibitory factor levels correlate with fatal outcome in sepsis. *Shock* **22**: 309–313.
30. Lehmann LE, Weber SU, Fuchs D *et al.* (2008) Oxidoreductase macrophage migration inhibitory factor is simultaneously increased in leukocyte subsets of patients with severe sepsis. *Biofactors* **33**: 281–291.
31. Chuang CC, Wang ST, Chen WC *et al.* (2007) Increases in serum macrophage migration inhibitory factor in patients with severe sepsis predict early mortality. *Shock* **27**: 503–506.
32. Gando S, Sawamura A, Hayakawa M *et al.* (2007) High macrophage migration inhibitory factor levels in disseminated intravascular coagulation patients with systemic inflammation. *Inflammation* **30**: 118–124.

33. Emonts M, Sweep FC, Grebenchtchikov N *et al.* (2007) Association between high levels of blood macrophage migration inhibitory factor, inappropriate adrenal response, and early death in patients with severe sepsis. *Clin Infect Dis* **44**: 1321–1328.
34. de Mendonca-Filho HT, Gomes RV, de Almeida Campos LA *et al.* (2004) Circulating levels of macrophage migration inhibitory factor are associated with mild pulmonary dysfunction after cardiopulmonary bypass. *Shock* **22**: 533–537.
35. Sprong T, Pickkers P, Geurts-Moespot A *et al.* (2007) Macrophage migration inhibitory factor (MIF) in meningococcal septic shock and experimental human endotoxemia. *Shock* **27**: 482–487.
36. Boeuf P, Vigan-Womas I, Jublot D *et al.* (2005) CyProQuant-PCR: A real time RT-PCR technique for profiling human cytokines, based on external RNA standards, readily automatable for clinical use. *BMC Immunol* **6**: 5.
37. Roger T, Ding X, Chanson AL *et al.* (2007) Regulation of constitutive and microbial pathogen-induced human macrophage migration inhibitory factor (MIF) gene expression. *Eur J Immunol* **37**: 3509–3521.
38. Temple SE, Cheong KY, Price P, Waterer GW. (2009) Pathogenic bacteria and TNF do not induce production of macrophage migration inhibitory factor (MIF) by human monocytes. *Cytokine* **46**: 316–318.
39. Maxime V, Fitting C, Annane D, Cavaillon JM. (2005) Corticoids normalize leukocyte production of macrophage migration inhibitory factor in septic shock. *J Infect Dis* **191**: 138–144.
40. Ostergaard C, Benfield T. (2009) Macrophage migration inhibitory factor in cerebrospinal fluid from patients with central nervous system infection. *Crit Care* **13**: R101.
41. Miyauchi T, Tsuruta R, Fujita M *et al.* (2009) Serum macrophage migration inhibitory factor reflects adrenal function in the hypothalamo-pituitary-adrenal axis of septic patients: An observational study. *BMC Infect Dis* **9**: 209.
42. Wang F, Shen X, Guo X *et al.* (2010) Spinal macrophage migration inhibitory factor contributes to the pathogenesis of inflammatory hyperalgesia in rats. *Pain* **148**: 275–283.
43. Iwaki T, Sugimura M, Nishihira J *et al.* (2003) Recombinant adenovirus vector bearing antisense macrophage migration inhibitory factor cDNA prevents acute lipopolysaccharide-induced liver failure in mice. *Lab Invest* **83**: 561–570.
44. Tohyama S, Onodera S, Tohyama H *et al.* (2008) A novel DNA vaccine-targeting macrophage migration inhibitory factor improves the survival of mice with sepsis. *Gene Ther* **15**: 1513–1522.
45. Wiersinga WJ, Calandra T, Kager LM *et al.* (2010) Expression and function of macrophage migration inhibitory factor (MIF) in melioidosis. *PLoS Negl Trop Dis* **4**: e605.

46. Chuang CC, Chuang YC, Chang WT *et al.* (2010) Macrophage migration inhibitory factor regulates interleukin-6 production by facilitating nuclear factor-kappa B activation during *Vibrio vulnificus* infection. *BMC Immunol* **11**: 50.
47. Benjamim CF, Hogaboam CM, Kunkel SL. (2004) The chronic consequences of severe sepsis. *J Leukoc Biol* **75**: 408–412.
48. Pollak N, Sterns T, Echtenacher B, Mannel DN. (2005) Improved resistance to bacterial superinfection in mice by treatment with macrophage migration inhibitory factor. *Infect Immun* **73**: 6488–6492.
49. Yende S, Angus DC, Kong L *et al.* (2009) The influence of macrophage migration inhibitory factor gene polymorphisms on outcome from community-acquired pneumonia. *FASEB J* **23**: 2403–2411.
50. Xia HH, Lam SK, Huang XR *et al.* (2004) *Helicobacter pylori* infection is associated with increased expression of macrophage migratory inhibitory factor — by epithelial cells, T cells, and macrophages — in gastric mucosa. *J Infect Dis* **190**: 293–302.
51. Beswick EJ, Pinchuk IV, Suarez G *et al.* (2006) *Helicobacter pylori* CagA-dependent macrophage migration inhibitory factor produced by gastric epithelial cells binds to CD74 and stimulates procarcinogenic events. *J Immunol* **176**: 6794–6801.
52. Wong BL, Zhu SL, Huang XR *et al.* (2009) Essential role for macrophage migration inhibitory factor in gastritis induced by *Helicobacter pylori*. *Am J Pathol* **174**: 1319–1328.
53. Beswick EJ, Reyes VE. (2008) Macrophage migration inhibitory factor and interleukin-8 produced by gastric epithelial cells during *Helicobacter pylori* exposure induce expression and activation of the epidermal growth factor receptor. *Infect Immun* **76**: 3233–3240.
54. Gadjeva M, Nagashima J, Zaidi T *et al.* (2010) Inhibition of macrophage migration inhibitory factor ameliorates ocular *Pseudomonas aeruginosa*-induced keratitis. *PLoS Pathog* **6**: e1000826.
55. Vera PL, Meyer-Siegler KL. (2003) Anatomical location of macrophage migration inhibitory factor in urogenital tissues, peripheral ganglia and lumbosacral spinal cord of the rat. *BMC Neurosci* **4**: 17.
56. Meyer-Siegler KL, Ordorica RC, Vera PL. (2004) Macrophage migration inhibitory factor is upregulated in an endotoxin-induced model of bladder inflammation in rats. *J Interferon Cytokine Res* **24**: 55–63.
57. Meyer-Siegler KL, Iczkowski KA, Vera PL. (2006) Macrophage migration inhibitory factor is increased in the urine of patients with urinary tract infection: Macrophage migration inhibitory factor-protein complexes in human urine. *J Urol* **175**: 1523–1528.

58. Otukesh H, Fereshtehnejad SM, Hoseini R *et al.* (2009) Urine macrophage migration inhibitory factor (MIF) in children with urinary tract infection: A possible predictor of acute pyelonephritis. *Pediatr Nephrol* **24**: 105–111.
59. Sevketoglu E, Yilmaz A, Gedikbasi A *et al.* (2010) Urinary macrophage migration inhibitory factor in children with urinary tract infection. *Pediatr Nephrol* **25**: 299–304.
60. Sundaramurthy V, Pieters J. (2007) Interactions of pathogenic mycobacteria with host macrophages. *Microbes Infect* **9**: 1671–1679.
61. Pieters J. (2008) *Mycobacterium tuberculosis* and the macrophage: Maintaining a balance. *Cell Host Microbe* **3**: 399–407.
62. Chan ED, Chan J, Schluger NW. (2001) What is the role of nitric oxide in murine and human host defense against tuberculosis? Current knowledge. *Am J Respir Cell Mol Biol* **25**: 606–612.
63. Saunders BM, Britton WJ. (2007) Life and death in the granuloma: Immunopathology of tuberculosis. *Immunol Cell Biol* **85**: 103–111.
64. Oddo M, Calandra T, Bucala R, Meylan PR. (2005) Macrophage migration inhibitory factor reduces the growth of virulent *Mycobacterium tuberculosis* in human macrophages. *Infect Immun* **73**: 3783–3786.
65. Yamada G, Shijubo N, Takagi-Takahashi Y *et al.* (2002) Elevated levels of serum macrophage migration inhibitory factor in patients with pulmonary tuberculosis. *Clin Immunol* **104**: 123–127.
66. Zenewicz LA, Shen H. (2007) Innate and adaptive immune responses to *Listeria monocytogenes*: A short overview. *Microbes Infect* **9**: 1208–1215.
67. Rayamajhi M, Humann J, Penheiter K *et al.* (2010) Induction of IFN-alphabeta enables *Listeria monocytogenes* to suppress macrophage activation by IFN-gamma. *J Exp Med* **207**: 327–337.
68. Sashinami H, Sakuraba H, Ishiguro Y *et al.* (2006) The role of macrophage migration inhibitory factor in lethal *Listeria monocytogenes* infection in mice. *Microb Pathog* **41**: 111–118.
69. Samsom JN, Annema A, Geertsma MF *et al.* (2000) Interleukin-10 has different effects on proliferation of *Listeria monocytogenes* in livers and spleens of mice. *Infect Immun* **68**: 4666–4672.
70. Hu X, Chakravarty SD, Ivashkiv LB. (2008) Regulation of interferon and Toll-like receptor signaling during macrophage activation by opposing feedforward and feedback inhibition mechanisms. *Immunol Rev* **226**: 41–56.
71. Nakane A, Okamoto M, Asano M *et al.* (1994) Protection by dexamethasone from a lethal infection with *Listeria monocytogenes* in mice. *FEMS Immunol Med Microbiol* **9**: 163–170.

72. Mastroeni P, Vazquez-Torres A, Fang FC *et al.* (2000) Antimicrobial actions of the NADPH phagocyte oxidase and inducible nitric oxide synthase in experimental salmonellosis. II. Effects on microbial proliferation and host survival *in vivo. J Exp Med* **192**: 237–248.
73. Vazquez-Torres A, Jones-Carson J, Mastroeni P *et al.* (2000) Antimicrobial actions of the NADPH phagocyte oxidase and inducible nitric oxide synthase in experimental salmonellosis. I. Effects on microbial killing by activated peritoneal macrophages *in vitro. J Exp Med* **192**: 227–236.
74. Stecher B, Robbiani R, Walker AW *et al.* (2007) *Salmonella enterica* serovar typhimurium exploits inflammation to compete with the intestinal microbiota. *PLoS Biol* **5**: 2177–2189.
75. Winter SE, Thiennimitr P, Winter MG *et al.* (2010) Gut inflammation provides a respiratory electron acceptor for *Salmonella. Nature* **467**: 426–429.
76. Koebernick H, Grode L, David JR *et al.* (2002) Macrophage migration inhibitory factor (MIF) plays a pivotal role in immunity against *Salmonella typhimurium. Proc Natl Acad Sci USA* **99**: 13681–13686.
77. MacFarlane AS, Schwacha MG, Eisenstein TK. (1999) *In vivo* blockage of nitric oxide with aminoguanidine inhibits immunosuppression induced by an attenuated strain of *Salmonella typhimurium*, potentiates *Salmonella* infection, and inhibits macrophage and polymorphonuclear leukocyte influx into the spleen. *Infect Immun* **67**: 891–898.
78. Buzoni-Gatel D, Werts C. (2006) *Toxoplasma gondii* and subversion of the immune system. *Trends Parasitol* **22**: 448–452.
79. Butcher BA, Greene RI, Henry SC *et al.* (2005) p47 GTPases regulate *Toxoplasma gondii* survival in activated macrophages. *Infect Immun* **73**: 3278–3286.
80. Flores M, Saavedra R, Bautista R *et al.* (2008) Macrophage migration inhibitory factor (MIF) is critical for the host resistance against *Toxoplasma gondii. FASEB J* **22**: 3661–3671.
81. Terrazas CA, Juarez I, Terrazas LI *et al.* (2010) *Toxoplasma gondii*: Impaired maturation and pro-inflammatory response of dendritic cells in MIF-deficient mice favors susceptibility to infection. *Exp Parasitol* **126**: 348–358.
82. Juttner S, Bernhagen J, Metz CN *et al.* (1998) Migration inhibitory factor induces killing of *Leishmania major* by macrophages: Dependence on reactive nitrogen intermediates and endogenous TNF-alpha. *J Immunol* **161**: 2383–2390.
83. Xu D, McSorley SJ, Tetley L *et al.* (1998) Protective effect on *Leishmania major* infection of migration inhibitory factor, TNF-alpha, and IFN-gamma administered orally via attenuated *Salmonella typhimurium. J Immunol* **160**: 1285–1289.

84. Satoskar AR, Bozza M, Rodriguez-Sosa M *et al.* (2001) Migration-inhibitory factor gene-deficient mice are susceptible to cutaneous *Leishmania major* infection. *Infect Immun* **69**: 906–911.
85. Junqueira C, Caetano B, Bartholomeu DC *et al.* (2010) The endless race between *Trypanosoma cruzi* and host immunity: Lessons for and beyond Chagas disease. *Expert Rev Mol Med* **12**: e29.
86. Reyes JL, Terrazas LI, Espinoza B *et al.* (2006) Macrophage migration inhibitory factor contributes to host defense against acute *Trypanosoma cruzi* infection. *Infect Immun* **74**: 3170–3179.
87. Zhang WJ, Sarawar S, Nguyen P *et al.* (1996) Lethal synergism between influenza infection and staphylococcal enterotoxin B in mice. *J Immunol* **157**: 5049–5060.
88. Nansen A, Randrup Thomsen A. (2001) Viral infection causes rapid sensitization to lipopolysaccharide: Central role of IFN-alpha beta. *J Immunol* **166**: 982–988.
89. Nansen A, Christensen JP, Marker O, Thomsen AR. (1997) Sensitization to lipopolysaccharide in mice with asymptomatic viral infection: Role of T cell-dependent production of interferon-gamma. *J Infect Dis* **176**: 151–157.
90. Jacobs F, Dubois C, Carlier Y, Goldman M. (1996) Administration of anti-CD3 monoclonal antibody during experimental Chagas' disease induces CD8+ cell-dependent lethal shock. *Clin Exp Immunol* **103**: 233–238.
91. Paiva CN, Pyrrho AS, Lannes-Vieira J *et al.* (2003) *Trypanosoma cruzi* sensitizes mice to fulminant SEB-induced shock: Overrelease of inflammatory cytokines and independence of Chagas' disease or TCR Vbeta-usage. *Shock* **19**: 163–168.
92. Paiva CN, Arras RH, Lessa LP *et al.* (2007) Unraveling the lethal synergism between *Trypanosoma cruzi* infection and LPS: A role for increased macrophage reactivity. *Eur J Immunol* **37**: 1355–1364.
93. Taylor-Robinson AW. (2010) Regulation of immunity to *Plasmodium*: Implications from mouse models for blood stage malaria vaccine design. *Exp Parasitol* **126**: 406–414.
94. Martiney JA, Sherry B, Metz CN *et al.* (2000) Macrophage migration inhibitory factor release by macrophages after ingestion of *Plasmodium chabaudi*-infected erythrocytes: Possible role in the pathogenesis of malarial anemia. *Infect Immun* **68**: 2259–2267.
95. McDevitt MA, Xie J, Shanmugasundaram G *et al.* (2006) A critical role for the host mediator macrophage migration inhibitory factor in the pathogenesis of malarial anemia. *J Exp Med* **203**: 1185–1196.
96. De Mast Q, Sweep FC, McCall M *et al.* (2008) A decrease of plasma macrophage migration inhibitory factor concentration is associated with lower numbers of circulating lymphocytes in experimental *Plasmodium falciparum* malaria. *Parasite Immunol* **30**: 133–138.

97. Awandare GA, Ouma Y, Ouma C *et al.* (2007) Role of monocyte-acquired hemozoin in suppression of macrophage migration inhibitory factor in children with severe malarial anemia. *Infect Immun* **75**: 201–210.
98. Haldar K, Mohandas N. (2009) Malaria, erythrocytic infection, and anemia. *Hematology Am Soc Hematol Educ Program*: 87–93.
99. Yap GS, Stevenson MM. (1994) Inhibition of *in vitro* erythropoiesis by soluble mediators in *Plasmodium chabaudi* AS malaria: Lack of a major role for interleukin 1, tumor necrosis factor alpha, and gamma interferon. *Infect Immun* **62**: 357–362.
100. Casals-Pascual C, Kai O, Cheung JO *et al.* (2006) Suppression of erythropoiesis in malarial anemia is associated with hemozoin *in vitro* and *in vivo*. *Blood* **108**: 2569–2577.
101. Lamikanra AA, Theron M, Kooij TW, Roberts DJ. (2009) Hemozoin (malarial pigment) directly promotes apoptosis of erythroid precursors. *PLoS One* **4**: e8446.
102. Skorokhod OA, Caione L, Marrocco T *et al.* (2010) Inhibition of erythropoiesis in malaria anemia: Role of hemozoin and hemozoin-generated 4-hydroxynonenal. *Blood* **116**(20): 4328–4337.
103. Awandare GA, Martinson JJ, Were T *et al.* (2009) MIF (macrophage migration inhibitory factor) promoter polymorphisms and susceptibility to severe malarial anemia. *J Infect Dis* **200**: 629–637.
104. Zhang W, Yue B, Wang GQ, Lu SL. (2002) Serum and ascites levels of macrophage migration inhibitory factor, TNF-alpha and IL-6 in patients with chronic virus hepatitis B and hepatitis cirrhosis. *Hepatobiliary Pancreat Dis Int* **1**: 577–580.
105. Chen LC, Lei HY, Liu CC *et al.* (2006) Correlation of serum levels of macrophage migration inhibitory factor with disease severity and clinical outcome in dengue patients. *Am J Trop Med Hyg* **74**: 142–147.
106. Arjona A, Foellmer HG, Town T *et al.* (2007) Abrogation of macrophage migration inhibitory factor decreases West Nile virus lethality by limiting viral neuroinvasion. *J Clin Invest* **117**: 3059–3066.
107. Assuncao-Miranda I, Amaral FA, Bozza FA *et al.* (2010) Contribution of macrophage migration inhibitory factor to the pathogenesis of dengue virus infection. *FASEB J* **24**: 218–228.
108. Wauquier N, Becquart P, Padilla C *et al.* (2010) Human fatal Zaire ebola virus infection is associated with an aberrant innate immunity and with massive lymphocyte apoptosis. *PLoS Negl Trop Dis* **4**(10): e837.
109. Arndt U, Wennemuth G, Barth P *et al.* (2002) Release of macrophage migration inhibitory factor and CXCL8/interleukin-8 from lung epithelial cells rendered necrotic by influenza A virus infection. *J Virol* **76**: 9298–9306.

110. Hou XQ, Gao YW, Yang ST *et al.* (2009) Role of macrophage migration inhibitory factor in influenza H5N1 virus pneumonia. *Acta Virol* **53**: 225–231.
111. Bacher M, Eickmann M, Schrader J *et al.* (2002) Human cytomegalovirus-mediated induction of MIF in fibroblasts. *Virology* **299**: 32–37.
112. Frascaroli G, Varani S, Blankenhorn N *et al.* (2009) Human cytomegalovirus paralyzes macrophage motility through down-regulation of chemokine receptors, reorganization of the cytoskeleton, and release of macrophage migration inhibitory factor. *J Immunol* **182**: 477–488.
113. Bacher M, Weihe E, Dietzschold B *et al.* (2002) Borna disease virus-induced accumulation of macrophage migration inhibitory factor in rat brain astrocytes is associated with inhibition of macrophage infiltration. *Glia* **37**: 291–306.
114. Kimura K, Nagaki M, Nishihira J *et al.* (2006) Role of macrophage migration inhibitory factor in hepatitis B virus-specific cytotoxic-T-lymphocyte-induced liver injury. *Clin Vaccine Immunol* **13**: 415–419.
115. Zhang S, Lin R, Zhou Z *et al.* (2006) Macrophage migration inhibitory factor interacts with HBx and inhibits its apoptotic activity. *Biochem Biophys Res Commun* **342**: 671–679.
116. Suzuki T, Ogata A, Tashiro K *et al.* (2000) Japanese encephalitis virus up-regulates expression of macrophage migration inhibitory factor (MIF) mRNA in the mouse brain. *Biochim Biophys Acta* **1517**: 100–106.
117. Sharma A, Bhattacharya B, Puri RK, Maheshwari RK. (2008) Venezuelan equine encephalitis virus infection causes modulation of inflammatory and immune response genes in mouse brain. *BMC Genomics* **9**: 289.
118. Assuncao-Miranda I, Bozza MT, Da Poian AT. (2009) Pro-inflammatory response resulting from sindbis virus infection of human macrophages: Implications for the pathogenesis of viral arthritis. *J Med Virol* **82**: 164–174.
119. Regis EG, Barreto-de-Souza V, Morgado MG *et al.* (2010) Elevated levels of macrophage migration inhibitory factor (MIF) in the plasma of HIV-1-infected patients and in HIV-1-infected cell cultures: A relevant role on viral replication. *Virology* **399**: 31–38.
120. Armstrong ME, Gantier M, Li L *et al.* (2008) Small interfering RNAs induce macrophage migration inhibitory factor production and proliferation in breast cancer cells via a double-stranded RNA-dependent protein kinase-dependent mechanism. *J Immunol* **180**: 7125–7133.
121. Rodriguez-Sosa M, Saavedra R, Tenorio EP *et al.* (2004) A STAT4-dependent Th1 response is required for resistance to the helminth parasite *Taenia crassiceps*. *Infect Immun* **72**: 4552–4560.
122. Rodriguez-Sosa M, Rosas LE, David JR *et al.* (2003) Macrophage migration inhibitory factor plays a critical role in mediating protection against the helminth parasite *Taenia crassiceps*. *Infect Immun* **71**: 1247–1254.

123. Magalhaes ES, Paiva CN, Souza HS *et al.* (2009) Macrophage migration inhibitory factor is critical to interleukin-5-driven eosinophilopoiesis and tissue eosinophilia triggered by *Schistosoma mansoni* infection. *FASEB J* **23**: 1262–1271.
124. Prieto-Lafuente L, Gregory WF, Allen JE, Maizels RM. (2009) MIF homologues from a filarial nematode parasite synergize with IL-4 to induce alternative activation of host macrophages. *J Leukoc Biol* **85**: 844–854.
125. Vieira-de-Abreu A, Calheiros AS, Mesquita-Santos FP *et al.* (2011) Crosstalk between MIF and eotaxin in allergic eosinophil activation forms LTC4-synthesizing lipid bodies. *Am J Respir Cell Mol Biol* **44**(4): 509–516.
126. Stavitsky AB, Metz C, Liu S *et al.* (2003) Blockade of macrophage migration inhibitory factor (MIF) in *Schistosoma japonicum*-infected mice results in an increased adult worm burden and reduced fecundity. *Parasite Immunol* **25**: 369–374.

III-2

Role of Macrophage Migration Inhibitory Factor (MIF) in Parasitic Diseases

Rashmi Tuladhar*,†, Ran Dong*,†, Sanjay Varikuti†, John R. David‡ and Abhay R. Satoskar*,†,§

1. Introduction

Macrophage migration inhibitory factor (MIF) is one of the first lymphokines to be discovered. Identified initially as a soluble factor produced by antigen stimulated lymphocytes which could inhibit macrophage migration in the delayed-type hypersensitivity reaction,[1] this lymphokine is now known to play a number of roles in a range of parasitic, bacterial, viral, and autoimmune diseases.[2]

MIF is produced by a number of cells such as those in the anterior pituitary gland, activated monocytes, macrophages,[3] and T cells.[4] MIF is elicited on stimulation with lipopolysaccharide (LPS), interferon-gamma (IFN-γ), and tumor necrosis factor-alpha (TNF-α).[5] MIF is also found circulating in serum.[6] Further orthologues of mammalian MIF have been found in many parasitic species such as *Leishmania, Plasmodium, Brugia, Eimeria, Trichinella,* and *Ancylostoma.*[7] Structurally MIF is characterized by the presence of a CXXC motif (in mammals) and functionally by its ability to tautomerize substrates as D-dopachrome and *p*-hydroxyphenyl pyruvate, and mediate oxido-reduction of protein thiols.[7,8]

MIF has been shown to be an important part of host innate immunity in circumventing sepsis,[9] and MIF can reverse the anti-inflammatory effects of glucocorticoids.[9] MIF has potent autocrine and paracrine effects that

*Department of Pathology, The Ohio State University, Columbus, OH, USA.
†Department of Microbiology, The Ohio State University, Columbus, OH, USA.
‡Department of Immunology and Infectious Diseases, Harvard School of Public Health, Boston, MA, USA.
§Corresponding author: Departments of Pathology and Microbiology, The Ohio State University, Columbus, OH 43210, USA. Email: abhay.satoksar@osumc.edu

promote cell growth and survival. MIF directly or indirectly promotes the induction of a number of cytokines, such as TNF, IFN-γ, IL-1β, IL-2, IL-6, IL-8, and macrophage inflammatory protein 2 (MIP-2).[4,10]

In this chapter, we take a closer look at the role of MIF in parasitic diseases.

2. MIF and Leishmaniasis

Leishmaniasis are a range of cutaneous and visceral infections affecting the reticuloendothelial cells by parasites of the genus *Leishmania*. It is a vector-borne disease transmitted to humans by the bite of sandflies.[11]

2.1 *Animal studies*

A mouse model of leishmaniasis caused by *L. major* has shown that the healing phenoype of this disease is associated with the induction of IL-12 leading to stimulation of a Th1 response and production of IFN-γ to activate macrophages. Such activated macrophages are highly leishmanicidal by virtue of increased nitric oxide (NO) synthesis.[12] A number of animal studies using this mouse model have implicated a critical role for MIF in this NO production along with IFN-γ and TNF-α.

A study done on peritoneal mouse macrophages treated with recombinant MIF secreted by COS cells showed an induction of NO synthesis via the enzyme NO synthase. This induction was shown to be enhanced when the macrophage was treated with IFN-γ along with MIF.[13] Pretreatment of BALB/c mouse macrophages with recombinant MIF showed a dose-dependent reduction in infection of macrophages by *L. major*. In this experiment, the leishmanicidal activity of MIF was shown to be enhanced when treated with IFN-γ along with MIF. MIF in a concentration greater than 1 μg/ml was shown to be necessary for significant parasite clearance. Physiologically, this is also the dose of MIF at which NO and TNF-α induction occurs. Endogenous TNF-α seems to play a critical role in the MIF-induced NO by macrophages, as it has been shown that mouse macrophages of TNF-R55$^{-/-}$ mice, which lack one TNF-α receptor, are incapable of clearing *L. major* infection even when pretreated with MIF. Further, addition of exogenous TNF-α does not compensate this MIF inactivity.[5] Oral inoculation of a vaccine strain of *Salmonella enterica* serovar Typhimurium expressing MIF along with TNF-α into BALB/c mice was shown to provide greater protection in the mice to subsequent *L. major* infection. In this study, it was the enhanced production of NO by the MIF inoculated mice that was shown to be the factor essential for the development of resistance.[14]

Mif$^{-/-}$ C57BL/6 mice infected with *L. major* developed larger lesions and had a greater parasite load than similarly infected wildtype (WT) C57BL/6 mice. The susceptible phenotype of the *Mif*$^{-/-}$ mice was associated with decreased NO and superoxide synthesis. Production of IFN-γ and IL-4 by the *Mif*$^{-/-}$ mouse lymph node cells on stimulation with *L. major* antigen was comparable to the WT mice, hence verifying it was not the skewing of Th1 response that led to susceptibility of *Mif*$^{-/-}$ mice, thus establishing the importance of MIF in leishmaniasis. *Mif*$^{-/-}$ mice were also associated with an increased IL-6 production. IL-6 is linked to superoxide production.[2]

2.2 *Homologue of MIF in Leishmania*

A number of *Leishmania* species such as *L. major*,[15] *L. braziliensis*, and *L. infantum* possess MIF orthologues in their genome. The *L. major* genome has two MIF orthologues, which have a 22% sequence identity with the mammalian MIF. While one of these proteins, Lm1740 MIF, has been shown to possess the ability to tautomerize, it does not have the thiol-protein oxidoreductase activity dependent on the CXXC motif in the mammalian MIF.[7,15] The parasite MIF orthologue is able to bind with the human MIF receptor CD74, and this binding has been shown to result in the internalization of MIF in an *in vitro* model using RAW cells. Further, bone marrow-derived murine macrophage stimulated with Lm1740 MIF activates the downstream ERK1/2 MAPK pathway.[7] MIF is known to regulate the levels of p53 in macrophages, which interferes with cell apoptosis.[16,17] Lm1740 MIF was also able to reduce the levels of macrophage p53 levels, suggesting that leishmanial MIF may have a role in increasing parasite survival by preventing apoptosis of infected macrophages.[7]

3. MIF and Malaria

The two major causes of death in human falciparum malaria are severe anemia and cerebral malaria. Various proinflammatory cytokines have been implicated in the pathophysiology of infections by malarial parasites. Anemia is characterized by the destruction of red blood cell (rbc) containing malaria parasites and suppression of red cell production.

3.1 *Animal Studies*

In the search for host factors that might be responsible for the inhibition of erythropoiesis in *Plasmodium chabaudi*-infected mice, IL-1β, TNF-α, and

IFN-γ were ruled out by antibody neutralization, whereas IL-12 protected susceptible mice. *P. chabaudi* infected red cells or malaria pigment (hemozoin) induced the release of MIF from macrophages. *P. chabaudi*-infected mice showed increased levels of MIF, which correlated with the severity of malaria, and increased hemozoin deposition, and MIF expression was elevated in the bone marrow and spleen in mice with active disease compared to those with early disease or controls. This was the first study that suggested that MIF might be the host factor involved in the development of anemia in malaria.[18]

In further studies on this model, concentrations of MIF found in patients with malaria suppressed erythropoietin-dependent erythroid colony formation. MIF synergized with TNF-α and IFN-γ, known inhibitors of red cell formation, even when these cytokines were at subinhibitory concentrations. MIF inhibited erythroid formation and hemoglobin production. Mitogen-activated protein kinase phosphorylation, which usually occurs during erythroid progenitor differentiation, also was inhibited by MIF. *Mif*$^{-/-}$ mice when infected with *P. chabaudi* showed less severe anemia, improved erythroid progenitor development, and increased survival compared to wildtype controls. Also reported was the finding that human mononuclear cells carrying highly expressed *MIF* alleles produced more MIF when stimulated by hemozoin compared to cells carrying low expression alleles (see further discussion below). The data further suggested that MIF could play an important role in the development of anemia in malaria and that polymorphisms in the *MIF* locus may influence the levels of MIF in the innate response to malaria infection.[19]

3.2 *MIF in the vasculature*

In exploring the possible role of MIF and NO in the pathogenesis of fatal falciparum and sepsis, MIF and inducible nitric oxide synthase (iNOS) in cerebral and systemic vasculature from fatal African pediatric cases were examined. Intense staining of MIF was observed in the systemic vascular endothelium of both malaria and sepsis deaths compared to death from non-infectious causes, as well as iNOS also was found. In contrast, MIF could not be detected in the vascular endothelium within the brain where staining of iNOS was less intense, suggesting that the suppression of the anti-inflammatory corticoids by MIF in the brain may be attenuated. These findings showed that vascular endothelial cells are the sites of intense inflammatory mediator activity during malaria and sepsis.[20,21]

3.3 *The Placenta and MIF*

There is a sequestration of infected red cells in the intervillous spaces of placenta in women infected by *P. falciparum* along with mononuclear cells causing maternal and fetal morbidity. Early studies showed that levels of MIF in intervillous blood (IVB) plasma was significantly elevated compared to peripheral and cord plasma, 500-fold and 4–6-fold, respectively. IVB mononuclear cells produce significantly more MIF than PBMC.[22]

In further studies, these authors compared the immunohistological characterization of MIF in placentas from *P. falciparum*-infected and those from uninfected women. Immunostaining was found in syncytiotrophoblasts, extravillous trophoblasts, IVB mononuclear cells, and amniotic epithelial cells whether they were malaria infected or not. Significantly higher levels of MIF were only found in amniotic epithelial cells and IVB mononuclear cells from *P. falciparum*-infected placentas compared to uninfected placentas. Falciparum-infected red blood cells stimulated increased secretion of MIF from syncytilized human trophoblast cells compared to controls. The findings suggested that placenta malaria modulates the expression of MIF in different placental compartments.[23]

To further understand the maternal/fetal morbidity caused by *P. falciparum*, the interaction of malaria-infected red blood cells with a receptor expressed on syncytiotrophoblasts was studied. Infected red blood cell-syncytiotrophoblast interaction increased secretion of MIF and MIP-1α and the chemotaxis of peripheral blood mononuclear cells to the syncytiotrophoblasts. It also increased JNK1 and the expression of TGF-β and IL-8. These studies showed that the syncytiotrophoblast participates by recruiting maternal immune cells to the maternal blood spaces during placental malarial infection.[24] In another study of *P. falciparum* infection during pregnancy, MIF levels were significantly higher in the placental plasma than in paired peripheral plasma.[25]

Whereas studies in mice suggested that increased MIF levels are associated with more severe malaria, studies of MIF levels in acute malaria infection have not provided such a clear-cut association. In one study, lower circulating lymphocyte levels were observed to parallel MIF levels, and in a study of 10 healthy human volunteers injected with *P. falciparum*, the levels of MIF were lowest on the eighth day post-infection.[26]

3.4 *Studies of MIF in malaria in Africa*

In some studies of African children with acute *P. falciparum* malaria, relatively low levels of MIF have been observed in comparison to healthy, malaria exposed children.[27–31] In one report, there was no association between blood MIF, parasitemia or hemoglobin concentration but IFN-γ levels were a predictor of circulating MIF concentrations.[27] Furthermore, children in one study with a prior history of mild malaria had higher plasma MIF levels than those with the same number of severe malaria episodes, suggesting that increased MIF expression may enhance the immune response and protect against developing severe malaria.[28] Another investigation showed that the number of circulating monocytes with ingested hemozoin was associated with both reduced hemoglobin and MIF levels.[29] In a genetic epidemiology study of 643 Kenyan children, however, an *MIF* promoter microsatellite repeat (–794 $CATT_{5-8}$) was associated with susceptibility to severe malarial anemia. The low expression –794 $CATT_6$/–173 G promoter haplotype was significantly associated with protection against severe malarial anemia, whereas the higher expression –794 $CATT_7$/–173 C and –794 $CATT_8$/–173 C promoter haplotypes were associated with a 70% increased risk of severe malaria anemia. A corresponding correlation with circulating MIF levels was not observed, however; this may reflect the limitations of plasma cytokine measurements as indicators of MIF expression in tissue sites, or the sensitivity of the particular ELISA methodology employed in these studies (R. Bucala, *personal communication*).

3.5 *Studies of MIF malaria in India and Brazil*

Studies in India have shown that elevated levels of plasma MIF are associated with fatal cerebral malaria.[32] Studies in the Brazilian Amazon showed that there was no difference in the frequency of anemia in *P. falciparum* and *P. vivax*, as well as no relation between the level of parasitemia and degree of anemia. MIF, as well as the proinflammatory cytokines TNF-α, IFN-γ, and MCP-1, were significantly increased in patients with both malaria types.[33]

3.6 *Plasmodium homologues of MIF*

Like human MIF, *P. berghei*-encoded MIF (PbMIF) has tautomerase and oxidoreductase activities, although with much less specific activity than mammalian MIF. PbMIF is produced by parasites both in the mammalian host and the mosquito vector, it is secreted into infected erythrocytes, and is released upon schizont rupture. *P. berghei* mutants lacking PbMIF completed their entire life cycle and showed no evident growth alterations, although rodents infected with the mutant *P. berghei* appeared to have higher reticulocyte counts.[34] The *P. falciparum* homologue (PfMIF) is expressed in asexual blood stage of the parasites, inhibits random migration of monocytes, and reduces surface expression of Toll-like receptors (TLR)-2 and -4, and CD68,[35] whereas mammalian MIF increases TLR-4 expression. The *Plasmodium*-derived MIF homologues were found to be conserved in both intrastrain and interspecies; while they lacked the C-X_X-C motif that is present in vertebrate MIFs, there was a chemokine-like C-C motif within the N-termini of these proteins. PfMIF also showed chemotactic activity toward human monocytes, similar to mammalian MIF, and the N-terminal proline was shown to be important for the tautomerase activity, as it is in mammalian MIF. PfMIF was found in the sera of patients infected with *P. falciparum*, using antibody detection.[36] *P. yoelii* MIF homologues (PyMIF) share a three-dimensional structure with mammalian MIF, have lower tautomerase activity, and inhibit macrophage apoptosis similar to mammalian MIF. Injection of recombinant PyMIF in mice during infection upregulated several proinflammatory cytokines and delayed the death of the mice.[37] The crystal structures of PfMIF and PbMIF were solved, revealing α/β folds and a homotrimer similar to mammalian MIF. Of special interest, the *Plasmodium* MIFs bound to CD74, the mammalian MIF receptor. This suggests that *Plasmodium* MIF has the ability to interfere or modulate host MIF activity through a competitive, receptor binding mechanism.[38] The epitope mapping of a monoclonal antibody against PfMIF showed a number of differences with mammalian MIF as well as other *Plamodium* MIFs.[39] The studies described above, taken together, suggest that *Plasmodium* MIFs can modulate and affect the immune response to the malaria parasite infection.

4. MIF and Toxoplasmosis

Toxoplasma gondii is an obligate intracellular protozoan parasite responsible for the disease toxoplasmosis. *Toxoplasma gondii* infects most warm-blooded animals, including humans. The definitive host of this single-celled parasite is the cat. *Toxoplasma gondii* infects a large proportion of the world's population but rarely causes clinically significant pathology. However, fetuses, newborns, and immunologically impaired patients are at high risk for severe or life-threatening disease due to this parasite.

4.1 *Immunity to toxoplasmosis*

Cell mediated immunity is thought to be the major host factor preventing chronic infection as well as determining the acquired resistance to reinfection.[40] Both CD4+ and CD8+ T cell responses are involved in the resolution of the infection.[41] Natural killer (NK) cells also appear to play a critical role during *T. gondii* infection by producing IFN-γ, which in turn activates macrophages into a microbicidal state. Many studies have implicated IFN-γ as a major cytokine involved in the immunological control of acute infection and the prevention of reactivation of latent infection.[42] The cytokines IL-2 and IL-12 also appear to be principal mediators of immunological resistance during *T. gondii* infection.[41]

There is growing evidence that MIF plays a very important role in providing resistance against *T. gondii* by promoting IL-12 and TNF-α production in macrophages as well as enhancing their microbicidal activity by increasing the production of NO.[5,17]

4.2 *Animal studies*

Studies done by different groups on *Mif* $^{-/-}$ mice with BALB/c or C57BL/6 genetic backgrounds have shown that *Mif* $^{-/-}$ mice are more susceptible to infection with the highly virulent RH as well as the moderately virulent ME49 strains of *T. gondii*.[43,44] *Mif* $^{-/-}$ mice showed greater liver damage and more brain cysts, and produced fewer proinflammatory cytokines.[44] Bone marrow-derived dendritic cells (BMDCs) from *Mif* $^{-/-}$ mice produced less interleukin-1β, IL-12, and TNF-α than WT BMDCs upon stimulation with *T. gondii* soluble antigen. Analysis of brains from patients with cerebral toxoplasmosis showed low levels of MIF expression.[44] Studies done by another group indicated that interaction with MIF during *T. gondii*-induced maturation is required for both the upregulation of surface MHC class II and co-stimulatory molecules on mature dendritic cells. Cytokine secretion also was found to be severely impaired in *Mif* $^{-/-}$ mice and was restored upon addition of exogenous MIF.[43] Together, these findings indicate that MIF plays a critical role in mediating protection against *T. gondii* infection by regulating proinflammatory responses.

5. MIF and Schistosomiasis

Schistosomiasis is caused by the pathogen *Schistosoma* species or blood flukes, and is considered one of the most prevalent parasitic diseases worldwide. Development of *Schistosoma* requires an intermediate snail host, and transmission into mammalian host is accomplished via direct penetration of the skin by the worm. Once inside the host, *schistosoma* migrates to the liver for sexual maturation. Eggs are passed out through feces and disseminated into the environment, where they infect snails and start another cycle.

Schistosomiasis has acute and chronic phases. Acute schistosomiasis is characterized by mild anemia, eosinophilia, fever and diarrhea. Chronic schistosomiasis manifests in formation of granuloma and fibrosis.

Schistosomiasis induces formation of hepatic granulomas which secrete MiF in an initial examination of MIF's role in the host response to this infection, *S. japonicum* infected mice were administered a neutralizing anti-MIF antibody 4–6 weeks postinfection, when granulomas formed and female worms produced eggs. Compared with controls, antibody-treated mice had more adult worms but half as many ova per worm pair in their livers. By contrast, immunoneutralization of MIF before infection or 6–8 weeks post-infection did not affect worm burden or fecundity. Splenocytes and granumola cells from antibody-treated mice showed reduced intracellular expression of TNF-α. These data are consistent with a model whereby a MIF-dependent granulomatous response enhances protection but promotes worm fecundity,[45] perhaps by up-regulating TNF-α expression.[46]

In another study using mice genetically deficient in *mif,* infection with *S. mansoni* caused similar counts of parasite eggs, worms, and granulomas as WT but with fewer eosinophils, which resulted in smaller granulomas. The decrease in eosinophils was due to impaired recruitment of eosinophils, which are chemoattracted to MIF. MIF also acts by protecting eosinophils from programmed cell death, which explains why *Mif*$^{-/-}$ mice had diminished precursors and mature eosinophils in bone marrow after infection compared with WT mice.[47] Therefore, MIF is positively involved in promoting granuloma formation in Schistosomiasis by recruiting eosinophils to inflamed site and protecting eosinophils from apoptosis in the bone marrow.

6. MIF and Taeniasis

Taenia, also known as tapeworm is an intestinal parasite. Larva or proglottids of *Taenia* are spread into the environment by feces from infected mammals. Infection is initiated by ingestion of contaminated food or water. The worm then travels from the small intestine to striated muscle and develop into cysticerci.

Helminthic infection normally induces a protective Th2 immunity by the host.[48] However in *T. crassiceps* infection, Th2 response plays a limited role and Th1 response is protective against the disease. This is supported by the finding that *Mif*$^{-/-}$ mice had reduced Th1 associated antibody IgG2a, while Th2 associated antibodies IgG1 and IgE were similar, and the mice were more susceptible compared to WT mice.[49] Increased NO production by activated macrophages is another defense mechanism by which Taenia is eliminated.[50] MIF has been proposed to activate macrophages by inducing IFN-γ and TNF-α. IL-6, which inhibits NO production by macrophages was down-regulated in WT mice compared to *Mif*$^{-/-}$.[49] Therefore, MIF is involved in eliciting Th1 response in resistance to *Taenia* by activating macrophages and enhancing NO production.

Contrary to other helminthic infection, increased amount of eosinophils were present in the peritoneum after *T. crassiceps* infection. This discrepancy might be due to preferred survival of eosinophils or specific parasite molecules, such as MIF ortholog in *Taenia,* which could alter cellular migration.[51]

7. MIF and Trypanosomiasis

Trypanosoma infects a broad range of hosts and causes a variety of diseases including sleeping sickness and Chagas disease, which are caused by *T. brucei* and *T. cruzi* respectively. Both forms of trypanosomiasis have acute and chronic stages. Chronic trypanosomiasis can remain dormant up to 10 to 20 years after initial infection and is fatal if left untreated.

Consistent with other diseases, MIF has been reported in resistance against *Trypanosoma* infection. *Mif*$^{-/-}$ mice infected with *T. cruzi* had higher parasite load, longer persistency, increased disease severity and mortality compared to WT mice.[52] A decrease in IL-12, IL-18, and TNF-α, was observed in infected *Mif*$^{-/-}$ mice, which consequently resulted in less activation of macrophages and less efficient clearing of *T. cruzi.* Moreover, *Mif*$^{-/-}$ mice had reduced iNOS mRNA which is involved in killing of the parasites. However, MIF seems to be also involved in immunopathology during trypanosomiasis, because high level of MIF in *T. cruzi* infected skeletal muscle was detected, and was correlated with disease severity and myopathology.

8. MIF and *Neospora* Infection

Neospora causes abortions in infected hosts such as dogs and cattle. Life cycle of *Neospora* is similar to that of *Toxoplasma,* in which the oocysts are ingested through contaminated food and water. *Neospora* can also be transmitted vertically.

Currently only one group has investigated the role of MIF in *Neospora.* Carvalho *et al.,* reported the ability of this parasite to induce MIF at the placental interface during infection. High level of MIF is associated with host Th1 response, which could be detrimental to fetus, and therefore MIF was indicated to attribute to abortions in infected animals.[53]

9. MIF Homologs in Some Other Parasites

9.1 *Trichinella MIF homolog*

Protein sequencing comparison revealed 25% identity between *Trichinella spiralis* MIF (TsMIF) and mammalian as well as parasitic *Brugia malayi* MIF. All highly conserved amino residues are present in TsMIF. Computational comparison of recombinant TsMIF (rTsMIF) and human MIF CD spectra indicated a similar composition of α-helixes, β-sheets, β-turns and unordered loop, which was supported by crystallographic data. Quaternary structure of rTsMIF is also similar to that of mammalian MIF (mMIF). TsMIF was tested for its tautomerase activity, and it had no lower catalytic activity and specificity than mMIFs. Studies further showed TsMIF inhibited random migration of human peripheral-blood mononuclear cells and reduced macrophage infiltration at inflamed tissues *in vivo*.[34] TsMIF has similar structure and functions compared to it mammalian molecules, which contributes to the suppressive effect on host macrophages, thus allowing more time for invasion and effectively establishing infection.

9.2 *Ancytostoma ceylanicum MiF (AceMIF) homolog*

AceMIF is 33% identical to mMIF at the amino acid level, and the secondary structure is superimposable to human MIF. However, most of the highly conserved residues are not present in AceMIF molecule, which gives AceMIF homolog unique protein surface residues, electrostatic potential and a broader enzymatic site. AceMIF is capable of interfering with macrophage migration, and such activity is not inhibited by ISO-1, a potent human MIF inhibitor, probably due to its unique active site. AceMIF has been shown to have some affinity to mMIF receptor CD74, but whether AceMIF exerts its function via CD74 remains to be elucidated. Human hookworm *A. ceylanicum* expresses MIF in a temporal dependent fashion; it is only expressed by adult worms and those at L3 stage, when the worms are inside mammalian host, but not by eggs or L1 worms, when they are more likely present in the environment. So it is likely that by modulating MIF expression, *A. ceylanicum* manipulates the host and evade immune response to favor disease establishment.[53]

References

1. David JR. (1966) Delayed hypersensitivity *in vitro*: Its mediation by cell-free substances formed by lymphoid cell-antigen interaction. *Proc Natl Acad Sci USA* **56**(1): 72–77.

2. Satoskar AR, Bozza M, Rodriguez-Sosa M *et al.* (2001) Migration-inhibitory factor gene-deficient mice are susceptible to cutaneous *Leishmania major* infection. Infect Immun **69**(2): 906–911.
3. Bernhagen J, Mitchell RA, Calandra T *et al.* (1994) Purification, bioactivity, and secondary structure analysis of mouse and human macrophage migration inhibitory factor (MIF). *Biochemistry* **33**(47): 14144–14155.
4. Calandra T, Bernhagen J, Metz CN *et al.* (1995) MIF as a glucocorticoid-induced modulator of cytokine production. *Nature* **377**(6544): 68–71.
5. Juttner S, Bernhagen J, Metz CN *et al.* (1998) Migration inhibitory factor induces killing of *Leishmania major* by macrophages: Dependence on reactive nitrogen intermediates and endogenous TNF-alpha. *J Immunol* **161**(5): 2383–2390.
6. Edwards KM, Bosch JA, Engeland CG *et al.* (2010) Elevated macrophage migration inhibitory factor (MIF) is associated with depressive symptoms, blunted cortisol reactivity to acute stress, and lowered morning cortisol. *Brain Behav Immun* **24**(7): 1202–1208.
7. Kamir D, Zierow S, Leng L *et al.* (2008) A *Leishmania* ortholog of macrophage migration inhibitory factor modulates host macrophage responses. *J Immunol* **180**(12): 8250–8261.
8. Kleemann R, Kapurniotu A, Frank RW *et al.* (1998) Disulfide analysis reveals a role for macrophage migration inhibitory factor (MIF) as thiol-protein oxidoreductase. *J Mol Biol* **280**(1): 85–102.
9. Bacher M, Metz CN, Calandra T *et al.* (1996) An essential regulatory role for macrophage migration inhibitory factor in T-cell activation. *Proc Natl Acad Sci USA* **93**(15): 7849–7854.
10. Makita H, Nishimura M, Miyamoto K *et al.* (1998) Effect of anti-macrophage migration inhibitory factor antibody on lipopolysaccharide-induced pulmonary neutrophil accumulation. *Am J Respir Crit Care Med* **158**(2): 573–579.
11. Kato H, Gomez EA, Caceres AG *et al.* (2010) Molecular epidemiology for vector research on leishmaniasis. *Int J Environ Res Public Health* **7**(3): 814–826.
12. Green SJ, Meltzer MS, Hibbs JB Jr, Nacy CA. (1990) Activated macrophages destroy intracellular *Leishmania major* amastigotes by an L-arginine-dependent killing mechanism. *J Immunol* **144**(1): 278–283.
13. Cunha FQ, Weiser WY, David JR *et al.* (1993) Recombinant migration inhibitory factor induces nitric oxide synthase in murine macrophages. *J Immunol* **150**(5): 1908–1912.
14. Xu D, McSorley SJ, Tetley L *et al.* (1998) Protective effect on *Leishmania major* infection of migration inhibitory factor, TNF-alpha, and IFN-gamma administered orally via attenuated *Salmonella typhimurium. J Immunol* **160**(3): 1285–1289.
15. Richardson JM, Morrison LS, Bland ND *et al.* (2009) Structures of *Leishmania major* orthologues of macrophage migration inhibitory factor. *Biochem Biophys Res Commun* **380**(3): 442–448.

16. Mitchell RA, Liao H, Chesney J *et al.* (2002) Macrophage migration inhibitory factor (MIF) sustains macrophage proinflammatory function by inhibiting p53: Regulatory role in the innate immune response. *Proc Natl Acad Sci USA* **99**(1): 345–350.
17. Calandra T, Roger T. (2003) Macrophage migration inhibitory factor: A regulator of innate immunity. *Nat Rev Immunol* **3**(10): 791–800.
18. Martiney JA, Sherry B, Metz CN *et al.* (2000) Macrophage migration inhibitory factor release by macrophages after ingestion of *Plasmodium chabaudi*-infected erythrocytes: Possible role in the pathogenesis of malarial anemia. *Infect Immun* **68**(4): 2259–2267.
19. McDevitt MA, Xie J, Shanmugasundaram G *et al.* (2006) A critical role for the host mediator macrophage migration inhibitory factor in the pathogenesis of malarial anemia. *J Exp Med* **203**(5): 1185–1196.
20. Clark IA, Awburn MM, Whitten RO *et al.* (2003) Tissue distribution of migration inhibitory factor and inducible nitric oxide synthase in falciparum malaria and sepsis in African children. *Malar J* **2**: 6.
21. Clark IA, Cowden WB. (2003) The pathophysiology of falciparum malaria. *Pharmacol Ther* **99**(2): 221–260.
22. Chaisavaneeyakorn S, Moore JM, Othoro C *et al.* (2002) Immunity to placental malaria. IV. Placental malaria is associated with up-regulation of macrophage migration inhibitory factor in intervillous blood. *J Infect Dis* **186**(9): 1371–1375.
23. Chaisavaneeyakorn S, Lucchi N, Abramowsky C *et al.* (2005) Immunohistological characterization of macrophage migration inhibitory factor expression in *Plasmodium falciparum*-infected placentas. *Infect Immun* **73**(6): 3287–3293.
24. Lucchi NW, Peterson DS, Moore JM. (2008) Immunologic activation of human syncytiotrophoblast by *Plasmodium falciparum. Malar J* **7**: 42.
25. Chaiyaroj SC, Rutta AS, Muenthaisong K *et al.* (2004) Reduced levels of transforming growth factor-beta1, interleukin-12 and increased migration inhibitory factor are associated with severe malaria. *Acta Trop* **89**(3): 319–327.
26. De Mast Q, Sweep FC, McCall M *et al.* (2008) A decrease of plasma macrophage migration inhibitory factor concentration is associated with lower numbers of circulating lymphocytes in experimental *Plasmodium falciparum* malaria. *Parasite Immunol* **30**(3): 133–138.
27. Awandare GA, Hittner JB, Kremsner PG *et al.* (2006) Decreased circulating macrophage migration inhibitory factor (MIF) protein and blood mononuclear cell MIF transcripts in children with *Plasmodium falciparum* malaria. *Clin Immunol* **119**(2): 219–225.
28. Awandare GA, Kremsner PG, Hittner JB *et al.* (2007) Higher production of peripheral blood macrophage migration inhibitory factor in healthy children with a history of mild malaria relative to children with a history of severe malaria. *Am J Trop Med Hyg* **76**(6): 1033–1036.

29. Awandare GA, Ouma Y, Ouma C *et al.* (2007) Role of monocyte-acquired hemozoin in suppression of macrophage migration inhibitory factor in children with severe malarial anemia. *Infect Immun* **75**(1): 201–210.
30. Awandare GA, Martinson JJ, Were T *et al.* (2009) MIF (macrophage migration inhibitory factor) promoter polymorphisms and susceptibility to severe malarial anemia. *J Infect Dis* **200**(4): 629–637.
31. Awandare GA, Ouma C, Keller CC *et al.* (2006) A macrophage migration inhibitory factor promoter polymorphism is associated with high-density parasitemia in children with malaria. *Genes Immun* **7**(7): 568–575.
32. Jain V, McClintock S, Nagpal AC *et al.* (2009) Macrophage migration inhibitory factor is associated with mortality in cerebral malaria patients in India. *BMC Res Notes* **2**: 36.
33. Fernandes AA, Carvalho LJ, Zanini GM *et al.* (2008) Similar cytokine responses and degrees of anemia in patients with *Plasmodium falciparum* and *Plasmodium vivax* infections in the Brazilian Amazon region. *Clin Vaccine Immunol* **15**(4): 650–658.
34. Augustijn KD, Kleemann R, Thompson J *et al.* (2007) Functional characterization of the *Plasmodium falciparum* and *P. berghei* homologues of macrophage migration inhibitory factor. *Infect Immun* **75**(3): 1116–1128.
35. Cordery DV, Kishore U, Kyes S *et al.* (2007) Characterization of a *Plasmodium falciparum* macrophage-migration inhibitory factor homologue. *J Infect Dis* **195**(6): 905–912.
36. Shao D, Han Z, Lin Y *et al.* (2008) Detection of *Plasmodium falciparum* derived macrophage migration inhibitory factor homologue in the sera of malaria patients. *Acta Trop* **106**(1): 9–15.
37. Shao D, Zhong X, Zhou YF *et al.* (2010) Structural and functional comparison of MIF ortholog from *Plasmodium yoelii* with MIF from its rodent host. *Mol Immunol* **47**(4): 726–737.
38. Dobson SE, Augustijn KD, Brannigan JA *et al.* The crystal structures of macrophage migration inhibitory factor from *Plasmodium falciparum* and *Plasmodium berghei*. *Protein Sci* **18**(12): 2578–2591.
39. Wang Z, Shao D, Zhong X *et al.* (2009) Epitope mapping of monoclonal antibody 1B9 against *Plasmodium falciparum*-derived macrophage migration inhibitory factor. *Immunol Invest* **38**(5): 422–433.
40. Gazzinelli RT, Hakim FT, Hieny S *et al.* (1991) Synergistic role of CD4+ and CD8+ T lymphocytes in IFN-gamma production and protective immunity induced by an attenuated *Toxoplasma gondii* vaccine. *J Immunol* **146**(1): 286–292.
41. Khan IA, Matsuura T, Kasper LH. (1994) Interleukin-12 enhances murine survival against acute toxoplasmosis. *Infect Immun* **62**(5): 1639–1642.

42. Araujo FG. Depletion of CD4+ T cells but not inhibition of the protective activity of IFN-gamma prevents cure of toxoplasmosis mediated by drug therapy in mice. *J Immunol* **149**(9): 3003–3007.
43. Terrazas CA, Juarez I, Terrazas LI *et al.* (2010) *Toxoplasma gondii*: Impaired maturation and pro-inflammatory response of dendritic cells in MIF-deficient mice favors susceptibility to infection. *Exp Parasitol* **126**(3): 348–358.
44. Flores M, Saavedra R, Bautista R *et al.* (2008) Macrophage migration inhibitory factor (MIF) is critical for the host resistance against *Toxoplasma gondii. FASEB J* **22**(10): 3661–3671.
45. Stavitsky AB, Metz C, Liu S *et al.* (2003) Blockade of macrophage migration inhibitory factor (MIF) in *Schistosoma japonicum*-infected mice results in an increased adult worm burden and reduced fecundity. *Parasite Immunol* **25**: 369–374.
46. Amiri P, Locksley RM, Parslow TG *et al.* (1992) Tumour necrosis factor alpha restores granulomas and induces parasite egg-laying in schistosome-infected SCID mice. *Nature* **356**: 604–607.
47. Magalhaes ES, Paiva CN, Souza HS *et al.* (2009) Macrophage migration inhibitory factor is critical to interleukin-5-driven eosinophilopoiesis and tissue eosinophilia triggered by *Schistosoma mansoni* infection. *FASEB J* **23**(4): 1262–1271.
48. Finkelman FD, Shea-Donohue T, Goldhill J *et al.* (1997) Cytokine regulation of host defense against parasitic gastrointestinal nematodes: Lessons from studies with rodent models. *Annu Rev Immunol* **15**: 505–533.
49. Rodriguez-Sosa M, Rosas LE, David JR *et al.* (2003) Macrophage migration inhibitory factor plays a critical role in mediating protection against the helminth parasite *Taenia crassiceps. Infect Immun* **71**(3): 1247–1254.
50. Ahmed SF, Oswald IP, Caspar P *et al.* (1997) Developmental differences determine larval susceptibility to nitric oxide-mediated killing in a murine model of vaccination against *Schistosoma mansoni. Infect Immun* **65**(1): 219–226.
51. Ohnmacht C, Pullner A, van Rooijen N, Voehringer D. (2007) Analysis of eosinophil turnover *in vivo* reveals their active recruitment to and prolonged survival in the peritoneal cavity. *J Immunol* **179**(7): 4766–4774.
52. Reyes JL, Terrazas LI, Espinoza B *et al.* (2006) Macrophage migration inhibitory factor contributes to host defense against acute *Trypanosoma cruzi* infection. *Infect Immun* **74**(6): 3170–3179.
53. Carvalho JV, Alves CM, Cardoso MR *et al.* (2010) Differential susceptibility of human trophoblastic (BeWo) and uterine cervical (HeLa) cells to *Neospora caninum* infection. *Int J Parasitol* **40**(14): 1629–1637.

54. Wu Z, Boonmars T, Nagano I *et al.* (2003) Molecular expression and characterization of a homologue of host cytokine macrophage migration inhibitory factor from *Trichinella* spp. *J Parasitol* **89**(3): 507–515.
55. Cho Y, Jones BF, Vermeire JJ *et al.* (2007) Structural and functional characterization of a secreted hookworm Macrophage Migration Inhibitory Factor (MIF) that interacts with the human MIF receptor CD74. *J Biol Chem* **282**(32): 23447–23456.

III-3

MIF and Pulmonary Disease

Gordon Cooke, Michelle E. Armstrong,
Helen Conroy and Seamas C. Donnelly*

1. Introduction

The lungs represent a classical end organ capable of mounting a rapid protective inflammatory response. To support its primary role in gas diffusion, it relies on a relatively narrow alveolar membrane that potentially could allow access systemically to a variety of external toxins and pathogens. Macrophage migration inhibitory factor (MIF) represents a classical key cytokine in the body's defensive armory within the lung.

While pulmonary inflammation evolved as primarily a protective defensive response, there are classical acute and chronic diseases where exaggerated persistent inflammation results in progressive injurious pulmonary disease. In this chapter, we will focus on specific diseases in which MIF has been implicated as a significant proinflammatory cytokine contributing to disease pathogenesis.

2. MIF and Acute Pulmonary Inflammatory Disease

The acute respiratory distress syndrome (ARDS) represents the classical example of catastrophic acute inflammatory disease taking out the lung. This disorder results in patients requiring mechanical ventilation and critical care support with an associate mortality of 25–30%.[1] It is well recognized in animal models that attenuating MIF activity is associated with downregulation of the pulmonary inflammatory response and less severe end-organ injury.

Makita *et al.* used an anti-MIF antibody to attenuate neutrophil migration to the lungs in response to intraperitoneally administered LPS.[2] A role for MIF

*Corresponding author: Medicine, Education & Research Centre, St. Vincent's University Hospital, Elm Park, Dublin 4, Ireland. Email: seamas.donnelly@ucd.ie

and its receptor CD74 in ALI has also been proposed by Takahashi *et al.*[3] Rittirsch *et al.* showed they could generate ALI experimentally via administration of LPS. By neutralizing MIF they were able to protect against LPS induced ALI.[4] Functional genomics has also confirmed what others have found in relation to ALI but also reported similar findings in ventilator ALI (ALI caused by mechanical stresses).[5,6] Yin *et al.* showed a role for heme oxygenase-1 (HO-1) in the negative regulation of lung MIF and TLR4-induced inflammatory response that was induced by LPS.[7] Their experiments indicated that following induction of HO-1 expression by administration of cobalt protoporphyrins (CoPP), there was a significant reduction in inflammatory cell infiltration, MIF expression, and MPO activity in lung tissues.

In seminal work, the role of MIF in human ARDS was highlighted when the authors showed that anti-MIF had the capacity to significantly attenuate proinflammatory cytokine production in *ex vivo* human pulmonary cells derived from ARDS patients. In addition, they revealed MIF's capacity to override glucocorticoid anti-inflammatory activity in these human cells. Thus, MIF had the capacity to drive this injurious inflammatory response by both directly augmenting pulmonary proinflammatory cytokine activity and indirectly by antagonizing the host's immunosuppressive glucocortioid response, thus potentiating this progressive injurious inflammatory response.[8]

3. Systemic Sepsis and the Lung

Virulent pathogens are a primary cause of an acute inflammatory response that results in significant end-organ pulmonary injury. Bozza *et al.* were among the first to describe a role for MIF in lung sepsis. They generated MIF knock-out mice, which when injected intraperitoneally to bacterial lipopolysaccharide, or *Staphylococcus aureus* enterotoxin B with D-galactosamine, showed lower production levels of TNF-α, normal levels of IL-6 and IL-12, and increased production of nitric oxide when compared with wildtype controls. They also instilled intratracheally *Pseudomonas aeruginosa* and found that *Mif*$^{-/-}$ animals cleared these Gram-negative bacteria better than the wildtype mice and had diminished neutrophil accumulation in their bronchoalveolar fluid when compared to the wildtype mice.[9] Bozza hypothesized that the adverse effect of MIF in sepsis could be overcome by counteracting the neutralization of the cytokine. A vaccination approach for neutralizing MIF was developed by Onodera *et al.*[10] Called MIF/TTX DNA vaccine, it was purported to be useful for ameliorating the symptoms of rheumatoid arthritis. This vaccine was also tested in mice, and it attenuated sepsis-related inflammation when compared to control vaccinated mice.[11] There may be concerns with the use of a global vaccine to MIF, as the authors did report wound healing deficiencies, but they hypothesized that this was due to the types of mice used and not to the side effects of the vaccine.

In important work by Lin *et al.* it was shown that lung-derived MIF significantly contributed to myocardial dysfunction in systemic sepsis.[12] A more recent paper by Wiersinga *et al.*[13] showed that in mice suffering from melioidosis, a severe disease caused by the bacterium *Burkholderia pseudomallei* that in itself is the major cause of community-acquired septicemia, anti-MIF treatment resulted in lower bacterial counts in the lungs when compared to untreated mice. The same study showed a 10-fold increase in bacterial load in mice given 50 μg of recombinant MIF in the liver and a trend towards higher bacterial counts in the lung and blood.[13] Although the anti-MIF treatment did not reverse the immune response to the induced melioidosis, the evidence overall pointed to an important role for MIF in the pathogenesis of this disease.

4. MIF and Chronic Pulmonary Inflammatory Disease

4.1 *Asthma*

Asthma is a heterogeneous syndrome characterized by variable, reversible airway obstruction and abnormally increased responsiveness (hyper reactivity) of the airways to various stimuli. It is characterized by wheezing, chest tightness, dyspnea, and/or cough, and results from widespread contraction of tracheobronchial smooth muscle (bronchoconstriction), hyper secretion of mucus, and mucosal edema, all of which result in the narrowing of the airways.[14]

Donnelly *et al.* first proposed a link between MIF and asthma when they showed that unstimulated human circulating eosinophils were found to contain pre-formed MIF and that stimulation of these cells with physiological relevant stimuli for asthma, namely IL-5 and C5α, *in vitro,* yielded significant release of MIF protein. Higher levels of MIF were also found to be present in BAL samples from clinically stable asthma patients as opposed to control patients.[15] Corticosteroids, either in inhaler, intravenous or tablet form, are the cornerstone of anti-inflammatory therapy in asthma. This treatment potently downregulates proinflammatory cytokine production but also contributes to the resolution of inflammation in asthma by significantly augmenting eosinophil apoptosis. It is intriguing to speculate that in the context of MIF's role in asthma, blocking this key cytokine would be directly anti-inflammatory but also, potentially, may override corticosteroids' ability to promote eosinophil apoptosis.[15,16] In 2000, Kawakami *et al.* found increased levels of MIF in the serum of asthma patients and following staining they found the MIF was co-localized with eosinophil peroxidase staining in the cytoplasm of sputum cells.[17] In 2002, Baugh *et al.*[18] described a functional promoter polymorphisim in the MIF gene that was associated with more aggressive disease in rheumatoid arthritis. They identified 5-, 6-, 7-, and 8-CATT repeat units and showed the 6-, 7-, and 8-CATT alleles had the highest levels of basal and stimulated MIF promoter activity.[18] When they assessed expression of these CATT alleles in mild, moderate, and severe asthma patients, they found a significant association between more brittle and severe asthma and individual patients genetically primed to be high MIF producers (i.e., 6-, 7-, and 8-CATT repeat units).[19]

4.2 *Cystic fibrosis*

Cystic fibrosis is the most common fatal inherited disease in Caucasians and is caused by mutations in the cystic fibrosis transmembrane regulator (*CFTR*) gene with a frequency of 1 in 2,500 life births.[20] It is characterized by obstruction of the airways with a thick mucous, which is followed by infection with *Pseudomonas* sp. and *Burkholderia* sp. Gram-negative bacteria. In 1999, Bozza *et al,* showed that MIF knock-out mice where better able to clear bacterial infections then wildtype mice,[9] suggesting a possible role for neutralizing MIF in sepsis. In 2003, Simon *et al.* showed that MIF was factor in the delayed apoptosis of neutrophils. Thus, the presence of enhanced MIF activity would promote the survival of inflamed leukocyte contributing to persistent chronic inflammation.[21] Neutrophils are a major source of MIF in the body,[22] and Donnelly *et al.* found a significant clinical association between the MIF CATT polymorphism and cystic fibrosis. Those patients genetically primed to be high MIF producers had earlier pseudomonas colonization and earlier end-organ dysfunction.[23] This study described a 5-CATT repeat polymorphism in the promoter region of the *MIF* gene that was associated with a significant decrease in *Pseudomonas* sp. colonization. A unique feature of the cytokine MIF is that it possesses tautomerase enzymatic activity. In the context of acute inflammation, investigators have shown that this unique enzymatic activity contributes to this cytokine's significant proinflammatory activity. A transgenic mouse has been generated that produces a mutant MIF lacking enzymatic activity (P1G).[24] In an animal model of chronic pulmonary pseudomonas infection, this mutant mouse exhibited significantly less pulmonary inflammation compared to wildtype mice.[25]

4.3 *MIF and lung cancer*

Lung cancer is the commonest fatal malignancy in the developed world, causing over 1 million deaths annually. With the increasing prevalence of smoking worldwide, particularly in India and China, mortality figures are predicted to substantially increase. Chronic inflammatory diseases are well recognized as being associated with enhanced risk of developing cancer. This ranges from chronic hepatitis infection and the development of liver cancer, peptic ulcer diseases, and the development of gastric cancer to idiopathic pulmonary fibrosis (IPF) and the enhanced risk of developing lung cancer. Recent review articles have highlighted the potential for MIF being a key link between chronic inflammatory diseases and the subsequent development of cancer.[26,27]

The initial descriptive association between lung cancer and MIF was made by Kamimura and colleagues. They found that in normal lung tissue, MIF, mRNA, and MIF protein were observed throughout the bronchial and alveolar epithelium, vascular smooth muscle, and alveolar macrophages, whereas in tumor tissue derived from patients with primary lung adenocarcinoma, they were found in significantly higher levels in the alveolar epithelium than in normal patients.[28] These authors also observed that patients who had no MIF protein in the nuclei of the tumor cells had a statistically significant worse prognosis compared with those patients who had MIF expression in their tumor nuclei. Tomiyasu *et al.*[29] demonstrated that NSCLC cells produce MIF and that the high levels of expression of MIF mRNA in lung cancer tissues were significantly associated with heavy smoking.

In 2001, White *et al.*[30] found that non-small cell lung cancer cells promote angiogenesis by inducing host peripheral blood monocytes (PBM) to upregulate CXC chemokine-dependant angiogenic activity. They observed that MIF was responsible in part for the induction of PBM-derived CXC chemokine-dependant angiogenic activity by NSCLC cells. They suggested that their findings indicate that tumors recruit macrophages to promote a proangiogenic process. Subsequently, they found a link between high levels of tumor associated CXC chemokine, MIF or VEGF and the risk of recurrence after resection of lung cancer.[31] McClelland *et al.* showed that the co-expression of MIF and its putative receptor CD74 in NSCLC is also associated with greater tumor vascularity and greater levels of angiogenic CXC chemokines.[32]

More recent research has shown that MIF can act together with its homologue, D-dopachrome tautomerase, to promote the expression of the proangiogenic factors CXCL8 and VEGF in NSCLC cells.[33] Arenberg *et al.* showed that in response to increased MIF expression due to lung injury, orthotopic tumors (Lewis lung carcinoma) injected after the acute phase were protected from apoptosis and showed increased proliferation.[34] Further evidence of the role of MIF in lung cancer metastasis, invasion, anchorage, and independent growth has been shown.[35] Rendon *et al.* found that the Rho GTPase family member, Rac1, is a likely mediator of MIF actions on cell migration and metastatic invasion in human NSCLC cells and that the small molecular weight inhibitor of MIF, ISO-1, was an effective inhibitor of MIF in blocking NSCLC invasive and anchorage-independent disease phenotypes.[35] ISO-1 was shown to be as effective as MIF siRNA but without some of its side effects, as noted by Armstrong *et al.*,[36] who found that siRNA had off-target, nonspecific effects including the activation of a PKR-dependent MIF in response to siRNA. This undesired side effect was associated with enhanced cell proliferation in MCF-7 cells as well as primary breast cancer cells. More recently, novel small molecular weight inhibitors of MIF's enzymatic activity have been shown to inhibit motility and growth of lung cancer cells such as 4-iodo-6-phenylpyrimidine (4-IPP) in a similar manner to ISO-1 but with 5–10 times greater potency.[37]

5. Conclusion

There is now substantial and persuasive evidence highlighting MIF as a key driver of the abnormal perpetuation of the inflammatory response and the development of chronic inflammatory pulmonary disease. In the context of a broader spectrum of lung disease and in particular lung cancer, there are key seminal observations underlining the importance of MIF in promoting an *in vivo* microenvironment favoring cancer growth and metastases. Taken together this evidence highlights targeting MIF as a valid therapeutic target in lung disease.

Acknowledgment

S. C. Donnelly is supported by Science Foundation Ireland (SFI) (08/IN1/B2035).

References

1. Bernard GR. (2005) Acute respiratory distress syndrome: A historical perspective. *Am J Resp Crit Care Med* **172**(7): 798–806.
2. Makita H *et al.* (1998) Effect of anti-macrophage migration inhibitory factor antibody on lipopolysaccharide-induced pulmonary neutrophil accumulation. *Am J Resp Crit Care Med* **158**(2): 573–579.

3. Takahashi K *et al.* (2009) Macrophage CD74 contributes to MIF-induced pulmonary inflammation. *Resp Res* **10**: 33.
4. Rittirsch D *et al.* (2008) Acute lung injury induced by lipopolysaccharide is independent of complement activation. *J Immunol* **180**(11): 7664–7672.
5. Nonas SA *et al.* (2005) Functional genomic insights into acute lung injury: Role of ventilators and mechanical stress. *Proc Am Thorac Soc* **2**(3): 188–194.
6. Meyer NJ, Garcia JG. (2007) Wading into the genomic pool to unravel acute lung injury genetics. *Proc Am Thorac Soc* **4**(1): 69–76.
7. Yin H *et al.* (2010) Heme oxygenase-1 upregulation improves lipopolysaccharide-induced acute lung injury involving suppression of macrophage migration inhibitory factor. *Mol Immunol* **47**(15): 2443–2449.
8. Donnelly SC *et al.* (1997) Regulatory role for macrophage migration inhibitory factor in acute respiratory distress syndrome. *Nat Med* **3**(3): 320–323.
9. Bozza M *et al.* (1999) Targeted disruption of migration inhibitory factor gene reveals its critical role in sepsis. *J Exp Med* **189**(2): 341–346.
10. Onodera S *et al.* (2007) A novel DNA vaccine targeting macrophage migration inhibitory factor protects joints from inflammation and destruction in murine models of arthritis. *Arthritis Rheum* **56**(2): 521–530.
11. Tohyama S *et al.* (2008) A novel DNA vaccine-targeting macrophage migration inhibitory factor improves the survival of mice with sepsis. *Gene Ther* **15**(23): 1513–1522.
12. Lin X *et al.* (2005) Macrophage migration inhibitory factor within the alveolar spaces induces changes in the heart during late experimental sepsis. *Shock* **24**(6): 556–563.
13. Wiersinga WJ *et al.* (2010) Expression and function of macrophage migration inhibitory factor (MIF) in melioidosis. *PLoS Negl Trop Dis* **4**(2): e605.
14. Gong H. (1990). Wheezing and asthma. In: Walker HK, Hall WD, Hurst JW (eds), *Clinical Methods: The History, Physical, and Laboratory Examinations.* Butterworth Publishers (Reed Publishing), Boston.
15. Rossi AG, *et al.* (1998) Human circulating eosinophils secrete macrophage migration inhibitory factor (MIF). Potential role in asthma. *J Clin Invest* **101**(12): 2869–2874.
16. Roger T *et al.* (2005) Macrophage migration inhibitory factor promotes innate immune responses by suppressing glucocorticoid-induced expression of mitogen-activated protein kinase phosphatase-1. *Euro J Immunol* **35**(12): 3405–3413.
17. Yamaguchi E *et al.* (2000) Macrophage migration inhibitory factor (MIF) in bronchial asthma. *Clin Exp Allergy* **30**(9): 1244–1249.
18. Baugh JA *et al.* (2002) A functional promoter polymorphism in the macrophage migration inhibitory factor (MIF) gene associated with disease severity in rheumatoid arthritis. *Genes Immun* **3**(3): 170–176.

19. Mizue Y *et al.* (2005) Role for macrophage migration inhibitory factor in asthma. *Proc Natl Acad Sci USA* **102**(40): 14410–14415.
20. Collins FS. (1992) Cystic fibrosis: Molecular biology and therapeutic implications. *Science* **256**(5058): 774–779.
21. Baumann R *et al.* (2003) Macrophage migration inhibitory factor delays apoptosis in neutrophils by inhibiting the mitochondria-dependent death pathway. *FASEB J* **17**(15): 2221–30.
22. Riedemann NC *et al.* (2004) Regulatory role of C5a on macrophage migration inhibitory factor release from neutrophils. *J Immunol* **173**(2): 1355–1359.
23. Plant BJ *et al.* (2005) Cystic fibrosis, disease severity, and a macrophage migration inhibitory factor polymorphism. *Am J Resp Crit Care Med* **172**(11): 1412–1415.
24. Fingerle-Rowson G *et al.* (2009) A tautomerase-null macrophage migration-inhibitory factor (MIF) gene knock-in mouse model reveals that protein interactions and not enzymatic activity mediate MIF-dependent growth regulation. *Mol Cell Biol* **29**(7): 1922–1932.
25. McLaughlin AM, Donnelly SC. (2006) Macrophage-migration inhibitory factor (MIF) drives growth of *Pseudomonas aeruginosa* via its isomerase enzymatic activity: Potential therapeutic role in cystic fibrosis. *Proc Am Thorac Soc*
26. Bucala R, Donnelly SC. (2007) Macrophage migration inhibitory factor: A probable link between inflammation and cancer. *Immunity* **26**(3): 281–285.
27. Conroy H, Mawhinney L, Donnelly SC. (2010) Inflammation and cancer: Macrophage migration inhibitory factor (MIF) — The potential missing link. *QJM Mon J Assoc Phys* **103**(11): 831–836.
28. Kamimura A *et al.* (2000) Intracellular distribution of macrophage migration inhibitory factor predicts the prognosis of patients with adenocarcinoma of the lung. *Cancer* **89**(2): 334–341.
29. Tomiyasu M *et al.* (2002) Quantification of macrophage migration inhibitory factor mRNA expression in non-small cell lung cancer tissues and its clinical significance. *Clin Cancer Res* **8**(12): 3755–3760.
30. White ES *et al.* (2001) Non-small cell lung cancer cells induce monocytes to increase expression of angiogenic activity. *J Immunol* **166**(12): 7549–7555.
31. White ES *et al.* (2003) Macrophage migration inhibitory factor and CXC chemokine expression in non-small cell lung cancer: Role in angiogenesis and prognosis. *Clin Cancer Res* **9**(2): 853–860.
32. McClelland M *et al.* (2009) Expression of CD74, the receptor for macrophage migration inhibitory factor, in non-small cell lung cancer. *Am J Pathol* **174**(2): 638–646.
33. Coleman AM *et al.* (2008) Cooperative regulation of non-small cell lung carcinoma angiogenic potential by macrophage migration inhibitory factor and its homolog, D-dopachrome tautomerase. *J Immunol* **181**(4): 2330–2337.

34. Arenberg D *et al.* (2010) Macrophage migration inhibitory factor (MIF) promotes tumor growth in the context of lung injury and repair. *Am J Resp Crit Care Med* **182**: 1030–1037.
35. Rendon BE *et al.* (2007) Regulation of human lung adenocarcinoma cell migration and invasion by macrophage migration inhibitory factor. *J Biol Chem* **282**(41): 29910–29918.
36. Armstrong ME *et al.* (2008) Small interfering RNAs induce macrophage migration inhibitory factor production and proliferation in breast cancer cells via a double-stranded RNA-dependent protein kinase-dependent mechanism. *J Immunol* **180**(11): 7125–7133.
37. Winner M *et al.* (2008) A novel, macrophage migration inhibitory factor suicide substrate inhibits motility and growth of lung cancer cells. *Cancer Res* **68**(18): 7253–7257.

III-4

The Role of MIF in Neurogenic Inflammation

Pedro L. Vera*,†,‡ and Katherine L. Meyer-Siegler†,§

1. Introduction

Neurogenic inflammation occurs when neuropeptides (mainly substance P and calcitonin in gene related peptide) are released from afferent nerve terminals at a peripheral site as a result of trauma, injury or axon reflexes. This chapter will review evidence for a role of macrophage migration inhibitory factor (MIF) in neurogenic inflammation using the bladder as a model system. Substance P elicits MIF release from urothelial cells *in vivo* and induces expression of MIF's receptor on the cell surface of urothelial cells. Once released from the urothelium, MIF binds to cell surface receptors in urothelial cells and mediates (i) inflammatory changes in the bladder, (ii) upregulation of urothelial cytokine production, and (iii) increased bladder motility. These results indicate that MIF released as a result of neurogenic inflammation is responsible, at least in part, for the inflammatory changes due to neurogenic inflammation in the bladder. Therefore, MIF is a possible target in chronic pelvic pain conditions where neurogenic inflammation may have a causative role, such as painful bladder syndrome/interstitial cystitis.

2. Neurogenic Inflammation: Description and Overview

Activation of afferent nerve terminals at a peripheral site (either as a result of trauma and injury or as a result of axon reflexes) elicits release of neurotransmitter from these afferent nerve terminals, mainly calcitonin gene-related

*Corresponding author: The Bay Pines VA Healthcare System, Research and Development, 10000 Bay Pines Boulevard, Bay Pines, FL 33744, USA. Email: pvera@health.usf.edu
†The Bay Pines VA Healthcare System, Bay Pines, FL, USA.
‡University of South Florida, Department of Surgery, Urology Division, Tampa, FL, USA.
§University of South Florida, Department of Molecular Medicine, Tampa, FL, USA.

peptide (CGRP) and substance P. The effect of these released neuropeptides is to produce vasodilation (perceived as warmth and redness; "calor","rubor"), plasma extravasation (swelling, "tumor") and hypersensitivity (due to hyper-excitability of terminal fibers of sensory neurons and central release of neuro-peptides, "dolor"), a condition that is referred to as "neurogenic inflammation."

The first description of neurogenic inflammation may be the findings of Goltz and Freusberg,[1] who reported increased vasodilation (measured as temperature changes) in the hind limbs of dogs mediated by the sciatic nerve. These observations were expanded by Bayliss,[2] who examined the effect of electrical stimulation of sciatic nerves in dogs, cats, and rabbits on hind limb vasodilation, and using ablation experiments, identified dorsal root ganglia neurons as responsible for the response. Subsequently, ample evidence has established release of neuropeptides from small diameter, capsaicin-sensitive afferent fibers as the mechanism of neurogenic inflammation.[3]

Currently, neurogenic inflammation is recognized as not only initiating or maintaining inflammation resulting from trauma, injury or pathogen,[4,5] but it is also recognized that neurogenic inflammation directly modulates immune cells (e.g., mast cells, macrophages) to induce cytokine expression and release that promotes inflammation.[4–6] Neurogenic inflammation is implicated in a number of diseases, including arthritis,[7] sepsis,[8] complex regional pain syndrome,[9] asthma,[10] cardiac dysfunction due to hypomagnesemia,[11] and chronic pelvic pain conditions such as chronic prostatitis and painful bladder syndrome/interstitial cystitis.[12–14]

Neurogenic inflammation is likely an important mechanism of pelvic organ dysfunction. Inflammation of one pelvic organ has been documented to elicit inflammatory changes in another pelvic organ through activation of spinal reflexes triggering neurogenic inflammation in the organ not initially affected ("visceral cross-talk").[15–18] The bladder receives afferent innervation from substance P and CGRP containing nerve fibers, and such innervation is responsible for neurogenic inflammation in this organ, which has been documented in a number of animal species.[19–21] This chapter will review experimental evidence supporting the conclusion that macrophage migration inhibitory factor (MIF) mediates, at least in part, the effects of neurogenic inflammation in the bladder. These findings may offer a novel target to guide further research into treatment options for chronic pelvic pain conditions such as chronic prostatitis and painful bladder syndrome/interstitial cystitis, two diseases with poorly understood etiologies and with only modestly successful therapeutic options. Since MIF is a central regulator of inflammation and is associated with several inflammatory conditions,[22,23] findings and mechanisms reviewed here may also apply in other inflammatory diseases where MIF is involved.

3. MIF in the Bladder

In the bladder, MIF is constitutively expressed by smooth muscle cells in the detrusor and by urothelial cells.[24] The urothelium generally consists of three layers: basal, intermediate, and apical. The apical layer consists of large, terminally differentiated epithelial cells, also termed umbrella cells. The urothelium contains rich stores of pre-formed MIF, predominantly in basal and intermediate cells, while apical cells (umbrella cells) are either devoid of MIF or only very lightly immunostained[24] (Fig. 1).

4. Substance P Induces MIF Release

4.1 *Substance P induces urothelial MIF release and upregulation in vivo*

We examined the effect of neurogenic inflammation on MIF release and upregulation in the bladder *in vivo*. In this model of neurogenic inflammation, administration of substance P results in plasma extravasation in the bladder

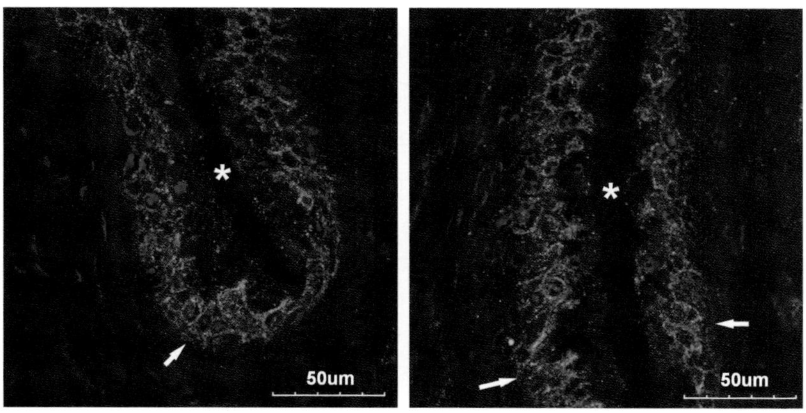

(a) (b)

Fig. 1. MIF immunostaining in urothelium. Urothelial cells in basal and intermediate layers of rat urothelium show prominent MIF immunofluorescence. (a, b) MIF immunofluorescence is shown in green; nuclei are counterstained with DAPI and appear blue; arrows indicate cells in basal layer. Apical cells lining the bladder lumen (indicated by asterisk) show little or no MIF immunostaining.

and bladder hypermotility.[25,26] We reported that treatment with substance P resulted in increased bladder edema (as expected) and also decreased levels in bladder MIF while increasing levels of urinary MIF (Fig. 2).[27] These results showed that substance P evoked MIF release from the bladder and into the urine as early as 30 min after substance P administration (Fig. 2b). In addition, we also reported that MIF is upregulated in the bladder after substance P treatment.[27] Moreover, substance P also increased production of nerve growth factor (NGF) by the bladder (an urinary marker correlated

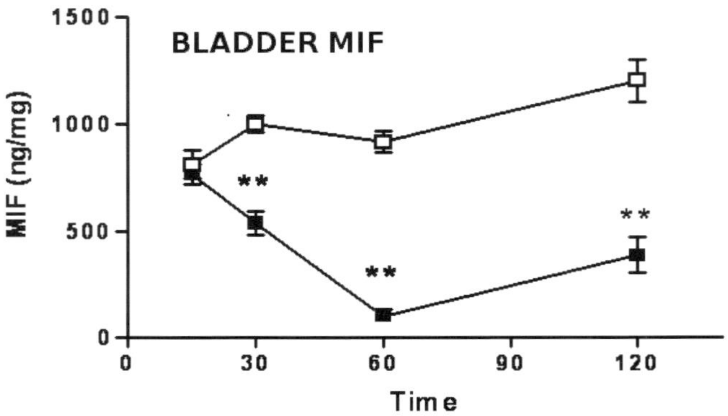

(a)

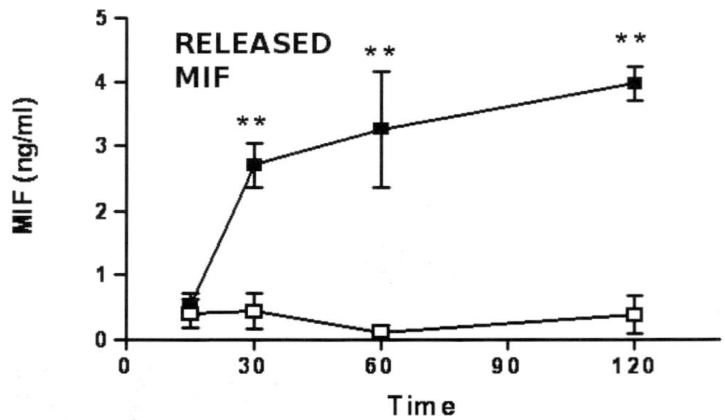

(b)

Fig. 2. Substance P induces MIF release in the bladder. Substance P administered systemically in a model of neurogenic inflammation elicited a decrease in the amount of MIF in bladder homogenates (a: ng MIF/mg of homogenate, as determined by ELISA) while concomitantly increasing the levels of MIF released into the urine (b: ng MIF/ml urine). Significant changes were first observed 30 min after substance P treatment. ** = $p < 0.01$, compared to control at equivalent time point. Open squares = saline treatment (control); filled squares = substance P treatment.

with bladder pain[28–30]) and *cox-2* and *c-fos* upregulation.[27] These findings established that MIF is released from the bladder in a model of neurogenic inflammation.

MIF is constitutively expressed in urothelial and also in smooth muscle cells.[24,31] Substance P treatment causes plasma extravasation, which adds serum MIF as a possible source of MIF in the urine. Hence, the exact source of increased urinary MIF in response to substance P treatment remained to be determined. In a separate series of experiments, we isolated the bladder from the kidneys by cutting both ureters, emptied the bladder of urine, and filled it with intravesical saline, collecting both intravesical fluid and urothelial cells (obtained by scraping the bladder gently at the end of the experiment to collect urothelial cells).[32] Under these conditions, substance P treatment decreased the levels of MIF in isolated urothelial cells while increasing the levels of MIF released into the bladder lumen, and upregulated MIF mRNA in urothelial cells.[33,34] Since we developed an MIF ELISA that detected rodent MIF from bladder homogenates, urine or intravesical fluid but not serum MIF, other possible sources of MIF were ruled out, and our experiments conclusively demonstrated that substance P caused urothelial MIF release and upregulation.

4.2 *Substance P induces MIF release from macrophages in vitro*

As part of neurogenic inflammation's modulation of the immune system, substance P modulates immune responses by acting directly on inflammatory cells (such as macrophages) by binding to the neurokinin-1(NK-1) receptor and activating signaling pathways leading to chemokine and cytokine production.[35,36] Therefore, since macrophages are an important source of MIF and secrete MIF when activated,[37] and also have functional NK-1 receptors,[35] we examined whether substance P stimulation can also elicit MIF release from macrophages.

Our results show that *in vitro* stimulation of RAW macrophages with substance P elicits MIF release, as early as 30 min after exposure (Figs. 3a and 3b). In addition, preincubation with a selective NK-1 antagonist (L-732,138)[38] greatly diminished substance P's ability to elicit MIF release from macrophages (Fig. 3c). These data suggested that activation of macrophage NK-1 receptors by substance P was responsible for eliciting MIF release. In fact, when RAW macrophages were exposed to a selective NK-1 agonist (GR73632),[39] MIF was released from macrophages in a dose-dependent manner (Fig.3d). Finally, substance P induces extracellularly regulated kinase (ERK) activation in RAW macrophages (Fig. 4). Thus, our findings demonstrate that substance P, through

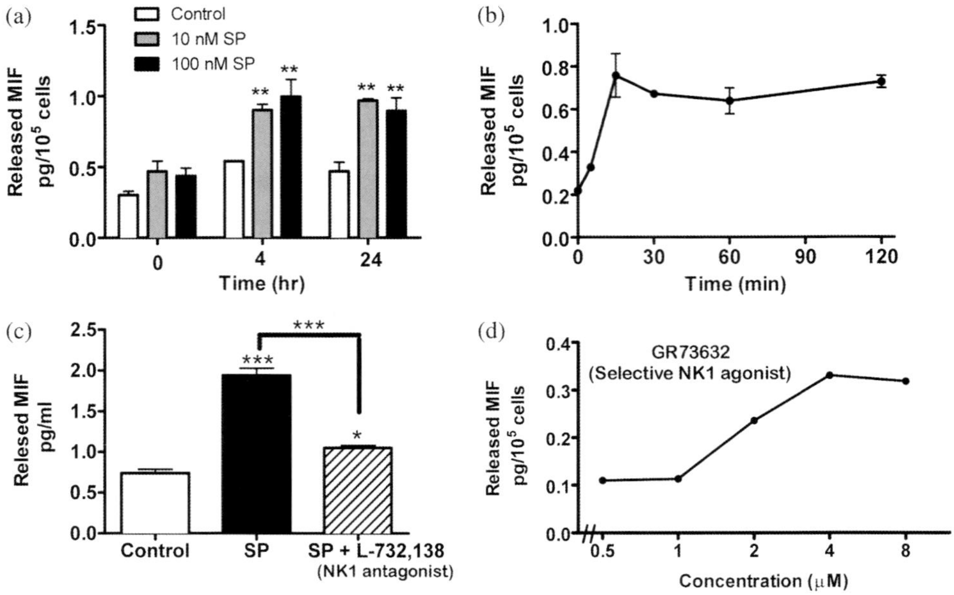

Fig. 3. Substance P induces MIF release from RAW macrophages *in vitro* via activation of NK-1 receptors. (a) Stimulation of RAW macrophages with substance P at 10 and 100 nM increased MIF released into the culture media at 4 and 24 hrs. (b) Time course of MIF release after substance P stimulation. (c) Pre-treatment with NK-1 antagonist (L-732,138) greatly decreased substance P-induced MIF release from macrophages. (d) Stimulation with selective NK-1 agonist (GR73632) induced MIF release in a concentration-dependent manner. *, **, *** = $p < 0.05, 0.01, 0.001$, respectively.

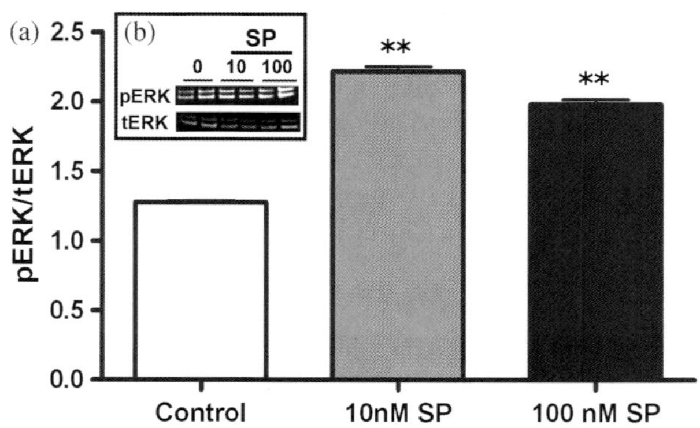

Fig. 4. Substance P-induced ERK activation in RAW macrophages *in vitro*. Ratio of phosphorylated ERK (pERK) to total ERK (tERK) in RAW macrophages increased after substance P (SP) treatment when compared to control treatment. Representative western blots of pERK and tERK are shown. ** = $p < 0.01$.

activation of NK-1 receptors, elicits MIF release from macrophages and thus links neurogenic inflammation, both at the urothelial cell and immune cell level, to MIF release.

5. Substance P-Induced Bladder MIF Release and Upregulation are Mediated by Peripheral Nerve Activation

We examined the physiological mechanism responsible for urothelial MIF release in response to substance P stimulation. In anesthetized rats that received substance P treatment to mimic neurogenic inflammation, the effect of urothelial MIF release by substance P could be blocked either by instilling lidocaine into the bladder (to block all nerve activity) or by cutting the pelvic nerves that supply innervation to the bladder (Fig. 5). Hence, these results suggest that activation of nerve terminals at the end-organ level are responsible for urothelial MIF release elicited by substance P. Further experiments established that urothelial MIF release is controlled by activation of post ganglionic fibers and mediated by muscarinic (via the parasympathetic nervous system) and/or adrenergic (viathe sympathetic nervous system) receptors at the bladder.[40]

Capsaicin is the active component of plants in the genus *Capsicum* responsible for the burning sensation associated with ingestion of hot peppers. Capsaicin also stimulates sensory fibers in the bladder to elicit neuropeptide release[41] and

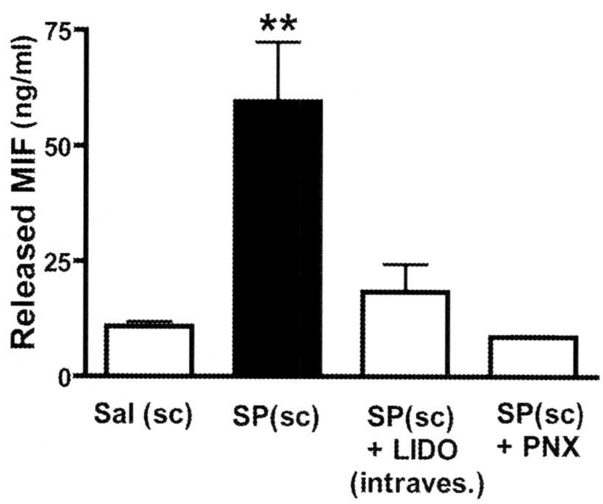

Fig. 5. Substance P (SP, subcutaneous) induced MIF release from the bladder that is blocked by intravesical treatment with lidocaine (LIDO intraves., to stop all nerve activity in the bladder) or by bladder denervation (PNX, cutting pelvic nerves bilaterally). ** = $p < 0.01$.

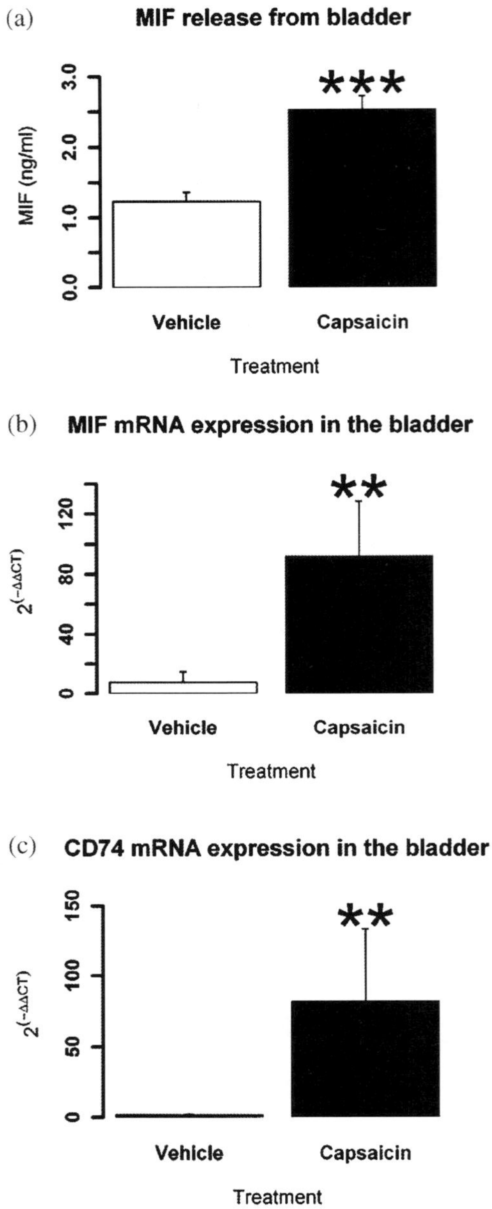

Fig. 6. Intravesical capsaicin, which activates afferent fibers in the bladder, also elicits MIF release from the bladder (a) as well as upregulation of MIF (b) and CD74 (c). **, *** = $p < 0.01$, 0.001, respectively.

has been investigated as a possible treatment for painful bladder syndrome/interstitial cystitis.[42] We showed that activation of C-fiber afferent terminals innervating the ladder by using intravesical capsaicin was sufficient to elicit urothelial MIF release,[43] bladder MIF upregulation, and bladder CD74 upregulation

(Fig. 6). Thus, these findings demonstrate that while afferent stimulation is sufficient to elicit urothelial MIF release and upregulation, central nervous system reflexes contribute to neurogenic inflammation in the bladder through activation of postganglionic elements involving both the parasympathetic (via muscarinic receptors) and the sympathetic nervous system (via adrenergic receptors).

6. MIF Mediates Inflammatory Changes due to Neurogenic Inflammation in the Bladder

Substance P elicits not only MIF release from urothelial cells but also upregulation and cell-surface expression of MIF's receptor, CD74.[33,44] In addition, substance P also induces production of proinflammatory cytokines in the bladder and more specifically, the urothelium.[34,45] We examined the role of released MIF during neurogenic inflammation by comparing histological changes in the bladder and inflammatory cytokine production in the bladder and specifically in the urothelium, in animals that received substance P treatment but also received intravesical antibodies to MIF or to CD74. Our results showed that intravesical treatment with MIF or CD74 antibodies (to block either released MIF or MIF-mediated signal transduction through binding to CD74) markedly decreased bladder histological changes caused by substance P (e.g., increased edema) while also markedly decreasing inflammatory cytokine production in the bladder and the urothelium, in particular.[34,35]

In addition to inducing morphological and molecular changes, substance P also increases bladder excitability, as measured by increased frequency of bladder contractions.[26] We examined the effect of intravesical treatment either with phosphate buffered saline (PBS) or antibodies to MIF (MIF antibodies, in order to block released MIF) on the frequency of bladder contractions due to substance P treatment (40 μg/kg in sterile PBS, subcutaneous). In this model, a cystometrogram is performed starting with an empty bladder, isolated from the ureters (so that urine does not flow into the bladder) and under isovolumetric conditions (closed urethra, transvesical catheter). The intercontraction interval (ICI) is defined as the time to contraction under constant filling conditions (Fig. 7a). Substance P treatment decreased the intercontraction interval, a measure of increased bladder excitability, in animals that received intravesical phosphate-buffered saline (PBS) as treatment (Fig. 7a, bottom panel). On the other hand, in animals treated with intravesical antibodies to MIF to block released MIF, substance P had no effect on the intercontraction interval (Figs. 7b and 7c). Animals treated with intravesical non specific immunoglobulin, IgG (as a control), still showed decreased ICI after substance P treatment (Fig. 7).

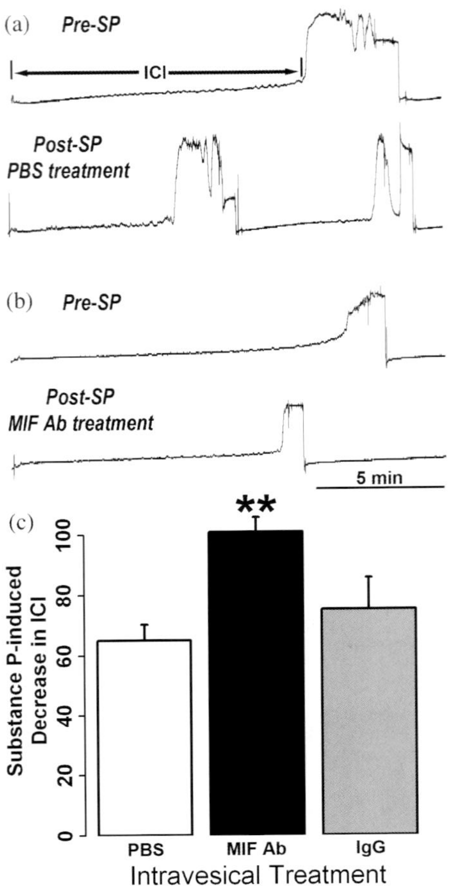

Fig. 7. Treatment with intravesical antibodies (Ab) to MIF prevents substance P-induced bladder hyperreflexia. The intercontraction interval (ICI) was measured in anesthetized rats by recording the time to a bladder contraction during filling through a transvesical catheter. (a) Changes in ICI before substance P (SP) and after SP treatment in animals treated with intravesical PBS showed a decrease in ICI. (b) Intravesical MIF Ab treatment prevented substance P's decrease in ICI. (c) Group data ($N = 6$) showed that substance P decreased ICI approximately 60% (compared to Pre-SP) in animals that received intravesical PBS (control), and this effect was blocked by intravesical MIF Ab, while intravesical nonspecific IgG had no effect (** = $p < 0.01$).

Thus, these findings indicate that many of the inflammatory changes produced by substance P during neurogenic inflammation in the bladder may be due to increased MIF release and increased MIF proinflammatory signaling in the bladder.

7. Model of MIF's Role in Neurogenic Inflammation in the Bladder

Our experimental findings of MIF's role in neurogenic inflammation in the bladder can be synthesized into a model for neurogenic inflammation (Fig. 8).

In this model, inflammatory stimuli (such as trauma, injury or other noxious stimuli) activate C-fiber afferent fibers in the bladder to trigger nerve-mediated release of constitutively expressed MIF from the urothelium into the bladder lumen (a local effect) while also eliciting pain and viscerovisceral reflexes (a central effect). Activation of viscerovisceral reflexes (as part of the central effect) will contribute to MIF release through activation of parasympathetic and/or sympathetic postganglionic pathways (mediated by the major pelvic ganglia, MPG; sympathetic ganglia, SG; respectively) activating muscarinic and/or adrenergic receptors at the bladder.

Tachykinins released from afferent terminals in the bladder will elicit neurogenic inflammation and urothelial MIF release, both effects mediated via NK-1 receptors. Released MIF binds to urothelial receptors (CD74, CXCR4;

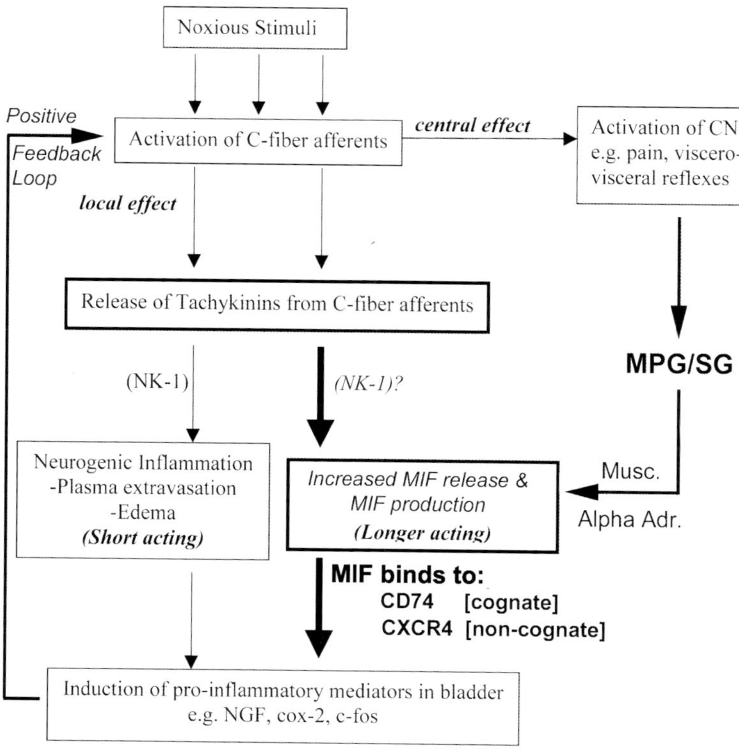

Fig. 8. Model of MIF and neurogenic inflammation. Adapted from Vera *et al.*[40]

which are also upregulated during inflammation) to mediate signal transduction, leading to production of proinflammatory cytokines by the urothelium. MIF-mediated proinflammatory signaling cascade then maintains and/or enhances bladder inflammation (positive feedback loop).

Understanding the basic mechanisms involved in urothelial MIF release, MIF-mediated signaling in the urothelium and bladder and MIF-mediated inflammatory cytokine production in the bladder is important in identifying targets that disrupt bladder inflammation and/or enhance urothelial proliferation and repair. Blocking MIF and/or receptors for MIF in the urothelium may prevent or reduce bladder inflammation.

8. Conclusions

Experimental evidence supports the conclusion that neurogenic inflammation in the bladder results in MIF release that acts to enhance or maintain bladder inflammation. Since there is experimental and clinical evidence suggesting a role for tachykinins (and specifically substance P) in chronic pelvic pain conditions such as painful bladder syndrome/interstitial cystitis and chronic prostatitis,[13,46–49] it is reasonable to investigate MIF's role in these diseases. Finally, substance P-induced MIF release may also be involved in other diseases where MIF has been implicated, including arthritis and sepsis. Our model provides an explanatory framework for neuromodulation of immune and local responses while providing attractive targets for therapeutic intervention.

Acknowledgments

This material is based upon work supported (or supported in part) by the Department of Veterans Affairs, Veterans Health Administration, Office of Research and Development, Biomedical Laboratory Research and Development (PLV, KLMS). This work was also supported by the National Institute of Diabetes and Digestive and Kidney Disorders DK075059 (PLV) and the Bay Pines Foundation. The funders had no role in study design, data collection, and analysis, decision to publish, or preparation of the manuscript. The contents of this article do not represent the views of the Department of Veterans Affairs or the United States Government.

Gary A. Smith, Jr. and Warren Cowart provided excellent technical assistance. Xihai Wang, M.D., and Lixian Jiang, Ph.D., contributed to the experiments described.

References

1. Goltz F, Freusberg A. (1874) Ueber gefässerweiternde Nerven. *Pflügers Archiv: Eur J Physiol* **9**: 174–197.
2. Bayliss WM. (1901) On the origin from the spinal cord of the vaso-dilator fibres of the hind-limb, and on the nature of these fibres. *J Physiol* **26**(3–4): 173–209.
3. Holzer P. (1988) Local effector functions of capsaicin-sensitive sensory nerve endings: Involvement of tachykinins, calcitonin gene-related peptide and other neuropeptides. *Neuroscience* **24**(3): 739–768.
4. Black PH. (2002) Stress and the inflammatory response: A review of neurogenic inflammation. *Brain Behav Immun* **16**(6): 622–653.
5. Richardson JD, Vasko MR. (2002) Cellular mechanisms of neurogenic inflammation. *J Pharmacol Exp Ther* **302**(3): 839–845.
6. Fernandes ES, Schmidhuber SM, Brain SD. (2009) Sensory-nerve-derived neuropeptides: Possible therapeutic targets. *Handb Exp Pharmacol* **194**: 393–416.
7. Campos MM, Calixto JB. (2000) Neurokinin mediation of edema and inflammation. *Neuropeptides* **34**(5): 314–322.
8. Hegde A, Zhang H, Moochhala SM, Bhatia M. (2007) Neurokinin-1receptor antagonist treatment protects mice against lung injury in polymicrobial sepsis. *J Leukoc Biol* **82**(3): 678–685.
9. Birklein F, Schmelz M. (2008) Neuropeptides, neurogenic inflammation and complex regional pain syndrome (CRPS). *Neurosci Lett* **437**(3): 199–202.
10. Pisi G, Olivieri D, Chetta A. (2009) The airway neurogenic inflammation: Clinical and pharmacological implications. *Inflamm Allergy Drug Targets* **8**(3): 176–181.
11. Kramer JH, Spurney C, Iantorno M *et al.* (2009) Neurogenic inflammation and cardiac dysfunction due to hypomagnesemia. *Am J Med Sci* **338**(1): 22–27.
12. Pontari MA, Ruggieri MR. (2004) Mechanisms in prostatitis/chronic pelvic pain syndrome. *J Urol* **172**(3): 839–845.
13. Wesselmann U. (2001) Neurogenic inflammation and chronic pelvic pain. *World J Urol* **19**(3): 180–185.
14. Theoharides TC, Whitmore K, Stanford E *et al.* (2008) Interstitial cystitis: Bladder pain and beyond. *Expert Opin Pharmacother* **9**(17): 2979–2994.
15. Morrison TC, Dmitrieva N, Winnard KP, Berkley KJ. (2006) Opposing viscerovisceral effects of surgically induced endometriosis and a control abdominal surgery on the rat bladder. *Fertil Steril* **86**(Suppl 4): 1067–1073.
16. Pan X-Q, Gonzalez JA, Chang S *et al.* (2010) Experimental colitis triggers the release of substance P and calcitonin gene-related peptide in the urinary bladder via TRPV1 signaling pathways. *Exp Neurol* **225**(2): 262–273.
17. Ustinova EE, Fraser MO, Pezzone MA. (2006) Colonic irritation in the rat sensitizes urinary bladder afferents to mechanical and chemical stimuli: An afferent origin of pelvic organ cross-sensitization. *Am J Physiol* **290**(6): F1478–F1487.

18. Vera PL, Meyer-Siegler KL. (2004) Inflammation of the rat prostate evokes release of macrophage migration inhibitory factor in the bladder: Evidence for a viscerovisceral reflex. *J Urol* **172**(6 Pt 1): 2440–2445.
19. Bjorling DE, Beckman M, Saban R. (2003) Neurogenic inflammation of the bladder. *Adv Exp Med Biol* **539**(Pt B): 551–583.
20. Candenas L, Lecci A, Pinto FM *et al.* (2005) Tachykinins and tachykinin receptors: Effects in the genitourinary tract. *Life Sci* **76**(8): 835–862.
21. Geppetti P, Nassini R, Materazzi S, Benemei S. (2008) The concept of neurogenic inflammation. *BJU Int* **101**(Suppl 3): 2–6.
22. Bach J-P, Rinn B, Meyer B *et al.* (2008) Role of MIF in inflammation and tumorigenesis. *Oncology* **75**(3-4): 127–133.
23. Hoi AY, Iskander MN, Morand EF. (2007) Macrophage migration inhibitory factor: A therapeutic target across inflammatory diseases. *Inflamm Allergy Drug Targets* **6**(3): 183–190.
24. Vera PL, Meyer-Siegler, KL. (2003) Anatomical location of macrophage migration inhibitory factor in urogenital tissues, peripheral ganglia and lumbosacral spinal cord of the rat. *BMC Neurosci* **4**: 17.
25. Gao GC, Wei ET. (1995) Inhibition of substance P-induced vascular leakage in rat by N-acetyl-neurotensin-(8-13). *Regul Pept* **58**(3): 117–121.
26. Chen WC, Hayakawa S, Shimizu K *et al.* (2004) Catechins prevents substance P-induced hyperactive bladder in rats via the downregulation of ICAM and ROS. *Neurosci Lett* **367**: 213–217.
27. Meyer-Siegler KL, Vera PL. (2004) Substance P induced release of macrophage migration inhibitory factor from rat bladder epithelium. *J Urol* **171**(4): 1698–1703.
28. Dmitrieva N, McMahon SB. (1996) Sensitisation of visceral afferents by nerve growth factor in the adult rat. *Pain* **66**(1): 87–97.
29. Lowe EM, Anand P, Terenghi G *et al.* (1997) Increased nerve growth factor levels in the urinary bladder of women with idiopathic sensory urgency and interstitial cystitis. *Br J Urol* **79**(4): 572–577.
30. Oddiah D, Anand P, McMahon SB, Rattray M. (1998) Rapid increase of NGF, BDNF and NT-3 mRNAs in inflamed bladder. *Neuroreport* **9**(7): 1455–1458.
31. Vera PL, Ordorica RC, Meyer-Siegler KL. (2003) Hydrochloric acid induced changes in macrophage migration inhibitory factor in the bladder, peripheral and central nervous system of the rat. *J Urol* **170**: 623–627.
32. Maitra A, Wistuba I, Virmani AK *et al.* (1999) Enrichment of epithelial cells for molecular studies. *Nat Med* **5**(4): 459–463.
33. Meyer-Siegler KL, Xia S-L,Vera PL. (2009) Substance P increases cell-surface expression of CD74 (receptor for macrophage migration inhibitory factor): *In vivo* biotinylation of urothelial cell-surface proteins. *Mediators Inflamm* **2009**: 535348.

34. Vera PL, Wang X, Bucala RJ, Meyer-Siegler KL. (2009) Intraluminal blockade of cell-surface CD74 and glucose regulated protein 78 prevents substance P-induced bladder inflammatory changes in the rat. *PLoS One* **4**(6): e5835.
35. Sun J, Ramnath RD, Zhi L *et al.* (2008) Substance P enhances NF-kappa B transactivation and chemokine response in murine macrophages via ERK1/2 and p38 MAPK signaling pathways. *Am J Physiol Cell Physiol* **294**(6): C1586–C1596.
36. O'Connor TM, O'Connell J, O'Brien DI *et al.* (2004) The role of substance P in inflammatory disease. *J Cell Physiol* **201**(2): 167–180.
37. Calandra T, Bernhagen J, Mitchell RA, Bucala R. (1994) The macrophage is an important and previously unrecognized source of macrophage migration inhibitory factor. *J Exp Med* **179**(6): 1895–1902.
38. Kuo HP, Hwang KH, Lin HC *et al.* (1998) Lipopolysaccharide enhances neurogenic plasma exudation in guinea-pig airways. *Br J Pharmacol* **125**(4): 711–716.
39. Sakurada C, Watanabe C, Inoue M *et al.* (1999) Spinal actions of GR73632, a novel tachykinin NK1 receptor agonist. *Peptides* **20**(2): 301–304.
40. Vera PL, Wang X, Meyer-Siegler KL. (2008) Neural control of substance P induced up-regulation and release of macrophage migration inhibitory factor in the rat bladder. *J Urol* **180**(1): 373–378.
41. Maggi CA, Meli A. (1988) Thesensory-efferent function of capsaicin-sensitive sensory neurons. *Gen Pharmacol* **19**(1): 1–43.
42. Payne CK, Mosbaugh PG, Forrest JB *et al.* (2005) Intravesical resiniferatoxin for the treatment of interstitial cystitis: A randomized, double-blind, placebo controlled trial. *J Urol* **173**(5): 1590–1594.
43. Vera PL, Iczkowski KA, Leng L *et al.* (2005) Macrophage migration inhibitory factor is released as a complex with alpha1-inhibitor-3 in the intraluminal fluid during bladder inflammation in the rat. *J Urol* **174**: 338–343.
44. Meyer-Siegler KL, Vera PL. (2005) Substance P induced changes in CD74 and CD44 in the rat bladder. *J Urol* **173**(2): 615–620.
45. Meyer-Siegler KL,Vera PL. (2004) Intraluminal antibodies to macrophage migration inhibitory factor decrease substance P induced inflammatory changes in the rat bladder and prostate. *J Urol* **172**(4 Pt 1): 1504–1509.
46. Pang X, Marchand J, Sant GR *et al.* (1995) Increased number of substance P positive nerve fibres in interstitial cystitis. *Br J Urol* **75**(6): 744–750.
47. Kushner L, Chiu PY, Brettschneider N *et al.* (2001) Urinary substance P concentration correlates with urinary frequency and urgency in interstitial cystitis patients treated with intravesical dimethyl sulfoxide and not intravesical anesthetic cocktail. *Urology* **57**(6 Suppl 1): 129.
48. Nazif O, Teichman JM, Gebhart GF. (2007) Neural upregulation in interstitial cystitis. *Urology* **69**(4 Suppl): 24–33.
49. Rudick CN, Chen MC, Mongiu AK, Klumpp DJ. (2007) Organ cross talk modulates pelvic pain. *Am J Physiol* **293**(3): R1191–R1198.

III-5

MIF and Lower Urinary Tract Disorders

Anthony DeAngelis*, Phillip P. Smith*,† and George A. Kuchel*,‡

1. Introduction

The lower urinary tract plays an essential role in human survival, health, and quality of life by permitting the elimination of toxic metabolic wastes through the storage and socially acceptable evacuation of urine. Moreover, bladder and prostate cancers represent extremely common contributors to early disability and death. Inflammatory pathways, including those involving MIF, have been implicated in many lower urinary tract diseases and disorders.

The goal of this chapter will be to first review lower urinary tract structure and function. A majority of cells comprising the lower urinary tract produce MIF and/or are sensitive to its effects. With these considerations in mind, MIF biology will be reviewed within the context of the lower urinary tract, emphasizing both similarities and differences when compared to other tissues. The chapter will conclude with a discussion of the evidence implicating MIF in specific diseases and disorders involving the lower urinary tract, emphasizing the ultimate potential for MIF-based diagnostics and therapeutics.

2. Lower Urinary Tract Structure and Function

The lower urinary tract is composed of the urinary bladder, the urethra, and directly related pelvic structures. It receives, stores, and evacuates urinary fluid volumes resulting from renal output. In humans, this process is normally under semiconscious control, while most of the conscious nature of its

*UConn Center on Aging, University of Connecticut Health Center, USA.
†Division of Urology, University of Connecticut Health Center, USA.
‡Corresponding author: UConn Center on Aging, University of Connecticut, Health Center, Farmington, CT 06030, USA. Email: kuchel@uchc.edu

function involves recognition and management of the full bladder. Normal function is dependent upon complex interactions of sensory, motor, structural, and cognitive functions. Deficiencies in the ability to unconsciously store urine without leakage as well as the conscious capacity to periodically evacuate the stored volume in an efficient manner may result from either lower urinary tract or systemic disorders. Several recent reviews provide summary details of current knowledge of the regulation of the sensorimotor functions of the lower urinary tract.[1–5]

The bladder is composed of two primary functional layers, an outer interlacing network of smooth muscle cells (detrusor) and a watertight and highly specialized inner epithelial lining (urothelium). Interlaced within the detrusor and also placed between the detrusor and the urothelium is a layer of myofibroblast cells functionally analogous to the interstitial cells of Cajal (ICC) in the gut.[5] The urothelium protects deeper tissues from infection and chemical damage caused by urine.[2] Beyond these barrier functions, urothelial cells are also a rich source of neuroactive and inflammatory molecules, which are thought to be prime mediators of sensory function and host defense in the bladder.[2] The robust capacity of normally quiescent urothelial cells to proliferate in response to injury[6] may contribute to their great propensity towards malignant transformation resulting in bladder carcinomas.[7] Ultimately, the detrusor functions as a muscular bag with an outlet, generating expulsive forces upon contraction. The ICCs are likely mediators of both sensory and muscular functions. The detrusor is normally relaxed during urine storage, thus permitting urine to drain from the kidneys into the bladder with minimal resistive force.[3]

The urethra is a tube through which voiding takes place. It consists of smooth muscle and connective tissue lined by urothelial cells. In addition to being the normal expulsive conduit for urine during voiding, it must remain closed during urine storage in order to maintain continence. Its sphincteric functions required for continence are not the result of an external annular muscular structure, as for example is the case for the external anal sphincter. Rather continence is maintained through a complex interaction involving active and passive features of the urethral tube itself, as it passes through the fibromuscular tissues of the pelvic floor. These interactions provide a watertight seal at rest, which may be quickly augmented by both neural inputs and pelvic floor structures at times of sudden increases in bladder pressure.

The requirements of successful urine storage include a bladder reservoir, which maintains low pressure during filling and has a socially adequate volume capacity. The sphincter mechanism keeps the outlet (urethra) watertight.

Volume sensory information is relayed to brain control areas, where it is integrated with systemic, environmental, and experiential information, allowing planning for socially appropriate bladder emptying.[4] The requirements for successful emptying begin with the recognition of a normally full bladder, as well as the ability to plan and prepare for bladder emptying. Upon the conscious desire to initiate voiding, central suppression of relatively hard-wired voiding reflexes is released, permitting urethral sphincteric relaxation and subsequent detrusor contraction, thus expelling urine. Bladder and urethral sensory feedback maintains the voiding state until the bladder has been emptied, at which time central control systems switch back to the urine storage state.

Both motor and sensory aspects of bladder and urethral function are subject to active neuropharmacologic regulation. Normal end-organ function in humans is primarily mediated through cholinergic and adrenergic pathways. The role of purinergic signaling within the lower urinary tract is not well defined but appears to include regulation of sensory and motor functions. These pathways are subject to modulation in response to trauma, disease, and potentially aging.

The prostate is an exocrine gland that surrounds the proximal portion of the urethra in men. Its function is reproductive resulting in the production of an alkaline fluid that enhances sperm survival and function. Nevertheless, with advancing age, prostatic tissues typically undergo hypertrophy with lower urinary tract voiding symptoms that may include obstruction. Finally, the propensity of prostatic cells to undergo malignant transformation contributes to the very high prevalence of prostate cancer in older men.

3. MIF Biology in the Lower Urinary Tract

3.1 *Cellular sites of MIF expression*

Macrophage migration inhibitory factor (MIF) is a pleiotropic protein that has been shown to be expressed in a broad variety of different cell types including lymphocytes,[8,9] pituicytes,[10] macrophages,[11,12] neurons,[13] fibroblasts,[14] adipocytes,[15] skeletal myocytes,[16] cardiomyocytes,[17] and smooth muscle cells.[18,19] Many of these cells represent key lower urinary tract structural components. In addition, several reports have shown that the MIF protein is present in various structures of the lower urinary tract including the urinary bladder,[19–21] prostate,[22] and the neurons responsible for innervating these pelvic organs.[22] Thus, it should not be surprising that in recent years information has emerged implicating MIF in a variety of related disease processes.

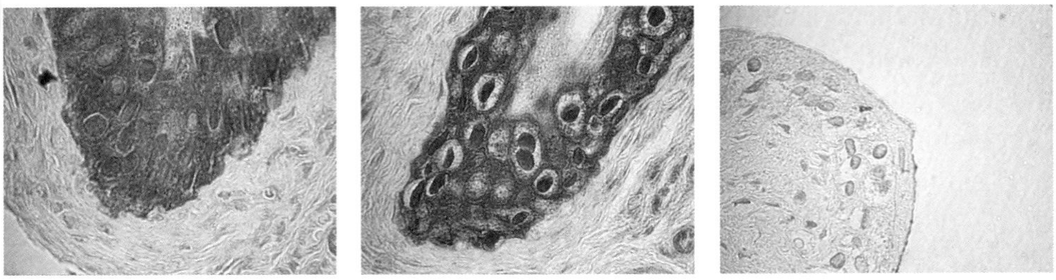

(a) (b) (c)

Fig 1. Images of MIF immunoreactivity in urothelial cells obtained from (a) sham-operated wildtype female mice, (b) wildtype female mice studies 3 weeks after partial bladder outlet obstruction surgery, and (c) in bladders from *Mif*$^{-/-}$ mice.

In rat[22] and mouse[19] urinary bladders, immunohistochemical analysis revealed very strong MIF protein expression within urothelial cells (Fig. 1). In these cells, MIF expression is primarily localized to basal and intermediate layers[19,22] as there is minimal staining of the superficial layers of the urothelium.[19,22] In the detrusor, far less intense MIF immunoreactivity can also be localized to smooth muscle cells involving both cytoplasmic and nuclear regions of the muscle fibers.[22] Based on ELISA measurements, overall MIF concentrations are 200.8 (± 5.75) ng/mg total protein and 67.3 (± 13) ng/mg total protein in extracts obtained from male and female rat bladders, respectively.[22]

The ventral lobe of the rat prostate contains 144.1 ng MIF/mg total protein, with MIF expression within the prostate localized mostly to acinar epithelial cells.[22] MIF is also expressed in various peripheral nerve cells that innervate and regulate lower urinary tract tissues. For example, MIF is expressed in both sympathetic and parasympathetic neurons located within the major pelvic ganglia that provide efferent innervations to the bladder and other major pelvic viscera.[22] MIF is also expressed in sympathetic neurons located within inferior mesenteric neurons and the lumbar sympathetic chain.[22] Finally, many sensory neurons within the dorsal root ganglia, including substance P-positive cells, also express MIF, as do many subpopulations of cells within the lumbosacral spinal cord.[22]

In summary, many cellular populations that constitute or innervate the lower urinary tract demonstrate clear evidence of MIF immunoreactivity. In some cases (e.g., bladder epithelial cells), MIF protein expression is unusually high for a cytokine, rivaled only by the type of MIF expression observed in pituitary cells. In spite of MIF's broad expression, minimal or no

MIF immunoreactivity is observed in other cellular populations within the lower urinary tract. MIF expression in these tissues may be influenced by stimuli as diverse as chemical cystitis,[21] substance P administration,[23] partial bladder obstruction,[19] viscerovisceral reflexes mediating pelvic visceral inflammation,[24] urinary tract infection,[25] and bilateral ovariectomy (Kuchel *et al., personal observations*), suggesting a broad range of biological functions within the lower urinary tract for this cytokine.

3.2 *Regulation of MIF release*

Pre-formed MIF is present in the urinary bladder[21] and prostate,[22] with the most abundant expression being found in the transitional epithelial layers of the bladder wall.[19,22] MIF is released from urothelial stores into the urine in response to potent inflammatory molecules such as substance P or cyclophosphamide and serves as an important early mediator in bladder inflammatory processes.[23,26] The release of MIF may be direct from urothelial cells or via neurally mediated reflex pathways. Substance P is a neurotransmitter found in subsets of sensory nerve fibers innervating the bladder and prostate. This molecule has also been found to act as an initiator of the inflammatory process. Intravesical lidocaine inhibits substance P-induced MIF release, as does ganglionic blockade, suggesting that substance P activation of C-fiber afferents adjacent to urothelial cells initiates a reflexic post-ganglionic release of MIF. Alpha-adrenergic blockade as well as pan-cholinergic blockade inhibit the SP-induced MIF release, suggesting involvement of sympathetic and possibly parasympathetic efferent pathways. As a urothelial C-fiber mediated response, these findings suggest a mechanism by which noxious urothelial stimulation may induce MIF release. This in turn leads to an upregulation of proinflammatory mediators such as NGF, COX-2, and *c-fos*, all of which are important in mediating the inflammatory response.[23] Moreover, rat prostate inflammation induced using intraprostatically injected formalin has been shown to evoke release of MIF in the bladder via a viscerovisceral reflex mechanism.[24] Similar to the substance P-induced cystitis model, MIF release was shown to result in an increased production of other inflammatory mediators (COX-2 and NFG) in the bladder and lumbosacral spinal cord, resulting in a potentiation of the inflammatory process.[24]

MIF secretion can be induced by a number of other stimuli, including LPS,[11,28] IFN-γ,[11] and TNF-α,[11] upon activation of the innate immune system. In addition, other factors such as thrombin,[29] corticosteroids,[30] adrenocorticotropic hormone,[31] and angiotensin II[32] have also been shown to promote the release of MIF. A large number of soluble proteins follow the classical route of secretion that utilizes a signal sequence peptide to direct the protein to the endoplasmic reticulum. MIF, however, lacks the genetic sequence encoding this peptide. Instead, experiments investigating the release of MIF from LPS-stimulated macrophages have shown that MIF release occurs via a nonclassical route and involves the ABCA1 transporter protein.[33] In addition to promoting cytokine release from macrophages,[11] LPS also promotes MIF release from endothelial[28] and urothelial cells (Kuchel *et al., personal observations*) in a dose-dependent manner. Thus, not only is MIF release mediated via the LPS receptor TLR-4, but MIF is, in turn, required for optimal *TLR4* gene expression.[34]

The systemic administration of potent inflammatory stimuli such as substance P[23] and cyclophosphamide[26] resulted in increased MIF levels in the urine (Table 1). This was accompanied by decreases in MIF immunoreactivity within urothelial cells and decreases in overall MIF protein levels in the bladder, while overall bladder MIF mRNA levels increased.[23,26] The installation of hydrochloric acid within the bladder not only denuded the urothelial layer but also decreased total bladder MIF protein levels (Table 1).[21] Overall bladder MIF mRNA levels remained unaltered.[21] At the same time, MIF immunoreactivity was increased within selected detrusor layer cells, which were assumed to be bladder smooth muscle cells, and MIF immunoreactivity as well as MIF mRNA levels increased in neurons located in the major pelvic ganglia, L6/S1 dorsal root ganglia, and L6/S1 spinal cord regions that innervate the bladder.[21] Instillation of LPS (lipopolysaccharide) into the bladder upregulates MIF mRNA and protein expression in both the bladder and the same neural structures (Table 1).[18] Finally, it appears that similar changes take place following at least two other physiological interventions. For example,

Table 1. Summary of findings describing the impact of various interventions on MIF urine release and expression in the bladder and its innervations. N.D. = not determined.

Stimulus	Urine MIF	Uroepithelial expression	Total bladder MIF protein	Total bladder MIF mRNA	Peripheral nervous system MIF	Spinal cord MIF
Hydrochloric acid[21]	N.D.	Denudation	↓	—	↑	↑
Substance P[23,36]	↑	↓	↓	↑	N.D.	N.D.
Cyclophosphamide[83]	↑	↓	↓	↑	N.D.	N.D.
Lipopolysaccharide[18]	N.D.	—	↑	↑	↑	↑
Partial bladder outlet obstruction[19]	N.D.	↓	N.D.	—	N.D.	N.D.

both partial bladder outlet obstruction (Fig. 1, Table 1)[19] and bilateral ovariectomy (Table 1; Kuchel *et al., personal observations*) result in thinning of the urothelial layer with depletion of MIF immunoreactivity within remaining urothelial cells and no change in bladder MIF mRNA levels. In spite of important differences between various experimental paradigms, all of these studies highlight the importance of considering different MIF sources within the bladder, which include both pre-formed MIF protein (mostly from urothelial stores) and newly synthesized MIF (potentially originating from urothelial cells, neurons, smooth muscle cells, fibroblasts, and infiltrating macrophages). At the same time, further studies are needed to help distinguish between the distinct contributions made by MIF synthesis, release, and inactivation.

3.3 *Regulation of MIF synthesis*

Since the administration of powerful inflammatory stimuli such as substance P,[23] cyclophosphamide,[26] and LPS[18] results in significant increases in overall bladder MIF mRNA levels, increased MIF synthesis must contribute to increased MIF bioavailability under these conditions. Nevertheless, the cellular source of such new MIF synthesis remains to be delineated, as does the relative contribution of enhanced MIF synthesis following other interventions (e.g., partial bladder outlet obstruction and bilateral ovariectomy).

There is a relative paucity of information on the regulation of MIF synthesis in lower urinary tract cells. Human uroepithelial bladder cancer cell line (HT-1376) cultures contain and release MIF protein under basal conditions.[20] Treatment with HA (high molecular weight hyaluronate) resulted in decreased *Mif* mRNA levels, increased intracellular protein content and decreased MIF secretion, suggesting that HA inhibits both MIF synthesis and release.[20] In contrast, thrombin may induce MIF synthesis in urothelial cells since thrombin induced both *Mif* mRNA expression and MIF protein release in UROtsa (human immortalized uroepithelial cells).[29] Moreover, based on the ability of lidocaine to reduce similar thrombin[29] and substance P[27] mediated *in vivo* changes, these effects may be, at least in part, neurally mediated.

Once released, MIF promotes the synthesis of numerous other proinflammatory cytokines and transcription factors, including nerve growth factor (NGF), *c-fos*, and COX-2.[23,24] NGF is an important mediator of the inflammatory response. Upon release, NGF binds to its receptor and activates the Ras-Raf-MAPK signaling pathway, leading to subsequent downstream upregulation of the transcription factor *c-fos*. Together with the transcription factor *c-jun*, *c-fos* binds to the AP-1 response element and promotes the synthesis of several proteins.[35] The *Mif* promoter contains an AP-1 binding site. Therefore, it is possible that an NGF upregulation in *c-fos* may promote an increase in MIF synthesis.

3.4 *Localization of MIF receptors*

MIF signal transduction is initiated upon binding of MIF to the CD74 receptor,[37] leading to subsequent activation of the CD74/CD44 receptor complex.[38] This results in the phosphorylation and activation of the extracellular signal regulated kinase (ERK)-1/2 MAP kinase signaling cascade, which is responsible for mediating many of the effects of MIF.[37,39]

The CD74/CD44 MIF receptor complex is located on many different cell types including those which constitute the normal bladder, such as urothelial cells,[20,40] smooth muscle cells,[41,42] and fibroblasts,[38] as well as inflammatory cells as part of inflammatory processes such as macrophages,[38] B cells,[43] and monocytes,[38] which may infiltrate injured bladder tissues in significant numbers. Moreover, inflammatory changes in the bladder have been shown to increase the expression of CD74 and CD44 in the urothelium.[26,40] These changes facilitate the proinflammatory effects of MIF, as demonstrated by increased activation and phosphorylation of the downstream ERK-1/2 pathway,[40] leading to the upregulation of other inflammatory mediators.[23]

3.5 *Urinary tract infections and MIF-protein complexes*

Urinary tract infections (UTIs) represent the most common urologic disorder among men and women and are associated with significant financial and personal cost burdens.[44,45] UTIs vary in pathogenesis and severity. However, most are caused by ascending colonization of colonic bacteria into the urinary tract, with the most common organism being *E. coli*.[46] Gram-negative bacteria such as *E. coli* contain lipopolysaccharide (LPS) in their outer membrane. LPS is an endotoxin that potently activates the innate immune response by binding to the CD14/TLR4/MD2 receptor complex, leading to a release of proinflammatory cytokines.[47,48]

MIF has been shown to play a role in the development of bladder inflammation in animal models of induced cystitis,[23] with experimentally induced bladder inflammation resulting in increased expression and secretion of MIF into the bladder lumen.[23] This occurs alongside an upregulation in CD74/CD44 MIF receptor expression levels, resulting in increased MIF signaling and increased production of proinflammatory cytokines.[49]

Consistent with this is the finding that MIF levels in the urine of patients with UTIs are significantly higher than those without a urinary tract infection.[25] A more in-depth analysis reveals that MIF exists in a complex with other proteins in the urine including ceruloplasmin,[25] albumin,[25] uromodulin,[25] and α1-inhibitor-3[25,50] (a member of the α-2 macroglobin proteinase inhibitor family). α1-inhibitor-3 and the MIF-α1-inhibitor-3 complex are present in bladder interstitium, are released into the bladder lumen following a substance P-mediated inflammatory challenge, and then bind to urothelial cells.[51] Following substance P administration, glucose regulated protein 78 (GRP78), which functions as a receptor for both α-α1-inhibitor-3 and MIF-α1-inhibitor-3 complexes, is expressed on the surface of urothelial cells.[52] Blocking CD74 and/or GRP78 prevented substance P-mediated inflammatory events in the bladder.[52]

3.6 *Cellular effects of MIF*

MIF was originally discovered as a soluble T cell factor responsible for inhibiting random migration of guinea pig peritoneal macrophages.[8,9] Since then, many other effects of MIF have been identified. Among many others, these include its proinflammatory effects on macrophages[53] and proliferative effects on fibroblasts.[39] Unfortunately, few studies have evaluated the effects of MIF on cells derived from the lower urinary tract. The origin and nature of infiltrating macrophages most likely does not vary between different tissues. Nevertheless, their behavior may be profoundly influenced by different local tissue environments. Moreover, the lineage or behavior of bladder fibroblasts or bladder smooth muscle cells may be quite different from similar cells found in other tissues. For example, vascular smooth muscles within the proximal aorta are derived from a neural crest lineage, while those within abdominal or pelvic aorta are mesenchymal in origin.[54] As a result of this simple lineage difference, these two categories of vascular muscle cells have been shown to demonstrate important phenotypic differences,[54–57] including differential responses to growth factors such as TGF-β.[55] Thus, once MIF effects are demonstrated in primary cells or a cell line obtained from another tissue, such observations need to be ultimately confirmed in such cells obtained from the bladder.

3.6.1 *Inflammation*

MIF is well known for its role in promoting inflammatory processes.[58,59] Early studies identified MIF as a pituitary-derived cytokine involved in promoting lethal endotoxemia.[10] Subsequent studies found the macrophage to represent an important source of MIF and demonstrated that the proinflammatory cytokines TNF-α and IFN-γ, as well as LPS, promoted the release of MIF from these cells.[11]

MIF exerts proinflammatory effects on the macrophage. These effects are mediated through the binding of MIF to the CD74/CD44 receptor complex. This in turn promotes activation of the ERK-1/2 MAP kinase signaling pathway, leading to inhibition of p53-induced NO apoptosis.[53] Moreover, activation of this pathway ultimately upregulates the expression of several inflammatory mediators, including COX-2, IL-1β, and TNF-α. Taken together, MIF then promotes an inflammatory state by protecting the macrophage from NO-induced apoptosis while promoting the upregulation and release of several proinflammatory cytokines.[53]

3.6.2 *Fibroblast survival and fibrosis*

MIF also promotes fibroblast proliferation and survival[39] and has been implicated in promoting increased collagen deposition in the setting of renal focal glomerular sclerosis[60] and partial bladder outlet obstruction.[19] Studies investigating the effects of MIF on NIH3T3 fibroblasts observed that MIF stimulates the proliferation of these cells through activation of the Ras-Raf-ERK1/2 MAP kinase signaling pathway and that this activation was protein kinase A dependent.[39]

3.6.3 *Smooth muscle cell survival*

MIF has been found to promote apoptosis and decrease survival in primary bladder smooth muscle cultures.[19] Bladders from adult *Mif* $^{-/-}$ mice weigh nearly 37% more than bladders obtained from wildtype littermates, a change which can be attributed to increases in both muscle and collagen mass.[19] Not only do these findings provide *in vivo* evidence in support of the hypothesis that MIF reduces bladder smooth muscle survival, but the bladder also represents one of the few tissues that exhibit a phenotype in *Mif* $^{-/-}$ mice studied under basal conditions.

Partial bladder outlet obstruction (PBOO) is a common consequence of prolonged benign prostatic hyperplasia (BPH) in older men. In animal models, PBOO has been shown to be associated with three sequential phases — muscle hypertrophy, compensation and decompensation involving muscle loss and fibrosis.[61] In a recent study, adult female mice were studied at a 3-week time point when evidence of both muscle hypertrophy and fibrosis was evident.[19] Unbiased quantitative morphometry demonstrated a 22% decrease in numbers of nucleated smooth muscle cells and a 79% increase in collagen.[19] In mice rendered null for the *Mif* gene, PBOO did not result in significant changes in muscle or collagen, providing *in vivo* evidence in support of MIF's antisurvival effect on bladder smooth muscle cells, as well as a contrasting prosurvival effect on bladder fibroblasts.[19]

In contrast to the well-established effects of MIF on macrophage and fibroblast survival, its effects on smooth muscle cells in other tissues vary. For example, inhibition of MIF in an atherogenic mouse model of restenosis resulted in a reduced macrophage/foam cell content and an increased smooth muscle cell content.[62] In contrast, administration of the MIF inhibitor ISO-1 significantly diminished the development of ovalbumin-mediated asthma, pulmonary inflammation, and increased airway smooth muscle thickness.[63]

3.6.4 *Nerve cell survival*

MIF protein is expressed in many nerve cells located in peripheral ganglia and the spinal cord that innervate the bladder and other lower urinary tract structures.[22] However, the expression of the two MIF receptors (CD74 and CD44) or the effects of MIF on the survival or function of these peripheral nerve cells is not well understood. In addition to its role in microglial function, CD74 has been localized to neurofibrillary tangles and neurons in cases of Alzheimer's disease[64] with evidence of CD74 upregulation in the spinal cord in a model of neuropathic pain.[65] The CD44 receptor is expressed at the neuromuscular junction[66] and may play a role in neural responses following axotomy[67] or seizure[68] injury.

Spinal cord injury is attenuated and functional recovery enhanced in mice rendered null for the *Mif* gene.[69] At the same time, MIF promotes cell death and aggravates a functional deficit in an ischemic stroke model.[70] Thus, MIF may contribute to neuronal death and dysfunction in lower urinary tract models.

4. Potential Roles of MIF in Diagnosis and Management

A growing appreciation of the role played by MIF and related pathways in lower urinary tract development, physiology, and pathology has led to the hope that MIF-based approaches could be used to improve clinicians' capacities to diagnose or manage common conditions involving the bladder, prostate, and associated structures. Ability to easily measure MIF protein levels in the urine may offer important insights into the presence of inflammatory and other disease processes involving the kidney or bladder. Functional polymorphisms in the human *MIF* gene promoter[71] could also offer insights into the risk of developing a MIF-associated lower urinary tract condition or identify those individuals whose clinical outcome is likely to be worse. Finally, the therapeutic use of MIF antagonists[72,73] may alter the natural history and clinical outcomes of such conditions.

4.1 *Urinary tract infection*

Adults with lower urinary tract infection (UTI or cystitis) have been shown to have nearly four-fold higher MIF levels (adjusted for urinary creatinine) in the urine when compared to controls.[25] Since MIF can be detected even in the urine from individuals without infection, it is unclear whether urine MIF can distinguish between individuals with UTI and those without any infection or between individuals with UTI and those with asymptomatic bacteruria.[74] Nevertheless, in children with an elevated urinary MIF-to-creatinine ratio was reported as being able to differentiate pyelonephritis (renal infection) from cystitis (bladder infection) and to also identify individuals at high risk of permanent renal damage.[75]

Most individuals with acute UTIs respond rapidly to antibiotic therapy and MIF antagonists would not seem relevant. In contrast, in subjects with chronic and recurrent cystitis, MIF antagonism may be proven to be helpful especially if associated with urinary retention resulting from detrusor underactivity.[76] In these individuals, increased MIF may mediate a vicious cycle driving the relationship between urinary tract infection, inflammation, detrusor underactivity, and urinary retention.[76] Gram-negative bacteria and lipopolysaccharides promote MIF bladder synthesis and release,[18,25] MIF mediates bladder muscle loss and fibrosis,[19] and these changes may contribute to the development of detrusor underactivity.[77] Some patients with detrusor underactivity suffer urinary retention (even if partial), potentially raising the risk of chronic UTI while also impeding its treatment.[76]

4.2 *Bladder outlet obstruction and detrusor underactivity*

Benign prostatic hyperplasia (BPH) is a common condition that produces irritative lower urinary tract symptoms in many older men.[78] In a significant subset, the condition progresses into partial or complete bladder outlet obstruction.[78] For example, in the Olmsted County study the 5-year risk of acute urinary retention was 10% for men aged 70–79 with BPH.[79] Detrusor decompensation resulting from prolonged PBOO has been associated with the development of bladder muscle degeneration and fibrosis.[61,80] The observation that MIF is implicated in the development of muscle cell death and fibrosis in the setting of PBOO raises potential diagnostic and therapeutic roles for MIF. First, men with BPH and higher expressing *MIF* gene polymorphisms may be at greater risk of developing partial or complete urinary retention. Thus, such men may perhaps need to be considered for earlier and more aggressive BPH management (e.g., surgery), with less aggressive approaches (e.g., "watchful waiting" or medications) being proposed to the others.[78] Second, MIF antagonists may prove to be helpful in slowing or preventing the progression to complete retention in such higher risk men.

4.3 *Bladder cancer*

The vast majority of bladder cancers are epithelial in origin, with invasion of detrusor muscle representing a poor prognostic sign.[7] Human bladder cancer (HT-1736) cells express MIF, and strategies designed to decrease MIF expression in these cells resulted in decreased proliferation.[20] When bladder cancer was induced in mice using N-butyl-N-4-hydroxybutyl-nitrosamine (BBN), bladder cancer with invasion of muscle was associated with elevated *Mif* mRNA levels and was less common in $Mif^{-/-}$ mice, suggesting that MIF plays a role in the progression to invasive bladder cancer.[81] Moreover, serum MIF levels were elevated in bladder cancer patients, while circulating MIF-antithrombin III complexes were lower, suggesting that levels of bioactive or free MIF were higher in patients with bladder cancer.[82] It remains to be seen whether elevations in bioavailable serum MIF or higher expressing *MIF* gene polymorphisms are associated with a greater risk of developing bladder cancer or its progression to the stage of muscle invasion. Also, the role of MIF antagonists in preventing the development of bladder cancer or its dissemination via muscle invasion also needs to be evaluated in preclinical and clinical studies.

References

1. Andersson KE, Hedlund P. (2002) Pharmacologic perspective on the physiology of the lower urinary tract. *Urology* **60**: 13–20.
2. Birder LA. (2006) Urinary bladder urothelium: Molecular sensors of chemical/thermal/mechanical stimuli. *Vascul Pharmacol* **45**: 221–226.
3. Birder LA *et al.* (2010) Neural control of the lower urinary tract: Peripheral and spinal mechanisms. *Neurourol Urodyn* **29**: 128–139.

4. Fowler CJ, Griffiths DJ. (2010) A decade of functional brain imaging applied to bladder control. *Neurourol Urodyn* **29**: 49–55.
5. McCloskey KD. (2010) Interstitial cells in the urinary bladder — Localization and function. *Neurourol Urodyn* **29**: 82–87.
6. Shin K *et al.* (2011) Hedgehog/Wnt feedback supports regenerative proliferation of epithelial stem cells in bladder. *Nature* **472**: 110–114.
7. Taylor JA III, Kuchel GA. (2009) Bladder cancer in the elderly: Clinical outcomes, basic mechanisms, and future research direction. *Nat Clin Pract Urol* **6**: 135–144.
8. Bloom BR, Bennett B. (1966) Mechanism of a reaction *in vitro* associated with delayed-type hypersensitivity. *Science* **153**: 80–82.
9. David JR. (1966) Delayed hypersensitivity *in vitro*: Its mediation by cell-free substances formed by lymphoid cell-antigen interaction. *Proc Natl Acad Sci USA* **56**: 72–77.
10. Bernhagen J *et al.* (1993) MIF is a pituitary-derived cytokine that potentiates lethal endotoxaemia. *Nature* **365**: 756–759.
11. Calandra T, Bernhagen J, Mitchell RA, Bucala R. (1994) The macrophage is an important and previously unrecognized source of macrophage migration inhibitory factor. *J Exp Med* **179**: 1895–1902.
12. Ashcroft GS *et al.* (2003) Estrogen modulates cutaneous wound healing by downregulating macrophage migration inhibitory factor. *J Clin Invest* **111**: 1309–1318.
13. Bacher M *et al.* (1998) MIF expression in the rat brain: Implications for neuronal function. *Mol Med* **4**: 217–230.
14. Liao H, Bucala R, Mitchell RA. (2003) Adhesion-dependent signaling by macrophage migration inhibitory factor (MIF). *J Biol Chem* **278**: 76–81.
15. Hirokawa J *et al.* (1997) Identification of macrophage migration inhibitory factor in adipose tissue and its induction by tumor necrosis factor-alpha. *Biochem Biophys Res Commun* **235**: 94–98.
16. Benigni F *et al.* (2000) The proinflammatory mediator macrophage migration inhibitory factor induces glucose catabolism in muscle. *J Clin Invest* **106**: 1291–1300.
17. Garner LB *et al.* (2003) Macrophage migration inhibitory factor (MIF) is a cardiac-derived myocardial depressant factor. *Am J Physiol Heart Circ Physiol* **285**: H2500–H2509.
18. Meyer-Siegler KL, Ordorica RC, Vera PL. (2004) Macrophage migration inhibitory factor is upregulated in an endotoxin-induced model of bladder inflammation in rats. *J Interferon Cytokine Res* **24**: 55–63.
19. Taylor JA *et al.* (2006) Null mutation in macrophage migration inhibitory factor (MIF) prevents muscle cell loss and fibrosis in partial bladder outlet obstruction. *Am J Physiol* **291**: F1343–F1353.

20. Meyer-Siegler KL, Leifheit EC, Vera PL. (2004) Inhibition of macrophage migration inhibitory factor decreases proliferation and cytokine expression in bladder cancer cells. *BMC Cancer* **4**: 34.
21. Vera PL, Ordorica RC, Meyer-Siegler KL. (2003) Hydrochloric acid induced changes in macrophage migration inhibitory factor in the bladder, peripheral and central nervous system of the rat. *J Urol* **170**: 623–627.
22. Vera PL, Meyer-Siegler KL. (2003) Anatomical location of macrophage migration inhibitory factor in urogenital tissues, peripheral ganglia and lumbosacral spinal cord of the rat. *BMC Neurosci.* **4**: 17.
23. Meyer-Siegler KL, Vera PL. (2004) Substance P induced release of macrophage migration inhibitory factor from rat bladder epithelium. *J Urol* **171**: 1698–1703.
24. Vera PL, Meyer-Siegler KL. (2004) Inflammation of the rat prostate evokes release of macrophage migration inhibitory factor in the bladder: Evidence for a viscerovisceral reflex. *J Urol* **172**: 2440–2445.
25. Meyer-Siegler KL, Iczkowski KA, Vera PL. (2006) Macrophage migration inhibitory factor is increased in the urine of patients with urinary tract infection: Macrophage migration inhibitory factor-protein complexes in human urine. *J Urol* **175**: 1523–1528.
26. Vera PL, Wang X, Meyer-Siegler KL. (2008) Upregulation of macrophage migration inhibitory factor (MIF) and CD74, receptor for MIF, in rat bladder during persistent cyclophosphamide-induced inflammation. *Exp Biol Med (Maywood)* **233**: 620–626.
27. Vera PL, Wang X, Meyer-Siegler KL. (2008) Neural control of substance P induced up-regulation and release of macrophage migration inhibitory factor in the rat bladder. *J Urol* **180**: 373–378.
28. Nishihira J, Koyama Y, Mizue Y. (1998) Identification of macrophage migration inhibitory factor (MIF) in human vascular endothelial cells and its induction by lipopolysaccharide. *Cytokine* **10**: 199–205.
29. Vera PL, Wolfe TE, Braley AE, Meyer-Siegler KL. (2010) Thrombin induces macrophage migration inhibitory factor release and upregulation in urothelium: A possible contribution to bladder inflammation. *PLoS One* **5**: e15904.
30. Calandra T *et al.* (1995) MIF as a glucocorticoid-induced modulator of cytokine production. *Nature* **377**: 68–71.
31. Nishino T *et al.* (1995) Localization of macrophage migration inhibitory factor (MIF) to secretory granules within the corticotrophic and thyrotrophic cells of the pituitary gland. *Mol Med* **1**: 781–788.
32. Rice EK *et al.* (2003) Induction of MIF synthesis and secretion by tubular epithelial cells: A novel action of angiotensin II. *Kidney Int* **63**: 1265–1275.
33. Flieger O *et al.* (2003) Regulated secretion of macrophage migration inhibitory factor is mediated by a non-classical pathway involving an ABC transporter. *FEBS Lett* **551**: 78–86.

34. Roger T, David J, Glauser MP, Calandra T. (2001) MIF regulates innate immune responses through modulation of Toll-like receptor 4. *Nature* **414**: 920–924.
35. Onodera S *et al.* (2002) Macrophage migration inhibitory factor up-regulates matrix metalloproteinase-9 and -13 in rat osteoblasts. Relevance to intracellular signaling pathways. *J Biol Chem* **277**: 7865–7874.
36. Meyer-Siegler KL, Vera PL. (2004) Intraluminal antibodies to macrophage migration inhibitory factor decrease substance P induced inflammatory changes in the rat bladder and prostate. *J Urol* **172**: 1504–1509.
37. Leng L *et al.* (2003) MIF signal transduction initiated by binding to CD74. *J Exp Med* **197**: 1467–1476.
38. Shi X *et al.* (2006) CD44 is the signaling component of the macrophage migration inhibitory factor-CD74 receptor complex. *Immunity* **25**: 595–606.
39. Mitchell RA, Metz CN, Peng T, Bucala R. (1999) Sustained mitogen-activated protein kinase (MAPK) and cytoplasmic phospholipase A2 activation by macrophage migration inhibitory factor (MIF). Regulatory role in cell proliferation and glucocorticoid action. *J Biol Chem* **274**: 18100–18106.
40. Meyer-Siegler KL, Vera PL. (2005) Substance P induced changes in CD74 and CD44 in the rat bladder. *J Urol* **173**: 615–620.
41. Martin-Ventura JL *et al.* (2009) Increased CD74 expression in human atherosclerotic plaques: Contribution to inflammatory responses in vascular cells. *Cardiovasc Res* **83**: 586–594.
42. Arafat HA, Wein AJ, Chacko S. (2002) Osteopontin gene expression and immunolocalization in the rabbit urinary tract. *J Urol* **167**: 746–752.
43. Gore Y *et al.* (2008) Macrophage migration inhibitory factor induces B cell survival by activation of a CD74-CD44 receptor complex. *J Biol Chem* **283**: 2784–2792.
44. Griebling TL. (2005) Urologic diseases in America project: Trends in resource use for urinary tract infections in men. *J Urol* **173**: 1288–1294.
45. Griebling TL. (2005) Urologic diseases in America project: Trends in resource use for urinary tract infections in women. *J Urol* **173**: 1281–1287.
46. Nicolle LE. (2008) Uncomplicated urinary tract infection in adults including uncomplicated pyelonephritis. *Urol Clin North Am* **35**: 1–12, v.
47. Olsson LE, Wheeler MA, Sessa WC, Weiss, RM. (1998) Bladder instillation and intraperitoneal injection of *Escherichia coli* lipopolysaccharide up-regulate cytokines and iNOS in rat urinary bladder. *J Pharmacol Exp Ther* **284**: 1203–1208.
48. Song J, Abraham SN. (2008) TLR-mediated immune responses in the urinary tract. *Curr Opin Microbiol* **11**: 66–73.
49. Chapple C, Khullar V, Gabriel Z, Dooley JA. (2005) The effects of antimuscarinic treatments in overactive bladder: A systematic review and meta-analysis. *Eur Urol* **48**: 5–26.

50. Vera PL, Iczkowski KA, Leng L *et al.* (2005) Macrophage migration inhibitory factor is released as a complex with alpha1-inhibitor-3 in the intraluminal fluid during bladder inflammation in the rat. *J Urol* **174**: 338–343.
51. Vera PL, Meyer-Siegler KL. (2006) Substance P induces localization of MIF/alpha1-inhibitor-3 complexes to umbrella cells via paracellular transit through the urothelium in the rat bladder. *BMC Urol* **6**: 24.
52. Vera PL, Wang X, Bucala RJ, Meyer-Siegler KL. (2009) Intraluminal blockade of cell-surface CD74 and glucose regulated protein 78 prevents substance P-induced bladder inflammatory changes in the rat. *PLoS One* **4**: e5835.
53. Mitchell RA *et al.* (2002) Macrophage migration inhibitory factor (MIF) sustains macrophage proinflammatory function by inhibiting p53: Regulatory role in the innate immune response. *Proc Natl Acad Sci USA* **99**: 345–350.
54. Topouzis S, Catravas JD, Ryan JW, Rosenquist TH. (1992) Influence of vascular smooth muscle heterogeneity on angiotensin converting enzyme activity in chicken embryonic aorta and in endothelial cells in culture. *Circ Res* **71**: 923–931.
55. Gadson PF Jr *et al.* (1997) Differential response of mesoderm- and neural crest-derived smooth muscle to TGF-beta1: Regulation of c-myb and alpha1 (I) procollagen genes. *Exp Cell Res* **230**: 169–180.
56. Thieszen SL, Dalton M, Gadson PF *et al.* (1996) Embryonic lineage of vascular smooth muscle cells determines responses to collagen matrices and integrin receptor expression. *Exp Cell Res* **227**: 135–145.
57. Leroux-Berger M *et al.* (2011) Pathological calcification of adult vascular smooth muscle cells differs upon their crest or mesodermal embryonic origin. *J Bone Miner Res* **26**(7): 1543–1553.
58. Bozza M *et al.* (1999) Targeted disruption of migration inhibitory factor gene reveals its critical role in sepsis. *J Exp Med* **189**: 341–346.
59. Calandra T *et al.* (2000) Protection from septic shock by neutralization of macrophage migration inhibitory factor. *Nat Med* **6**: 164–170.
60. Matsumoto K *et al.* (2005) Elevated macrophage migration inhibitory factor (MIF) levels in the urine of patients with focal glomerular sclerosis. *Clin Exp Immunol* **139**: 338–347.
61. Buttyan R, Chen MW, Levin RM. (1997) Animal models of bladder outlet obstruction and molecular insights into the basis for the development of bladder dysfunction. *Eur Urol* **32**(Suppl 1): 32–39.
62. Schober A *et al.* (2004) Stabilization of atherosclerotic plaques by blockade of macrophage migration inhibitory factor after vascular injury in apolipoprotein E-deficient mice. *Circulation* **109**: 380–385.
63. Chen PF *et al.* (2010) ISO-1, a macrophage migration inhibitory factor antagonist, inhibits airway remodeling in a murine model of chronic asthma. *Mol Med* **16**: 400–408.

64. Bryan KJ *et al.* (2008) Expression of CD74 is increased in neurofibrillary tangles in Alzheimer's disease. *Mol Neurodegener* **3**: 13.
65. Wang F *et al.* (2011) Spinal macrophage migration inhibitory factor is a major contributor to rodent neuropathic pain-like hypersensitivity. *Anesthesiology* **114**: 643–659.
66. Gorlewicz A *et al.* (2009) CD44 is expressed in non-myelinating Schwann cells of the adult rat, and may play a role in neurodegeneration-induced glial plasticity at the neuromuscular junction. *Neurobiol Dis* **34**: 245–258.
67. Makwana M *et al.* (2010) Peripheral facial nerve axotomy in mice causes sprouting of motor axons into perineuronal central white matter: Time course and molecular characterization. *J Comp Neurol* **518**: 699–721.
68. Borges K, McDermott DL, Dingledine R. (2004) Reciprocal changes of CD44 and GAP-43 expression in the dentate gyrus inner molecular layer after status epilepticus in mice. *Exp Neurol* **188**: 1–10.
69. Nishio Y. *et al.* (2009) Deletion of macrophage migration inhibitory factor attenuates neuronal death and promotes functional recovery after compression-induced spinal cord injury in mice. *Acta Neuropathol* **117**: 321–328.
70. Inacio AR, Ruscher K, Leng L *et al.* (2011) Macrophage migration inhibitory factor promotes cell death and aggravates neurologic deficits after experimental stroke. *J Cereb Blood Flow Metab* **31**: 1093–1106.
71. Gregersen PK, Bucala R. (2003) Macrophage migration inhibitory factor, MIF alleles, and the genetics of inflammatory disorders: Incorporating disease outcome into the definition of phenotype. *Arthritis Rheum* **48**: 1171–1176.
72. Morand EF, Leech M, Bernhagen J. (2006) MIF: A new cytokine link between rheumatoid arthritis and atherosclerosis. *Nat Rev Drug Discov* **5**: 399–410.
73. Greven D, Leng L, Bucala R. (2010) Autoimmune diseases: MIF as a therapeutic target. *Expert Opin Ther Targets* **14**: 253–264.
74. Juthani-Mehta M. (2007) Asymptomatic bacteriuria and urinary tract infection in older adults. *Clin Geriatr Med* **23**: 585–594, vii.
75. Otukesh H *et al.* (2009) Urine macrophage migration inhibitory factor (MIF) in children with urinary tract infection: A possible predictor of acute pyelonephritis. *Pediatr Nephrol* **24**: 105–111.
76. Taylor JA, Kuchel GA. (2006) Detrusor underactivity: Clinical features and pathogenesis of an underdiagnosed geriatric condition. *J Am Geriatr Soc* **54**: 1920–1933.
77. Elbadawi A, Yalla SV, Resnick NM. (1993) Structural basis of geriatric voiding dysfunction. II. Aging detrusor: Normal versus impaired contractility. *J Urol* **150**: 1657–1667.
78. Roehrborn CG. (2011) Male lower urinary tract symptoms (LUTS) and benign prostatic hyperplasia (BPH). *Med Clin North Am* **95**: 87–100.

79. Jacobsen SJ *et al.* (1997) Natural history of prostatism: Risk factors for acute urinary retention. *J Urol* **158**: 481–487.
80. Elbadawi A, Yalla SV, Resnick NM. (1993) Structural basis of geriatric voiding dysfunction. IV. Bladder outlet obstruction. *J Urol* **150**: 1681–1695.
81. Taylor JA III *et al.* (2007) Null mutation for macrophage migration inhibitory factor (MIF) is associated with less aggressive bladder cancer in mice. *BMC Cancer* **7**: 135.
82. Meyer-Siegler KL, Cox J, Leng L *et al.* (2010) Macrophage migration inhibitory factor anti-thrombin III complexes are decreased in bladder cancer patient serum: Complex formation as a mechanism of inactivation. *Cancer Lett* **290**: 49–57.
83. Vera PL, Iczkowski KA, Wang X, Meyer-Siegler KL. (2008) Cyclophosphamide-induced cystitis increases bladder CXCR4 expression and CXCR4-macrophage migration inhibitory factor association. *PLoS One* **3**: e3898.

PART IV

Neoplasia

IV-1

MIF in the Pathogenesis of Urological Cancer

Katherine L. Meyer-Siegler[*,†,‡] and Pedro L. Vera[†,§]

1. Introduction

Emergent evidence suggests that the proinflammatory cytokine macrophage migration inhibitory factor (MIF) may serve as an important link between chronic inflammation and carcinogenesis as evidenced by the increase in serum MIF found in patients with various cancer types. MIF functional promoter polymorphisms are associated with prostate cancer recurrence. Prostate cancer patients with the −173 C/7 CATT carrier genotype had an increased risk of prostate recurrence at five years. Thus, these genotypes may serve as an independent prognostic marker. Aggressive urological cancer cells cultured *in vitro* require MIF activated signal transduction pathways for growth. Thus, blocking MIF either at the ligand (MIF) or receptor (CD74) may provide new, targeted specific therapies for urological cancer treatments.

Many studies support the paradigm that current or persistent inflammation plays a role in cancer development.[1–3] Inflammation-induced events such as growth factor production, release of reactive oxygen species, and activation of signal transduction pathways that promote cell survival and proliferation are considered risk factors for cancer development and progression.[4,5] Epidemiological support of this paradigm includes the observed relationship between Barrett's esophagus and esophageal cancer, chronic pancreatitis and pancreatic cancer, cirrhosis and liver cancer, and inflammatory bowel disease and colon carcinoma.[6–10] Inflammation is frequently observed in prostate biopsies and radical prostatectomy specimens, and is often a feature

*Corresponding author: The Bay Pines VA Healthcare System, Research and Development, 10000 Bay Pines Blvd., Bay Pines, FL 33744, USA. Email: katherine.siegler@va.gov
†The Bay Pines VA Healthcare System, Research and Development, Bay Pines, FL, USA.
‡University of South Florida, Department of Molecular Medicine, Tampa, FL, USA.
§University of South Florida, Department of Surgery, Urology Division, Tampa, FL, USA.

of atrophic foci that exhibit a characteristic increased proliferative index.[3,11] Bladder cancer is often associated with exposure to known chemicals that are irritants to the urinary tract.[12–14]

Studies over the last two decades have suggested that macrophage migration inhibitory factor (MIF) functions as a link between inflammation and cancer.[15] MIF is a distinctive cytokine in this regard not only because it acts as a key initiator of the innate immune response but because it exists pre-formed within cytoplasmic pools such that inflammatory insult can result in rapid MIF release and subsequent tissue response.[16] MIF is secreted not only by immune cells but also by the anterior pituitary gland and most other cell types, including epithelial cells.[17] Significant MIF amounts can be found in the blood, urine, cerebrospinal, and other body fluids.[18] In general, cells that facilitate acute inflammation in response to pathogens and other proinflammatory stimuli, produce, secrete, and respond to MIF.

Secreted MIF is a cognate ligand for its receptor cluster of differentiation-74 (CD74), which is expressed mainly by major histocompatibility complex class II positive cells, namely antigen-presenting cells such as macrophages, lymphocytes, dendritic cells, and endothelial cells.[19] Since the discovery of CD74 as the MIF receptor, it has been localized to the surface of the various cell types during inflammation and found to be upregulated on the surface of various cancer cells, providing a potential mechanism for these cells to respond to secreted MIF in an autocrine manner. CD74-MIF interaction thereby stimulates the expression and secretion of the proinflammatory cytokines tumor necrosis factor-alpha (TNF-α), interferon-gamma (IFN-γ), interleukin-1beta (IL-1β), interleukin-6 (IL-6), and interleukin-8 (IL-8), in addition to macrophage inflammatory protein-2 (MIP-2), cyclooxygenase-2 (COX-2), nitric oxide, and products of the arachidonic acid pathway.[20]

Various proinflammatory molecules stimulate MIF release, including lipopolysaccharide and thrombin.[21,22] Thrombin can affect tissues by activating a novel family of G protein-coupled proteinase activated receptors (PARs) by exposing a "tethered" receptor-triggering ligand.[23] PARs may play roles in many settings, including cancer, arthritis, asthma, inflammatory bowel disease, neurodegeneration, and cardiovascular disease, as well as in pathogen-induced inflammation.[24] Surprisingly, thrombin induces the secretion of MIF, which is required for activation of extracellular signal-related kinase-1/2 (ERK-1/2) via binding to newly expressed CD74 on the cell surface.[22]

With regard to signal transduction, MIF has two unusual effects: sustained ERK1/2 mitogen-activated protein kinase activation, which is a feature of transformation by oncogenic Ras mutation, as well as antagonism of the growth arrest and proapoptotic actions of p53 tumor suppressor.[25,26] The unique spectrum of MIF biological activities, especially its role in regulating tumor suppressor genes, angiogenesis, and cell senescence all support this cytokine as an important link between chronic inflammation and the development of cancer.

MIF activity is associated with inflammation, tumor cell proliferation, and cytokine production as exemplified by studies on glioblastoma, lymphoma, melanoma, prostate, breast, bladder, colon, lung, liver, ovarian, and gastric cancers.[15] More recent research in our laboratory has focused on the control of MIF release and activity with antithrombin III (ATIII) identified as an endogenous biological regulator of MIF activity.[27] The biological activities of both MIF and ATIII are blocked by this interaction, suggesting that binding of MIF to ATIII attenuates MIF bioactivity and may reduce MIF-dependent proinflammatory responses. Therefore, the binding of MIF to endogenous molecules may be an important feature in the control of MIF activity *in vivo* including its role in carcinogenesis.

Taking into account all of these features, a model for MIF in inflammation and carcinogensis can be proposed (Fig. 1). Normal epithelial cells secrete basal amounts of MIF. However, since CD74 is not available at the cell surface, little response to the MIF occurs. Under an acute inflammatory signal, CD74 is expressed on the cell surface, and thrombin entering from the exudate activates PAR receptors inducing an increase in released MIF and the subsequent binding of MIF to CD74. This results in sustained activation of ERK-1/2 generating nuclear factor kappa-light-chain-enhancer of B cells (NF-κB)-dependent transcription and the production of additional proinflammatory molecules. Eventually, the active thrombin and MIF remaining in the tissue is inactivated by complex formation with ATIII, resulting in decreased epithelial cell MIF secretion. Under normal conditions, MIF and thrombin activity are attenuated by ATIII and the cells return to their preinflammatory state (Fig. 1). Development of chronic inflammation impels a continuous supply of active thrombin, which drives the secretion of MIF. CD74 continues to be expressed on the cell surface resulting in continued ERK-1/2 activation and NF-κB-driven gene transcription. The amounts of ATIII present within the tissue are no longer sufficient to attenuate the inflammatory reaction. Highly invasive malignant cells will progress to continually express cell surface CD74 and additional changes that eventually result in dependency upon MIF for cell proliferation.

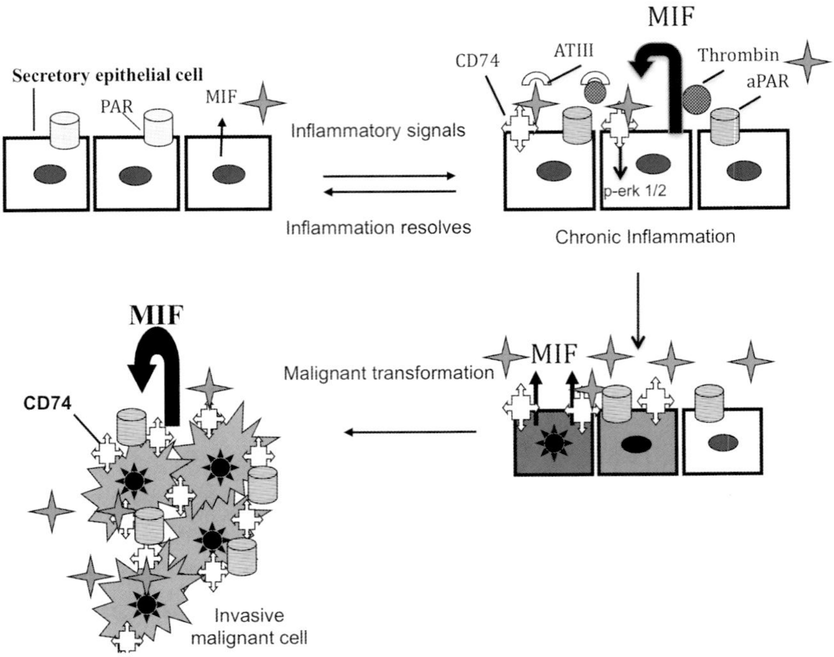

Fig. 1. A model for MIF in inflammation and carcinogenesis. Normal secretory epithelial cells synthesize and secrete basal MIF. During the inflammatory response, MIF expression and secretion are upregulated and PAR becomes activated (aPAR). Concomitantly cell surface expression of CD74 occurs. Binding of MIF to CD74 promotes cell growth and the production of other inflammatory molecules. Chronic inflammation provides stimuli to premalignant cells. Carcinogenesis results in cells that secrete increasing amounts of MIF. Highly invasive malignant cells progress to continually express cell surface CD74 resulting in dependency upon MIF for cell proliferation.

2. MIF as a Marker of Urological Cancer

One obstacle to the successful management of urological cancer treatment is associated with the extreme variation in cancer progression; some prostate and bladder cancers remain localized, whereas others metastasize quickly.[28,29] In addition, the current curative and palliative urological cancer treatments, which include radical prostatectomy, orchiectomy, and bladder resection, have significant deleterious effects on a patient's quality of life. Methods to delineate clinically significant tumors are required to clarify the risks and benefits to patients of the various treatment options, as well as to develop new clinical treatment strategies. Findings from numerous studies suggest a direct link between inflammation and urological carcinogenesis based upon clinical observations relating inflammatory lifestyles (smoking), exposure to repeated urinary irritants (chemotherapy treatments), multiple urinary tract infections, prostatitis, urinary stones, and indwelling catheters with bladder and prostate cancer.[30,31] These associations suggest that inflammatory cytokine amounts could be a relevant biomarker for cancer diagnosis. In addition, once-formed tumor cells continue to produce and utilize some of the signaling molecules of the inflammatory response, which aid in their invasion, migration and metastasis, suggesting that continued upregulation of inflammatory mediators might promote urological cancer progression.

2.1 *Increased MIF mRNA and protein in urological tumors*

An initial study undertaken by our laboratory compared the transcriptome (i.e., the complete collection of mRNAs) of benign prostate tissue to a prostate tumor lymph node metastasis.[32] These data revealed an upregulation in MIF mRNA in tumor compared with benign prostate tissue. Subsequent clinical studies demonstrated increased MIF protein within prostate tumor regions compared with normal control tissues from the same patient and these findings were subsequently confirmed by Del Vecchio.[33,34] Further studies documented an increase in CD74 staining intensity within the ductal epithelium with increases in CD74 staining associated with prostate cancer tissues (Fig. 2). The differential expression in normal compared with tumor tissue suggested that MIF and CD74 protein expression might provide useful diagnostic and/or prognostic information. We expanded upon our initial studies of prostate cancer to include bladder cancer patients by screening a bladder cancer tissue array for MIF staining intensity by immunohistochemistry. In agreement with the studies of prostate tumors, these studies also demonstrated an increase in MIF immunostaining in tumor tissue compared with histologically normal regions of the same tissue sample.[33]

2.2 *Serum MIF as a urological cancer biomarker*

Based upon the tumor immunostaining results, we proposed that a significant amount of the excess MIF produced by the tumors would be secreted and possibly reflected in elevated serum MIF amounts. In healthy human subjects, circulating MIF concentrations are on the order of 2–6 ng/ml with fluctuations associated with circadian rhythm.[35] Consistent with the concept established by analysis of tumor tissue, serum MIF amounts were significantly higher in prostate and bladder cancer patients than in controls. In general, prostate cancer patient serum MIF concentrations are 2.5 times that

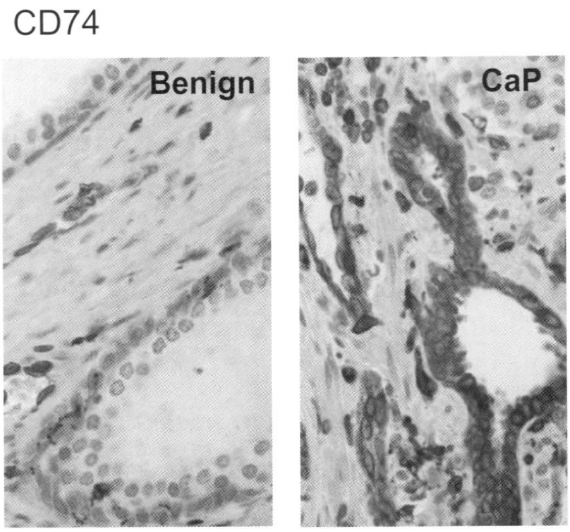

(a)

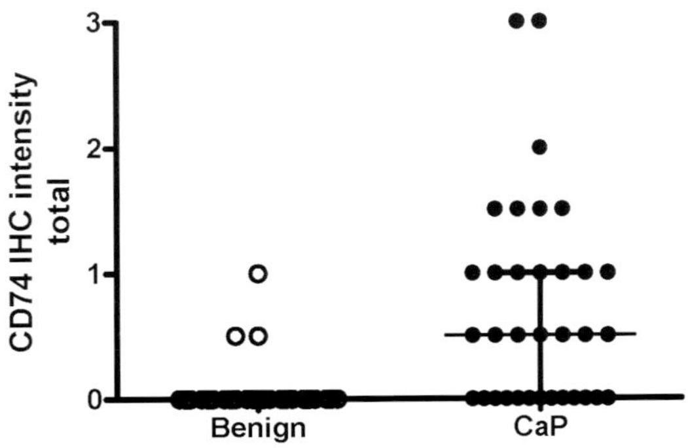

(b)

Fig. 2. CD74 immunostaining in benign and prostate cancer (CaP). (a) Prostate tissue arrays were stained for CD74 and staining intensity scored for the location and intensity of stain. Data shown are benign and CaP regions from the same prostate. (b) Total prostate staining scores were grouped according to benign and prostate cancer as defined by: 0 = no staining detected through 3 = very strong staining. There is a significant increase in CD74 epithelial staining in CaP compared with benign tissue ($p < 0.0001$; Wilcoxon signed rank test).

of age-matched controls (2.8 ± 0.3 ng/ml in controls, 6.9 ± 1.4 ng/ml prostate cancer patients; data expressed as mean plus/minus standard error, $p < 0.01$), while the serum MIF concentrations are 3.5 times higher than controls in bladder cancer patients (9.7 ± 1.1 ng/ml; data expressed as mean plus/minus standard error, $p < 0.001$). Investigators studying prostate as well as a variety other cancer types have confirmed these results.[27,36–39] These data suggest that serum MIF could serve as a generalized cancer biomarker.

2.3 *MIF* promoter polymorphisms as a marker of urological cancer

Various investigators are studying known genetic polymorphisms as disease biomarkers since these genetic variations within the DNA itself are not prone to the variability exhibited by serum protein determinations. It is likely that common polymorphisms (>1% frequency) within one or more pathways directly influence the expression of cancer. Often the expressed cancer phenotype is subtle or not apparent at all. Yet, in combination with and against a vulnerable background, the wrong set of variables could cumulatively undermine one or more biological functions and result in cancer.

Two known *MIF* promoter polymorphisms are functional in that they direct phenotypic variability in total MIF protein production.[40,41] Therefore, carriers of high expressing *MIF* polymorphisms would on average produce higher amounts of MIF protein compared with individuals carrying low expressing polymorphisms. These polymorphisms are persistent within the genome since individuals with high expressing *MIF* polymorphisms may be better able to control various infectious agents through efficient activation of the innate immune response. However, persistent *MIF* upregulation in the absence of infectious agents could result in a chronic inflammatory environment that may initiate and promote cancer development. Thus, the *MIF* polymorphisms may be an example of antagonistic pleiotropy, where beneficial mutations accumulated in one environment (ability to fight infection) prove deleterious in the second (chronic inflammation resulting in tumorigenesis). Analysis of *MIF* promoter polymorphisms defined a significant association between high producing *MIF* alleles (–173 C and/or –794 7-CATT repeat) and both the incidence and recurrence of prostate cancer.[42] Similar associations were found by other investigators studying prostate as well as numerous other cancer types.[43–47]

Based upon our preliminary data, *MIF* promoter mutations and loss of heterozygosity are uncommon in prostate cancer. Analyzed prostate cancer patients demonstrated the same genotype within tumor and benign regions of the prostate, as well as in DNA isolated from stored serum samples. These data strongly suggest that the observed genetic variability in the *MIF* promoter is present in germline cells.

Our studies and those of Ding report a significant association between the high expressing *MIF* polymorphisms and higher Gleason sum, which is the current pathological scaling system used to predict prostate cancer

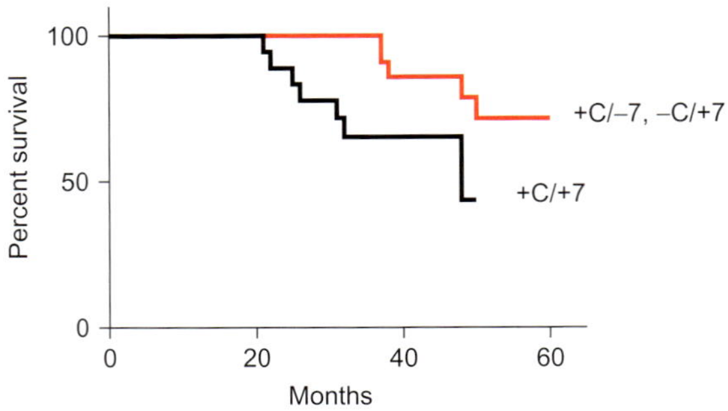

Fig. 3. Cox proportional hazard model estimates of remaining free from prostate specific antigen recurrence based upon macrophage migration inhibitory factor genotype. Biochemical recurrence was defined as two prostate specific antigen values of 0.4 ng/ml or greater and rising or the initiation of adjuvant therapy. Patients not experiencing recurrence by 60 months post-initial diagnosis were censored based upon the date of last observation. –173 C/7-CATT carrier genotype: 7 events, 11 censored versus remaining genotypes: 5 events, 30 censored. Adapted from Ref. 42.

recurrence.[42,44] The presence of high expressing *MIF* promoter polymorphisms (–173 C and/or –794, 7-CATT) appears to be prognostic and is independent of tumor Gleason evaluation. Patients (n = 18) with a –173 C/7-CATT carrier genotype had a significantly shorter time interval to prostate cancer recurrence compared with prostate cancer patients (n = 35) with any of the remaining genotypes (Fig. 4).[42] Based on the results of this preliminary study, determining *MIF* promoter genotype could prove useful as an early screening test for urological cancer prognosis.

Despite the strong correlation between elevated serum MIF and bladder or prostate cancer, serum MIF amounts alone have limited specificity in prostate or bladder cancer diagnosis, since elevated serum MIF is associated with several inflammatory conditions and other cancers. Additionally, serum MIF amounts within individuals are subject to diurnal variation, which would confound determination of baseline serum MIF amounts. The utility of these *MIF* polymorphisms in urological cancer prognosis awaits studies with larger patient populations.

3. MIF Inactivation via Binding to Endogenous Molecules

The findings of significant levels of MIF in blood, urine, and tissues suggest that circulating or extracellular MIF is readily present under basal ("normal") conditions.[18,35,48] Baseline levels of MIF are found in the serum and urine of normal patients without underlying inflammatory conditions. Several mechanisms could be proposed to explain why MIF is not proinflammatory under these circumstances: (i) Threshold levels are required to initiate proinflammatory signal transduction. (ii) CD74, the MIF receptor, is present in low amounts on cell surfaces under noninflammatory conditions, such that its availability to bind MIF is below the threshold required to initiate proinflammatory signaling. (iii) MIF exists in different forms, for example, active versus inactive. Binding to other proteins could regulate MIF activity.

We have previously identified MIF containing protein complexes in both urine and serum and therefore propose that the binding of MIF to endogenous molecules is an important feature in the control of *in vivo* MIF activity.[27,37,49,50] MIF was found to be bound to uromodulin, ceruloplasmin, and albumin in the urine of patients with urinary tract infections.[50]

We recently identified a biologically active MIF complex consisting of MIF and antithrombin III (ATIII) in human serum.[27] ATIII-MIF interaction abolishes MIF tautomerase activity. The precise role of this enzymatic activity has not been fully defined. However, blocking MIF tautomerase activity impairs MIF biological activity potentially by disrupting an important site required for MIF protein-protein interactions.[51] ATIII blocks MIF *in vitro* biological activity as determined by the loss of ERK-1/2 phosphorylation and decreased cell proliferation upon exogenous stimulation of mouse macrophages (RAW 264.7 cell line) with ATIII-MIF complexes.[27] Although the precise mechanism of ATIII-MIF interaction is not known, the loss of tautomerase activity suggests that the interaction of ATIII with MIF blocks a binding site required for MIF activity. These data provide evidence for our model, which proposes that shifts in the equilibrium between MIF, ATIII, and thrombin at the site of inflammation may regulate the local inflammatory response (Fig. 1). ATIII-MIF complexes may circulate in the blood to the liver for degradation without eliciting an inflammatory signal.

Interestingly, we found that the amount of MIF unbound to ATIII was increased in the serum of bladder cancer patients. By inference, bladder cancer patients should have more bioactive MIF. Therefore, ATIII may provide a potential therapy to cancer patients by reducing the availability of bioactive MIF.

4. MIF Production and Activity in Cultured Cells: Implications for Prostate and Bladder Cancer

Recent studies have demonstrated that thrombin can induce release of MIF from urothelial cells *in vitro*. Addition of thrombin to benign transformed

Table 1. Thrombin stimulated MIF secretion. Cells were exposed to thrombin with or without ISO-1 for the time intervals indicated. Secreted MIF was determined by ELISA and is expressed as ng/ml plus/minus standard error.

	UROtsa		RWPE-1	
	15 min	90 min	15 min	90 min
Control	44.1 ± 2.4	97.2 ± 5.0	22.3 ± 3.7	91.5 ± 4.4
ISO-1	42.8 ± 4.1	50.8 ± 4.9*	29.2 ± 6.1	42.6 ± 5.2*

*$p < 0.01$ compared to 90 min control.

urothelial cells (UROtsa) or prostate cells (RWPE-1) resulted in biphasic MIF release (Table 1). The initial release of MIF occurred within 15 minutes of thrombin exposure and could be initiated by the addition of a PAR-1 specific agonist (SFLLRN-NH_2).[52] Additional MIF was released 60 to 90 minutes later and was inhibited by treating the cells with the MIF specific antagonist (S,R)-3-(4-Hydroxyphenyl)-4,5-dihydro-5-isoazole acetic acid, methyl ester (ISO-1). The second peak of MIF release was preceded by an increase in MIF mRNA and was dependent upon the initial release of MIF induced by thrombin exposure, since ISO-1 treatment blocked both the second peak of MIF release into the culture medium and the upregulation of MIF mRNA.

The possible therapeutic value of MIF inhibition in cancer treatment has been established by studies that document decreased tumor cell proliferation and invasion with blocking of MIF.[53–57] Cancer cells have varying dependence upon MIF for survival, as evidenced by study of cell lines with varying abilities to form xenograft tumors.[56,57] In particular, there appears to be a noticeable shift in reliance on MIF for cell proliferation in androgen-independent prostate cancer cell lines DU-145 and PC-3 when compared with androgen-responsive and/or benign prostate cell lines.[57]

Key to the requirement for MIF in cell viability was the increased expression of cell surface CD74, which was only detected in the androgen-independent tumor cells. Treatments directed against MIF or CD74 reduced cell proliferation, MIF secretion, and tumor cell invasiveness.[57,58] More recent studies comparing MIF expression in transformed benign cells and cancer cells found increased intracellular and secreted MIF in highly invasive bladder and prostate cancer cell lines when compared to immortalized benign cell lines (Fig. 4).

UROtsa (SV40 large T-antigen immortalized urothelium) is a histologically normal benign urothelial cell line that exhibits basal MIF secretion (5.4 pg/10^5

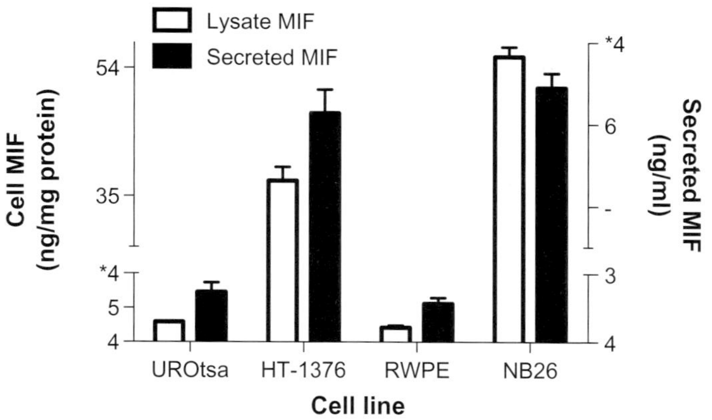

Fig. 4. MIF amounts. Cell lysates and cell culture medium from 80% confluent cells grown under standard conditions were assayed for MIF by ELISA. The urothelial cells exhibit increased cell lysate and secreted MIF amounts associated with the more invasive cell lines (HT-1376 and WPE1-NB26, highly invasive bladder cancer or prostate cancer cells, respectively). Data are expressed as ng MIF per mg total cell lysate protein and ng/ml culture medium.

cells in 24 hours), whereas the bladder cancer cell line HT-1376 secretes approximately 2.5 times more MIF per cell (14.4 pg/10^5 cells in 24 hours). The RWPE-1 cell line is derived from epithelial cells from the peripheral zone of a histologically normal adult human prostate, which were immortalized by transfection with HPV-18. This cell line exhibits basal MIF secretion equivalent to the UROtsa cell line (7.6 pg/10^5 cells in 24 hours), whereas the prostate cancer cell line WPE1-NB26 secretes approximately 3.5 times more MIF per cell (28.4 pg/10^5 cells). The WPE1-NB26 cell line was derived from RWPE-1 cells by exposure to N-methyl-N-nitrosourea.[59] The changes in MIF amounts associated with tumor aggressiveness in the WPE1-NB26 cell line are likely a consequence of epigenetic mechanisms associated with prostate cancer progression rather than genetic mutations present in unrelated cell lines.

The WEP1-NB26 cell line exhibits altered signaling pathways when compared to the parental RWPE-1 cell line. The WEP1-NB26 cell line exhibits persistent phosphorylation at threonine 308 in AKT-1, as well as ERK-1/2 phosphorylation, loss of PTEN and loss of phosphorylation of c-Raf. It should be noted that the WPE-1 NB-26 cell line is androgen-dependent, so the changes in signaling pathways in this cell line are likely related to those found in stage D2 prostate cancer.[60] Additional studies are in progress to determine the specific MIF associated changes that accompany tumor cell aggressiveness.

5. Conclusions and Outlook

The involvement of MIF in tumor cell proliferation makes it a potential treatment target. Small molecule or endogenous MIF inhibitors such as ISO-1, ATIII or uromodulin could be of potential therapeutic use for cancer treatment. The close association of MIF amounts in serum and tissues, as well as the association of high expressing MIF promoter polymorphisms with cancer diagnosis and prognosis, suggests that MIF genotyping may serve as an excellent urological cancer biomarker.

Acknowledgments

This material is based upon work supported (or supported in part) by the Department of Veterans Affairs, Veterans Health Administration, Office of Research and Development, Biomedical Laboratory Research and Development (KLMS, PLV). The funders had no role in study design, data collection and analysis, decision to publish or preparation of the manuscript. The contents of this article do not represent the views of the Department of Veterans Affairs or the United States Government. Gary A. Smith, Jr. and Warren Cowart provided excellent technical assistance. Alexander Braley, Mira Janjus, and Terra Wolfe contributed to the experiments described.

References

1. Cordon-Cardo C, Prives C. (1999) At the crossroads of inflammation and tumorigenesis. *J Exp Med* **190**(10): 1367–1370.
2. Karin M. (2009) NF-kappaB as a critical link between inflammation and cancer. *Cold Spring Harb Perspect Biol* **1**(5): a000141.
3. Stock D, Groome PA, Siemens DR. (2008) Inflammation and prostate cancer: A future target for prevention and therapy? *Urol Clin North Am* **35**(1): 117–130.
4. Coussens LM, Werb Z. (2002) Inflammation and cancer. *Nature* **420**(6917): 860–867.
5. Lin WW, Karin M. (2007) A cytokine-mediated link between innate immunity, inflammation, and cancer. *J Clin Invest* **117**(5): 1175–1183.
6. Kamisawa T, Horiguchi S, Hayashi S *et al.* (2007) K-ras mutation in the major duodenal papilla and gastric and colonic mucosa in patients with autoimmune pancreatitis. *J Gastroenterol* **45**(7): 771–778.

7. Li Y, Wo JM, Ray MB *et al.* (2006) Cyclooxygenase-2 and epithelial growth factor receptor up-regulation during progression of Barrett's esophagus to adenocarcinoma. *World J Gastroenterol* **12**(6): 928–934.
8. Krejs GJ. (2006) Pancreatic cancer: Epidemiology and risk factors. *Dig Dis* **28**(2): 355–358.
9. Uomo I, Miraglia S, Pastorello M. (2007) Inflammation and pancreatic ductal adenocarcinoma: A potential scenario for novel drug targets. *J Pancreatol* **11**(3): 199–202.
10. Terzic J, Grivennikov S, Karin E, Karin M. (2010) Inflammation and colon cancer. Gastroenterology **138**(6): 2101–2114, e5.
11. De Marzo AM, Platz EA, Sutcliffe S *et al.* (2007) Inflammation in prostate carcinogenesis. *Nat Rev Cancer* **7**(4): 256–269.
12. Marsh GM, Gula MJ, Youk AO, Cassidy LD. (2002) Bladder cancer among chemical workers exposed to nitrogen products and other substances. *Am J Ind Med* **42**(4): 286–295.
13. Greenberg RS, Mandel JS, Pastides H *et al.* (2001) A meta-analysis of cohort studies describing mortality and cancer incidence among chemical workers in the United States and western Europe. *Epidemiology* **12**(6): 727–740.
14. Ugnat AM, Luo W, Semenciw R, Mao Y. (2004) Occupational exposure to chemical and petrochemical industries and bladder cancer risk in four western Canadian provinces. *Chronic Dis Can* **25**(2): 7–15.
15. Bucala R, Donnelly SC. (2007) Macrophage migration inhibitory factor: A probable link between inflammation and cancer. *Immunity* **26**(3): 281–285.
16. Flaster H, Bernhagen J, Calandra T, Bucala R. (2007) The macrophage migration inhibitory factor-glucocorticoid dyad: Regulation of inflammation and immunity. *Mol Endocrinol* **21**(6): 1267–1280.
17. Bucala R. (1996) MIF re-discovered: Pituitary hormone and glucocorticoid-induced regulator of cytokine production. *Cytokine Growth Factor Rev* **7**(1): 19–24.
18. Lolis E, Bucala R. (2003) Macrophage migration inhibitory factor. *Expert Opin Ther Targets* **7**(2): 153–164.
19. Leng L, Metz CN, Fang Y *et al.* (2003) MIF signal transduction initiated by binding to CD74. *J Exp Med* **197**(11): 1467–1476.
20. Calandra T, Roger T. (2003) Macrophage migration inhibitory factor: A regulator of innate immunity. *Nat Rev Immunol* **3**(10): 791–800.
21. Shimizu T, Nishihira J, Watanabe H. (2004) Macrophage migration inhibitory factor is induced by thrombin and factor Xa in endothelial cells. *J Biol Chem* **279**(14): 13729–13737.
22. Wadgaonkar R, Somnay K, Garcia JG. (2008) Thrombin induced secretion of macrophage migration inhibitory factor (MIF) and its effect on nuclear signaling in endothelium. *J Cell Biochem* **105**(5): 1279–1288.

23. Gandhi PS, Chen Z, Di Cera E. (2010) Crystal structure of thrombin bound to the uncleaved extracellular fragment of PAR1. *J Biol Chem* **285**(20): 15393–15398.
24. Chen D, Dorling A. (2009) Critical roles for thrombin in acute and chronic inflammation. *J Thromb Haemost* **7**(Suppl 1): 122–126.
25. Mitchell RA, Metz CN, Peng T, Bucala R. (1999) Sustained mitogen-activated protein kinase (MAPK) and cytoplasmic phospholipase A2 activation by macrophage migration inhibitory factor (MIF). Regulatory role in cell proliferation and glucocorticoid action. *J Biol Chem* **274**(25): 18100–18106.
26. Hudson JD, Shoaibi MA, Maestro R *et al.* (1999) A proinflammatory cytokine inhibits p53 tumor suppressor activity. *J Exp Med* **190**(10): 1375–1382.
27. Meyer-Siegler KL, Cox J, Leng L *et al.* (2009) Macrophage migration inhibitory factor (MIF) anti-thrombin III complexes are decreased in bladder cancer patient serum: Complex formation as a possible mechanism of MIF inactivation. *Cancer Lett* **290**(1): 49–57.
28. Kompier LC, van Tilborg AA, Zwarthoff EC. (2010) Bladder cancer: Novel molecular characteristics, diagnostic, and therapeutic implications. *Urol Oncol* **28**(1): 91–96.
29. Coen JJ, Feldman AS, Smith MR, Zietman AL. (2010) Watchful waiting for localized prostate cancer in the PSA era: What have been the triggers for intervention? *BJU Int* **106**(8): 1401–1420.
30. Hirao Y, Kim WJ, Fujimoto K. (2009) Environmental factors promoting bladder cancer. *Curr Opin Urol* **19**(5): 494–499.
31. Bardia A, Platz EA, Yegnasubramanian S *et al.* (2009) Anti-inflammatory drugs, antioxidants, and prostate cancer prevention. *Curr Opin Pharmacol* **9**(4): 419–426.
32. Meyer-Siegler K, Hudson PB. (1996) Enhanced expression of macrophage migration inhibitory factor in prostatic adenocarcinoma metastases. *Urology* **48**(3): 448–452.
33. Meyer-Siegler K, Fattor RA, Hudson PB. (1998) Expression of macrophage migration inhibitory factor in the human prostate. *Diagn Mol Pathol* **7**(1): 44–50.
34. del Vecchio MT, Tripodi SA, Arcuri F *et al.* (2000) Macrophage migration inhibitory factor in prostatic adenocarcinoma: Correlation with tumor grading and combination endocrine treatment-related changes. *Prostate* **45**(1): 51–57.
35. Petrovsky N, Socha L, Silva D *et al.* (2003) Macrophage migration inhibitory factor exhibits a pronounced circadian rhythm relevant to its role as a glucocorticoid counter-regulator. *Immunol Cell Biol* **81**(2): 137–143.
36. Meyer-Siegler KL, Bellino MA, Tannenbaum M. (2002) Macrophage migration inhibitory factor evaluation compared with prostate specific antigen as a biomarker in patients with prostate carcinoma. *Cancer* **94**(5): 1449–1456.

37. Meyer-Siegler KL, Iczkowski KA, Vera PL. (2005) Further evidence for increased macrophage migration inhibitory factor expression in prostate cancer. *BMC Cancer* **5**(1): 73.
38. Muramaki M, Miyake H, Yamada Y, Hara I. (2006) Clinical utility of serum macrophage migration inhibitory factor in men with prostate cancer as a novel biomarker of detection and disease progression. *Oncol Rep* **15**(1): 253–257.
39. Stephan C, Xu C, Brown DA *et al.* (2006) Three new serum markers for prostate cancer detection within a percent free PSA-based artificial neural network. *Prostate* **66**(6): 651–659.
40. Baugh JA, Chitnis S, Donnelly SC *et al.* (2002) A functional promoter polymorphism in the macrophage migration inhibitory factor (MIF) gene associated with disease severity in rheumatoid arthritis. *Genes Immun* **3**(3): 170–176.
41. Donn RP, Alourfi Z, Zeggini E *et al.* (2001) A novel 5′-flanking region polymorphism of macrophage migration inhibitory factor is associated with systemic-onset juvenile idiopathic arthritis. *Arthritis Rheum* **44**(8): 1782–1785.
42. Meyer-Siegler KL, Vera PL, Iczkowski KA *et al.* (2007) Macrophage migration inhibitory factor (MIF) gene polymorphisms are associated with increased prostate cancer incidence. *Genes Immun* **8**(8): 646–652.
43. Arisawa T, Tahara T, Shibata T *et al.* (2008) Functional promoter polymorphisms of the macrophage migration inhibitory factor gene in gastric carcinogenesis. *Oncol Rep* **19**(1): 223–228.
44. Ding GX, Zhou SQ, Xu Z *et al.* (2009) The association between MIF-173 G>C polymorphism and prostate cancer in southern Chinese. *J Surg Oncol* **100**(2): 106–110.
45. Yasasever V, Camlica H, Duranyildiz D *et al.* (2007) Macrophage migration inhibitory factor in cancer. *Cancer Invest* **25**(8): 715–719.
46. Tahara T, Shibata T, Nakamura M *et al.* (2009) Effect of polymorphisms of IL-17A, −17F and MIF genes on CpG island hyper-methylation (CIHM) in the human gastric mucosa. *Int J Mol Med* **24**(4): 563–569.
47. Xue Y, Xu H, Rong L *et al.* (2010) The MIF −173G/C polymorphism and risk of childhood acute lymphoblastic leukemia in a Chinese population. *Leuk Res* **34**(10): 1282–1286.
48. Brown FG, Nikolic-Paterson DJ, Chadban SJ *et al.* (2001) Urine macrophage migration inhibitory factor concentrations as a diagnostic tool in human renal allograft rejection. *Transplantation* **71**(12): 1777–1783.
49. Vera PL, Iczkowski KA, Leng L *et al.* (2005) Macrophage migration inhibitory factor is released as a complex with alpha1-inhibitor-3 in the intraluminal fluid during bladder inflammation in the rat. *J Urol* **174**(1): 338–343.
50. Meyer-Siegler KL, Iczkowski KA, Vera PL (2006) Macrophage migration inhibitory factor is increased in the urine of patients with urinary tract

infection: Macrophage migration inhibitory factor-protein complexes in human urine. *J Urol* **175**(4): 1523–1528.

51. Fingerle-Rowson G, Kaleswarapu DR, Schlander C *et al.* (2009) A tautomerase-null MIF gene knock-in mouse reveals that protein interactions and not enzymatic activity mediate MIF-dependent growth regulation. *Mol Cell Biol* **29**(7): 1922–1932.
52. Asokananthan N, Graham PT, Fink J *et al.* (2002) Activation of protease-activated receptor (PAR)-1, PAR-2, and PAR-4 stimulates IL-6, IL-8, and prostaglandin E2 release from human respiratory epithelial cells. *J Immunol* **168**(7): 3577–3585.
53. Wilson JM, Coletta PL, Cuthbert RJ *et al.* (2005) Macrophage migration inhibitory factor promotes intestinal tumorigenesis. *Gastroenterology* **129**(5): 1485–1503.
54. Ren Y, Tsui HT, Poon RT *et al.* (2003) Macrophage migration inhibitory factor: Roles in regulating tumor cell migration and expression of angiogenic factors in hepatocellular carcinoma. *Int J Cancer* **107**(1): 22–29.
55. Mitchell RA, Bucala R. (2000) Tumor growth-promoting properties of macrophage migration inhibitory factor (MIF). *Semin Cancer Biol* **10**(5): 359–366.
56. Meyer-Siegler KL, Leifheit EC, Vera PL. (2004) Inhibition of macrophage migration inhibitory factor decreases proliferation and cytokine expression in bladder cancer cells. *BMC Cancer* **4**: 34.
57. Meyer-Siegler KL, Iczkowski KA, Leng L *et al.* (2006) Inhibition of macrophage migration inhibitory factor or its receptor (CD74) attenuates growth and invasion of DU-145 prostate cancer cells. *J Immunol* **177**(12): 8730–8739.
58. Meyer-Siegler K. (2000) Macrophage migration inhibitory factor increases MMP-2 activity in DU-145 prostate cells. *Cytokine* **12**(7): 914–921.
59. Webber MM, Quader ST, Kleinman HK *et al.* (2001) Human cell lines as an *in vitro/in vivo* model for prostate carcinogenesis and progression. *Prostate* **47**(1): 1–13.
60. Crawford ED, Blumenstein BA. (1997) Proposed substages for metastatic prostate cancer. *Urology* **50**(6): 1027–1028.

IV-2

MIF in Ovarian Cancer: Detection and Treatment

Guy Nadel, Ayesha B. Alvero and Gil Mor*

1. Introduction

Epithelial ovarian cancer (EOC) is the fourth leading cause of gynecologic cancer deaths in the Western world. In 2009, 21,550 new cases were diagnosed in the USA alone, and 14,600 are expected to die from the disease within five years.[1] The high death rate results from the high percentage of cases diagnosed at a relatively late stage where metastasis had already spread throughout the peritoneal cavity. Despite advances in surgery and chemotherapy, the five-year survival rate is less than 15%.[2]

The "cancer stem cell model" suggests that in many forms of tumors, cancer initiation and propagation is driven by a population of cells with the ability to self-renew and the capacity to give rise to heterogeneous lineages of cancer cells.[3–5] These cells, known as cancer stem cells or tumor initiating cells, are believed to be both the source of primary tumor as well as the cause of recurrence and chemoresistance.[3,6]

One of our earlier observations relating to the heterogeneity of ovarian cancer cells was the detection and isolation of cancer cells from fresh ovarian tumors with differential chemosensitivity patterns. Thus, from the same ovarian tumor, we were able to identify at least two types of EOC cells based on their chemo response: Type I chemo-resistant and Type II chemosensitive EOC cells.[7,8] Further characterization showed that these cells have additional differences in terms of their growth, cytokine production, and intracellular markers.[8] While Type II EOC cells represent the "classical" ovarian cancer cells characterized by fast growth and cell division, and lack of cell-to-cell contact inhibition, Type I cells are characterized by slower growth,

*Corresponding author: Department of Obstetrics, Gynecology and Reproductive Sciences, Reproductive Immunology Unit, Yale University School of Medicine, 333 Cedar St., LSOG 305A, New Haven, CT 06520, USA. Email: gil.mor@yale.edu

which is inhibited upon cell-to-cell contact.[6,7] In addition, Type I, but not Type II, EOC cells have constitutive NF-κB activity and constitutively secrete IL-6, IL-8, MCP-1, and GRO-α.[7,9] Gene expression microarray analysis obtained from these two types of cells further showed numerous differentially expressed genes including Cytokeratin 18, the TLR adapter protein, MyD88, and several genes that are associated with stem cell phenotype such as CD44, Oct-4, SSEA-4, and others, which were highly expressed in the Type I EOC cells.[6] These findings suggest that Type I EOC cells represent the population with stem-like properties.

Indeed, we later demonstrated that Type I EOC cells, sorted by the cell surface marker CD44, are able to establish xenografts in mice. This tumor initiating capacity was not observed with Type II EOC cells. Microscopic analysis of the xenografts obtained from Type I EOC cells showed recapitulation of the morphology of the original tumor as well as presence of both $CD44^+$ and $CD44^-$ cells.[6] Taken together, these data demonstrate that Type I EOC cells have the capacity to self-renew as well as differentiate.

The process of differentiation was also observed *in vitro,* wherein 100% of $CD44^+$ cells seeded at very low density eventually gave rise to cultures that looked morphologically different. Whereas the original $CD44^+$ culture doubles every 36 hrs, the resulting $CD44^-$ culture doubles faster, every 16 hrs. Moreover, while $CD44^+$ cells (slow dividing cells) are chemoresistant, the newly differentiated $CD44^-$ cell cultures (fast dividing cells) became responsive to chemotherapy.

Evaluation of the presence of $CD44^+$ cells in ovarian cancer tumors revealed their presence in clusters and in close proximity to the stroma surrounding the tumor. These cells are morphologically less differentiated with an immature appearance including larger size, higher nuclear-to-cytoplasm (N/C) ratio, vesicular chromatin pattern, and prominent nucleoli.[10]

2. Inflammation and Cancer

Numerous studies have provided convincing evidence supporting the notion that bacterial- and viral-induced inflammatory processes can mediate tumorigenesis.[11] Surgical removal of a primary tumor is often followed by rapid growth of previously dormant metastases as observed in *in vivo* studies, and the bacterial component, lipopolysaccharide (LPS), has been suggested to be responsible for this effect.[12] Indeed, BALB/c mice receiving a tail vein injection of 4T1 mouse mammary carcinoma cells showed an increase in lung metastases following LPS injection.[13] In humans, chronic infection and inflammation are considered two of the most important epigenetic and environmental factors contributing to tumorigenesis and tumor progression.[14,15]

In 1858, Rudolf Virchow noted that the generation of cancer often occurred at sites of chronic inflammation.[16,17] He also hypothesized that chronic inflammation could promote the proliferation of cells and thus, the development of cancer. In the past 15 years, numerous cancers have been shown to be associated with local chronic inflammation. Chronic inflammatory bowel diseases such as chronic ulcerative colitis and Crohn's disease have strong association with colon cancer.[15] Similarly, gastric cancer has a strong link with chronic *Helicobacter pylori* infection and the resulting inflammation.[18] Ovarian endometriosis is an established risk factor for certain types of epithelial ovarian cancers.[19–21] Other examples include chronic bronchitis and lung cancer, schistosomiasis and bladder cancer, papillomavirus infection and cervical cancer, chronic pancreatitis and pancreatic cancer, chronic cholecystitis and gall bladder cancer, and hepatitis virus B and C infection and liver cancer.[11,22,23] Moreover, epidemiological studies showed that regular intake of nonsteroidal anti-inflammatory drugs (NSAIDs) lowered the risk of developing some types of cancers.[15,24,25]

NF-κB is one of the key transcription factors in proinflammatory responses, and abundant evidence has linked NF-κB activation and cancer development.[26,27] Several cytokines and chemokines produced at the tumor microenvironment by immune cells, such as macrophages, are thought to drive the neoplastic process.[28] However, the contribution of cancer cells themselves in the maintenance of a proinflammatory environment that promotes cancer growth is largely overlooked and sometimes considered passive. As mentioned above, Type I EOC stem cells constitutively secrete the pro-growth cytokines IL-6, IL-8, MCP-1, and GRO-α. This may significantly contribute to the maintenance of an inflammatory environment, which may promote tissue repair and renewal.[8] The main trigger of constitutive NF-κB/cytokine production in Type I EOC stem cells is IKK-β.[7] Upon differentiation, Type I EOC stem cells lose IKK-β expression, NF-κB activity, and therefore the capacity to constitutively produce these cytokines. The regulation of IKK-β during the process of differentiation seems to be regulated by a cluster of microRNAs, including miR199a and miR214.[7,29]

Although the constitutive secretion of IL-6, IL-8, MCP-1, and GRO-α is only observed in the Type I EOC stem cells, the cytokine macrophage migration inhibitory factor (MIF) is highly expressed in both Type I and Type II EOC cells. MIF is detected in all ovarian cancer cell culture supernatants, in xenograft tumor lysates, as well as in patient samples including serum and malignant ovarian ascites. The demonstration that MIF is one of the few cytokines/chemokines produced by both types of ovarian cancer cells suggests that MIF may have an important role in the regulation of ovarian tumor growth and progression.

3. Ovarian Cancer and MIF

3.1 *Detection*

MIF is classically defined as an inflammatory cytokine that plays an important role in regulating innate and adaptive immune responses. It has been suggested that the potent proinflammatory effect of MIF may mediate some of the stimulatory effects of inflammation on cancer progression. Indeed, MIF expression has been observed in several forms of cancer, including gastric, prostate, breast, colon, brain, bladder, kidney, skin, and lung-derived tumors.[30–32] Recent studies have shown that MIF may have a key role in carcinogenesis, enhancement of angiogenesis, as well as tumor invasion.[33–35]

We initially evaluated MIF secretion by EOC cells and its levels in serum samples obtained from ovarian cancer patients. We found that EOC cell lines secrete high levels of MIF *in vitro*. These results, however, were not seen in normal ovarian surface epithelial cells.[36] Moreover, we showed that this *in vitro* finding correlates *in vivo*, where we observed high levels of MIF in the serum of patients with EOC but not in normal controls.[37] Other studies have shown a correlation between MIF expression and enhancement of tumor growth. MIF's effect on tumor growth has been associated with induction of angiogenesis through activation of the inflammatory cytokines, TNF-α and IL-6, as well as the proangiogenic cytokine, VEGF.[38,39] In another study, MIF was shown to be essential in regulating tumor-immune cell interaction by downregulation of the NK cell activator, NKG2D.[40]

These findings have three major implications: First, abnormal MIF expression may be related to the pathogenesis of ovarian cancer and therefore it may represent a potential target for treatment. Second, the constitutive expression of MIF by EOC cells suggests a specific signaling pathway that links inflammation, tumorigenesis, and tumor progression (see below). The demonstration that MIF, a proinflammatory cytokine, is upregulated in the serum of EOC patients but not in normal controls suggests that it may contribute to the inflammatory process associated with malignant neoplasia and may affect the antitumoral immune response. Third, MIF may serve as a specific marker for ovarian cancer.

Our initial studies identified MIF as a potential marker associated with ovarian cancer.[36] Using a microarray screening system, we identified MIF to be differentially expressed in newly diagnosed ovarian cancer patients compared to normal controls.[41] In this study, we showed that MIF could be used as a tumor marker with a sensitivity of 78% and specificity of 53%. In subsequent studies, we demonstrated that MIF combined with leptin, prolactin, osteopontin, and CA125 are able to detect ovarian cancer with a sensitivity of 95% and specificity of 99%. More importantly, the combination of these six markers is able to detect early stages (I and II) of ovarian cancer with a sensitivity that is significantly higher than CA125.[37]

3.2 *Treatment*

Although almost 80% of ovarian cancer patients respond to first line chemotherapy (paclitaxel and carboplatin), a significant number of these responders (60–80%) will recur within 2–5 years. In these cases, recurrent disease is characterized by chemoresistance. Therefore, there is a need for the identification and development of new target therapies that would overcome chemoresistance. Compounds that target elements within the pathways that drive cancer may help improve treatment efficacy and minimize side effects.

Since we observed MIF expression in both subtypes of EOC cells, we hypothesized that MIF inhibition could have an impact on the growth of these cells. To answer this hypothesis, we screened a series of small molecule MIF antagonists developed by William Jorgensen's lab at Yale. One of the tested MIF inhibitors had a significant effect on EOC cell growth (Figs. 1 and 2). The negative effect on growth was observed in the Type II EOC cells as well as in the chemoresistant Type I EOC stem cells (Fig. 1). The specificity of the MIF inhibitor was demonstrated by testing against normal human ovarian surface epithelial cells (OSE) and human endometrial stromal cells, which do not express MIF. The MIF inhibitor had no effect on the growth of these cells (Nadal *et al., in preparation*). Interestingly, neither a neutralizing anti-MIF receptor antibody (LN2) or an anti-MIF antibody (III.D.9) influenced EOC cell growth. These suggest that the antitumor effect of MIF inhibitors is

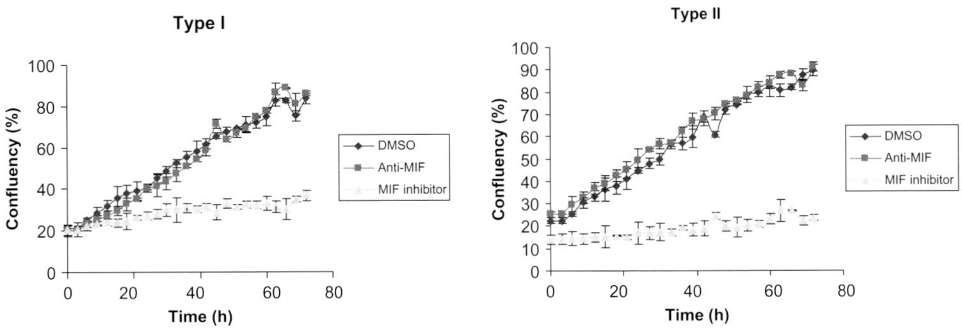

Fig. 1. Type I and II EOC cells were treated with MIF inhibitor or antibody against MIF (anti-MIF). Effect on cell growth was determined by plotting culture confluency through time using a real-time video imaging system (Incucyte).

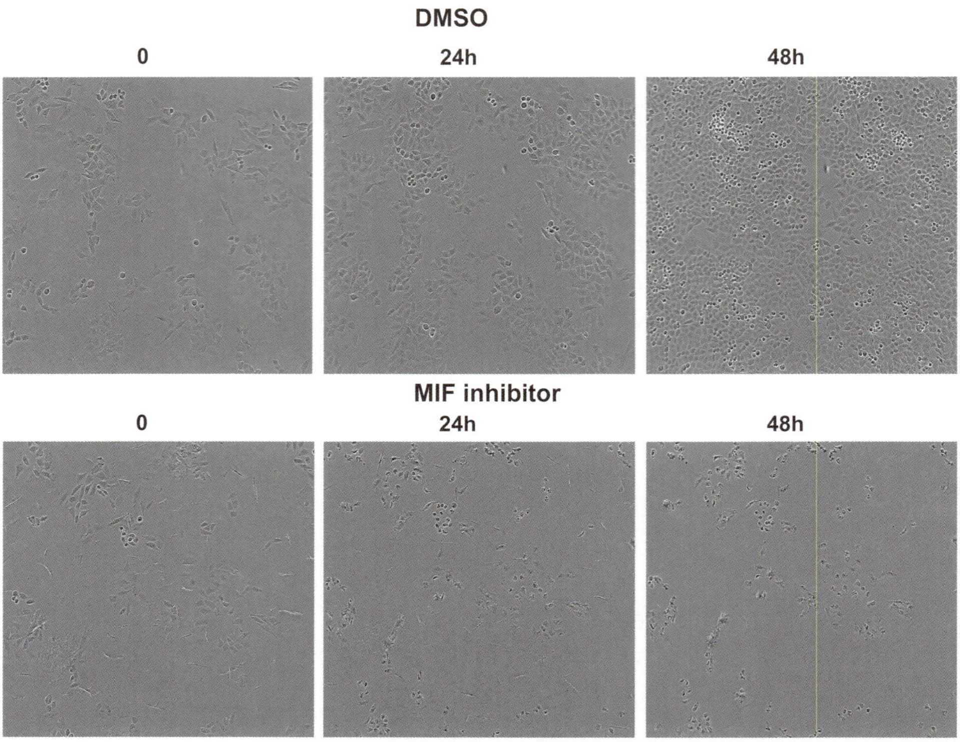

Fig. 2. A candidate small molecule MIF inhibitor but not anti-MIF antibody induces cell death in Type I EOC stem cells. Type I cells were treated with MIF inhibitor and cell morphology monitored after 24 hrs and 48 hrs.

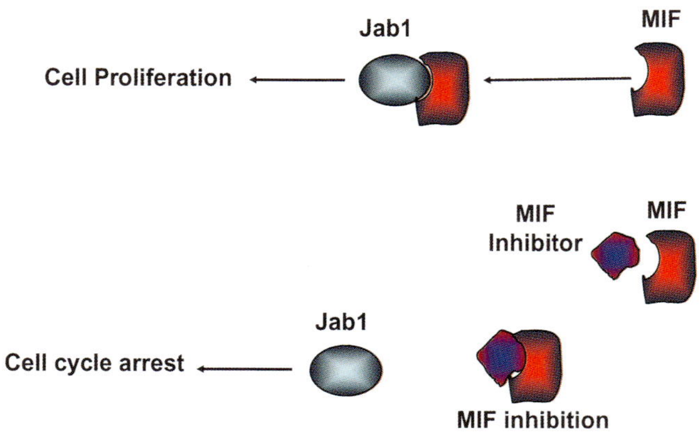

Fig 3. Proposed model for the mechanism of MIF inhibitor-induced cell death in EOC cells. MIF inhibitor releases Jab1 from interaction with MIF to induce cell death.

not associated with the secreted form of MIF nor MIF binding to the CD74 receptor complex, but possibly through an intracellular pathway that may involve Jab1. Jab1 has a dual role in the regulation of cell proliferation and death by activation of c-Jun and JunB, respectively.[42,43] MIF interacts with Jab1 and inhibits its activity, which results in stabilization of p27 and upregulation of cyclin D1 and ERK.[44]

Preliminary studies in our laboratory suggest that treatment of EOC cells with MIF inhibitors releases Jab1 from the inhibitory effect of MIF and results in degradation of p27 as well as downregulation of cyclin D1 and dephosphorylation of ERK, leading to cell death. These results imply that the survival of EOC cells rely partly on the inhibition of Jab1 by MIF (Figure 3).

4. Conclusion

MIF is highly expressed and secreted in the serum of EOC patients. Hence, MIF is currently used as part of a panel of biomarkers for the early detection of ovarian cancer.[37] There is a growing body of evidence suggesting an essential and specific role for MIF in EOC progression and maintenance. Our data demonstrate that *in vitro* inhibition of MIF activity in human EOC cells negated its proliferative effect and induced cell death in relatively low doses. Based on these results, it can be inferred that small MIF binding molecules may be of value in the treatment of EOC patients. The use of these molecules is advantageous given the simple and robust manufacturing procedures and relatively low cost. The demonstration that MIF inhibitors are specific against MIF-expressing cancer cells suggest a relevant therapeutic window.

Acknowledgment

These studies were supported in part by grants from NCI/NIH RO1CA127913, RO1CA118678, the Janet Burros Memorial Foundation, the Sands Family Foundation and the Discovery To Cure Research Program.

References

1. Jemal A, Siegel R, Ward E *et al.* (2009) Cancer statistics. *CA Cancer J Clin* **59**(4): 225–249.
2. Edwards BK, Ward E, Kohler BA *et al.* (2010) Annual report to the nation on the status of cancer, 1975–2006, featuring colorectal cancer trends and impact of

interventions (risk factors, screening, and treatment) to reduce future rates. *Cancer* **116**(3): 544–573.

3. Mor G, Yin G, Chefetz I *et al.* (2011) Ovarian cancer stem cells and inflammation. *Cancer Biol Ther* **11**(8): 708–713.
4. Clarke MF, Fuller M. (2006) Stem cells and cancer: Two faces of eve. *Cell* **124**(6): 1111–1115.
5. Alison MR, Murphy G, Leedham S. (2008) Stem cells and cancer: A deadly mix. Cell Tissue Res **331**(1): 109–124.
6. Alvero AB, Chen R, Fu HH *et al.* (2009) Molecular phenotyping of human ovarian cancer stem cells unravels the mechanisms for repair and chemoresistance. *Cell Cycle* **8**(1): 158–166.
7. Chen R, Alvero AB, Silasi DA *et al.* (2008) Regulation of IKKbeta by miR-199a affects NF-kappaB activity in ovarian cancer cells. *Oncogene* **27**(34): 4712–4723.
8. Kelly MG, Alvero AB, Chen R *et al.* (2006) TLR-4 signaling promotes tumor growth and paclitaxel chemoresistance in ovarian cancer. *Cancer Res* **66**(7): 3859–3868.
9. Silasi DA, Alvero AB, Illuzzi J *et al.* (2006) MyD88 predicts chemoresistance to paclitaxel in epithelial ovarian cancer. *Yale J Biol Med* **79**(3–4): 153–163.
10. Yin G, Chen R, Alvero AB *et al.* (2010) TWISTing stemness, inflammation and proliferation of epithelial ovarian cancer cells through MIR199A2/214. *Oncogene* **29**(24): 3545–3553.
11. Coussens LM, Werb Z. (2002) Inflammation and cancer. *Nature* **420**(6917): 860–867.
12. Pidgeon GP, Harmey JH, Kay E *et al.* (1999) The role of endotoxin/lipopolysaccharide in surgically induced tumour growth in a murine model of metastatic disease. *Br J Cancer* **81**(8): 1311–1317.
13. Harmey JH, Bucana CD, Lu W *et al.* (2002) Lipopolysaccharide-induced metastatic growth is associated with increased angiogenesis, vascular permeability and tumor cell invasion. *Int J Cancer* **101**(5): 415–422.
14. Beachy PA, Karhadkar SS, Berman DM. (2004) Mending and malignancy. *Nature* **431**(7007): 402.
15. Balkwill F, Coussens LM. (2004) Cancer: An inflammatory link. *Nature* **431**(7007): 405–406.
16. Virchow R. (1858) Reizung und Reizbarkeit. *Arch Pathol Anat Klein Med* **14**: 1–63.
17. Virchow R. (1863) *Aetiologie der neoplastischen Geschwulste/Pathogenie der neoplastischen Geschwulste. Die Krankhaften Geschwulste.* Verlag von August Hirschwald, Berlin, pp. 57–101.
18. Ernst PB, Takaishi H, Crowe SE. (2001) *Helicobacter pylori* infection as a model for gastrointestinal immunity and chronic inflammatory diseases. *Dig Dis* **19**(2): 104–111.

19. Riman T, Dickman PW, Nilsson S *et al.* (2004) Some life-style factors and the risk of invasive epithelial ovarian cancer in Swedish women. *Eur J Epidemiol* **19**(11): 1011–1019.
20. Sekizawa A, Amemiya S, Otsuka J *et al.* (2004) Malignant transformation of endometriosis: Application of laser microdissection for analysis of genetic alterations according to pathological changes. *Med Electron Microsc* **37**(2): 97–100.
21. Giudice LC, Kao LC. (2004) Endometriosis. *Lancet* **364**(9447): 1789–1799.
22. Balkwill F, Mantovani A. (2001) Inflammation and cancer: Back to Virchow? *Lancet* **357**(9255): 539–545.
23. Li Q, Withoff S, Verma IM. (2005) Inflammation-associated cancer: NF-kappaB is the lynchpin. *Trends Immunol* **26**(6): 318–325.
24. Gupta RA, Dubois RN. (2001) Colorectal cancer prevention and treatment by inhibition of cyclooxygenase-2. *Nat Rev Cancer* **1**(1): 11–21.
25. Marx J. (2004) Cancer research. Inflammation and cancer: The link grows stronger. *Science* **306**(5698): 966–968.
26. Chen R, Alvero AB, Silasi DA, Mor G. (2007) Inflammation, cancer and chemoresistance: Taking advantage of the toll-like receptor signaling pathway. *Am J Reprod Immunol* **57**(2): 93–107.
27. Karin M. (2006) Nuclear factor-kappaB in cancer development and progression. *Nature* **441**(7092): 431–436.
28. Luo JL, Kamata H, Karin M. (2005) The anti-death machinery in IKK/NF-kappaB signaling. *J Clin Immunol* **25**(6): 541–550.
29. Chen R, Alvero AB, Silasi DA *et al.* (2008) Cancers take their Toll — The function and regulation of Toll-like receptors in cancer cells. *Oncogene* **27**(2): 225–233.
30. Bucala R, Donnelly SC. (2007) Macrophage migration inhibitory factor: A probable link between inflammation and cancer. *Immunity* **26**: 281–285.
31. Wilson JM, Coletta PL, Cuthbert RJ *et al.* (2005) Macrophage migration inhibitory factor promotes intestinal tumorigenesis. *Gastroenterology* **129**(5): 1485–1503.
32. Hagemann T, Wilson J, Kulbe H *et al.* (2005) Macrophages induce invasiveness of epithelial cancer cells via NF-kappa B and JNK. *J Immunol* **175**(2): 1197–1205.
33. Hudson JD, Shoaibi MA, Maestro R *et al.* (1999) A proinflammatory cytokine inhibits p53 tumor suppressor activity. *J Exp Med* **190**(10): 1375–1382.
34. Petrenko O, Moll UM. (2005) Macrophage migration inhibitory factor MIF interferes with the Rb-E2F pathway. *Mol Cell* **17**(2): 225–236.
35. Mitchell RA, Liao H, Chesney J *et al.* (2002) Macrophage migration inhibitory factor (MIF) sustains macrophage proinflammatory function by inhibiting p53: Regulatory role in the innate immune response. *Proc Natl Acad Sci USA* **99**(1): 345–350.
36. Agarwal R, Whang DH, Alvero AB *et al.* (2007) Macrophage migration inhibitory factor expression in ovarian cancer. *Am J Obstet Gynecol* **196**(4): 348, e1–5.

37. Visintin I, Feng Z, Longton G *et al.* (2008) Diagnostic markers for early detection of ovarian cancer. *Clin Cancer Res* **14**(4): 1065–1072.
38. Hagemann T, Robinson SC, Thompson RG *et al.* (2007) Ovarian cancer cell-derived migration inhibitory factor enhances tumor growth, progression, and angiogenesis. *Mol Cancer Ther* **6**(7): 1993–2002.
39. Kulbe H, Thompson R, Wilson JL *et al.* (2007) The inflammatory cytokine tumor necrosis factor-alpha generates an autocrine tumor-promoting network in epithelial ovarian cancer cells. *Cancer Res* **67**(2): 585–592.
40. Krockenberger M, Dombrowski Y, Weidler C *et al.* (2008) Macrophage migration inhibitory factor contributes to the immune escape of ovarian cancer by down-regulating NKG2D. *J Immunol* **180**(11): 7338–7348.
41. Mor G, Visintin I, Lai Y *et al.* (2005) Serum protein markers for early detection of ovarian cancer. *Proc Natl Acad Sci USA* **102**(21): 7677–7682.
42. Shaulian E, Karin M. (2002) AP-1 as a regulator of cell life and death. *Nat Cell Biol* **4**(5): E131–E136.
43. Eferl R, Wagner EF. (2003) AP-1: A double-edged sword in tumorigenesis. *Nat Rev Cancer* **3**(11): 859–868.
44. Nemajerova A, Mena P, Fingerle-Rowson G *et al.* (2007) Impaired DNA damage checkpoint response in MIF-deficient mice. *EMBO J* **26**: 987–997.

IV-3

The Role of MIF on Tumorigenicity of Embryonic Stem Cells

Yi Ren*

1. Introduction

Embryonic stem cells (ESCs) have the potential to differentiate into all cell lineages. *In vitro* differentiation of many cell types has been achieved, such as cardiomyocytes, neurons, oligodendrocytes, pancreatic beta cells, and erythrocytes.[1] Induced pluripotent stem (iPS) cells are a novel stem cell population generated by reprogramming adult cells with several transcription factors (Oct3/4, Sox2, Klf4, and c-Myc).[2,3] ESCs and iPS cells offer great potential as a source of cells for regenerative medicine. However, ESCs and iPS cells are generally limited in their plasticity, as their pluripotency is also associated with teratoma formation after transplantation, which is a major obstacle for clinical application. Therefore, there is an urgent need to develop safe and effective approaches for treatment of various diseases using ESCs and iPS cells. Understanding the mechanisms of teratoma development by ESCs and iPS cells may lead to practical advances in the use of these cells in clinical medicine.

2. Risks of Teratoma Formation after ESC Transplantation

Teratomas are tumors that form by the differentiation of cells from all of the three germ layers (mesoderm, endoderm, and ectoderm). Transplantation of undifferentiated mouse ESCs into the striatum leads to the formation of teratomas in a high proportion of animals.[4] Although it has been proposed that teratoma formation can be reduced by *in vitro* predifferentiation of ESCs, recent observation revealed that not only undifferentiated human ESCs

*Corresponding author: W. M. Keck Center for Collaborative Neuroscience, Rutgers, The State University of New Jersey, Nelson Labs D-251, 604 Allison Road, Piscataway, NJ 08854, USA. Email: ren@dls.rutgers.edu

(hESCs) but also proliferating neural progenitors can generate tumors.[5,6] Dressel and colleagues systematically compared the tumorigenicity of mouse ESCs and *in vitro* differentiated neuronal cells in various recipients.[7] They showed that administration of undifferentiated ESCs and differentiated cells into syngeneic or allogeneic immunodeficient mice resulted in teratomas in about 95% of recipients. Both cell types did not give rise to tumors in immunocompetent allogeneic mice or xenogeneic rats. However, in 61% of cyclosporine A-treated rats, teratomas developed after injection of differentiated cells. Teratomas have also been observed after injection of ESCs or *in vitro* differentiated cells into different tissues, including liver and myocardium.[8–11] These results suggest that immunosuppression is required for the functional integration of grafted cells but is associated in many cases with teratoma formation. Human iPS cells are a potential source of patient-specific pluripotent stem cells that evade immune rejection and may offer a high therapeutic potential. However, it appears inevitable that iPS cells will develop teratomas because even *Myc*-deficient iPS cells form teratomas.[12,13] The potential tumorigenicity of stem cells thus must be carefully evaluated before the clinical application of any stem cells in regenerative medicine.

3. The Role of MIF in Teratoma Formation

Teratoma development requires several molecular processes, such as self-renewal, rapid proliferation, lack of contact inhibition, and telomerase activity.[14,15] The marker genes for hESCs and those required for the generation of iPS cells, including *Oct4, Sox2, Nanog, Lin28, Myc,* and *Klf4,* are linked to stem cell tumorigenesis.[15] Many tumor suppresser genes are hypermethylated and silenced in hESCs.[15,16] In addition, *Survivin,* an antiapoptotic gene, also contributes to hESC-induced teratoma formation.[17] However, the mechanisms of teratoma development are complex and may require the presence of supporting host cells in the ESC microenvironment. A study from our laboratory demonstrated that syngeneic ESC transplantation provokes an inflammatory response that involves the rapid recruitment of bone marrow-derived macrophages; these cells created a microenvironment that facilitates the initiation and progression of teratomas. Infiltrated macrophages were activated by ESCs and produced MIF and other angiogenic factors, thereby stimulating angiogenesis and teratoma development. Lack of host MIF was associated with reduced angiogenesis and teratoma growth without affecting the evident pluripotency of ESCs (Fig. 1). However, the genetic deletion of MIF from ESCs (intrinsic MIF deficiency) did not

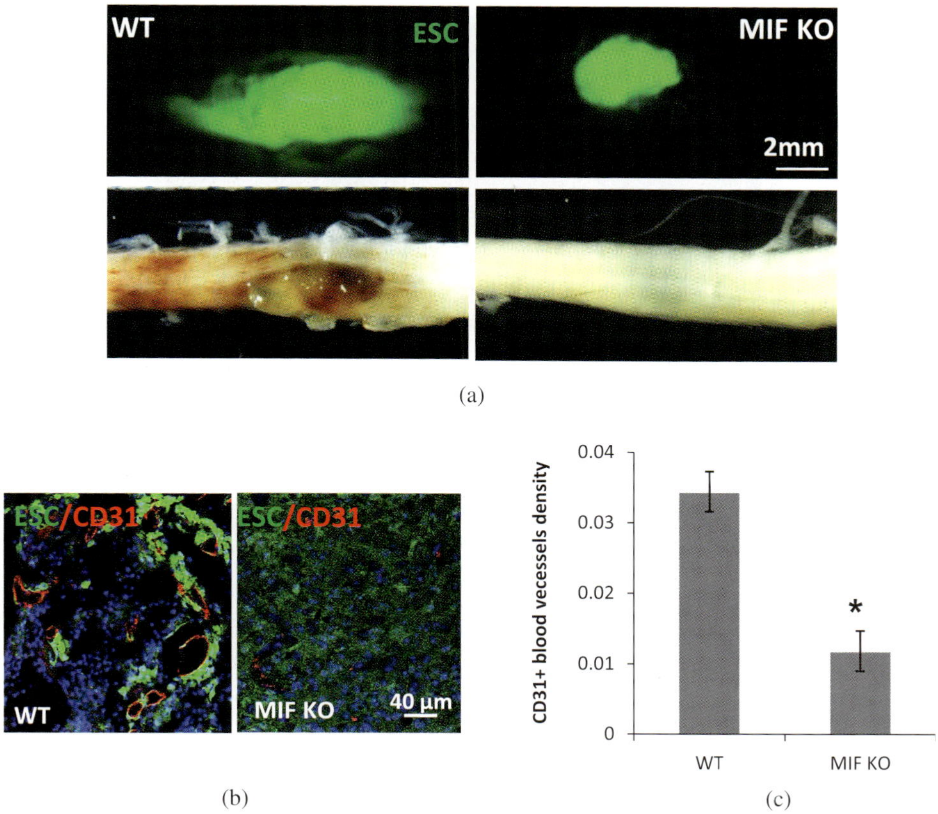

Fig. 1. MIF enhances teratoma angiogenesis and growth. Mouse EGFP-ESCs were stereotaxically injected into the mouse spinal cord, and histological examination was evaluated at 2 weeks after cell transplantation. (a) Micrographs showing strong EGFP expression in teratomas (upper panel) and the presence of associated blood vessels (lower panel) from wildtype (WT) mice (left panel) and MIF knockout (KO) mice (right panel). (b) Immunostaining of sections from WT (left panel) and MIF KO mice (right panel) for the endothelial marker, CD31 (red). (c) Quantification of CD31 positive endothelial cells in teratomas from WT and MIF KO mice (*$p < 0.05$).

influence teratoma development and ESC pluripotency. These data suggested that genetic deletion of MIF from the host but not from ESCs specifically reduced teratoma development.

3.1 *The role of MIF in ESC-induced macrophage activation and infiltration*

Macrophages, derived from peripheral blood monocytes, are a prominent component of the leukocyte infiltrate in inflamed tissues and malignant

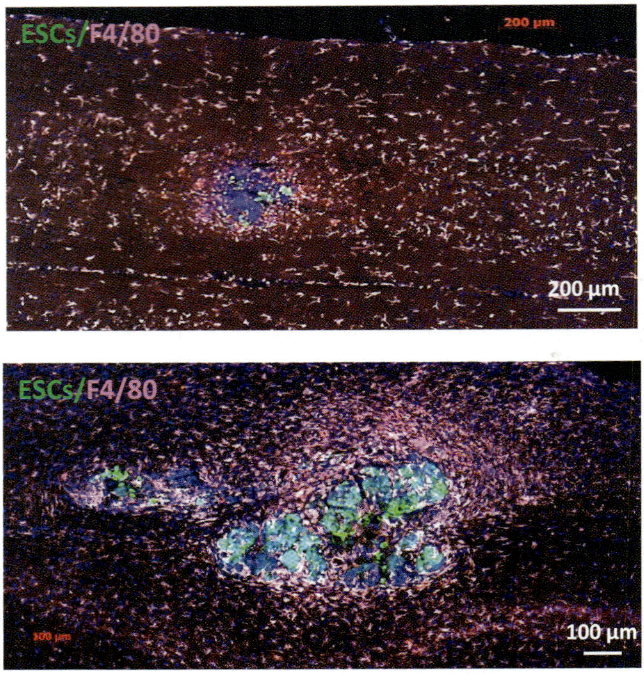

Fig. 2. Macrophage infiltration in longitudinal (coronal) section of spinal cord at 1 day (upper panel) and 1 week (lower panel) after EGFP-ESC injection. Macrophages/microglia cells were detected by Mac-2 (red) and IBA-1 (purple) antibodies.

tumors. There are two major pathways that induce macrophage activation *in vivo*: M1 (classical) and M2 (alternative) macrophage activation.[18–22] Tumor associated macrophages (TAMs) have mostly tumor promoting functions and display a polarized M2 phenotype.[22–24] M2 macrophages also support tumor survival, growth, and metastasis, and play an important role in tumor angiogenesis and immune evasion.[24–26] We studied macrophage infiltration during teratoma development after ESC injection in mouse spinal cord. Macrophages were present within the spinal cord as early as the first 24 hours after ESC injection, with peak macrophage infiltration occurring after 1 to 2 weeks (Fig. 2). Notably, there was a significant reduction in macrophage infiltration in MIF-deficient mice compared to wildtype mice (Fig. 3). Moreover, these infiltrated macrophages showed an M2 activation profile as some macrophages expressed YM1, a M2 marker. *In vitro* experiments confirmed that ESCs secrete products that induced M2 activation. ESCs not only induced M2 activation but also stimulated macrophages to produce MIF, VEGF, and metalloproteinase-9 (MMP-9). MIF released by infiltrating macrophages was both necessary and sufficient to drive angiogenesis and support teratoma progression, and led to the conclusion that MIF is a key regulator in the link

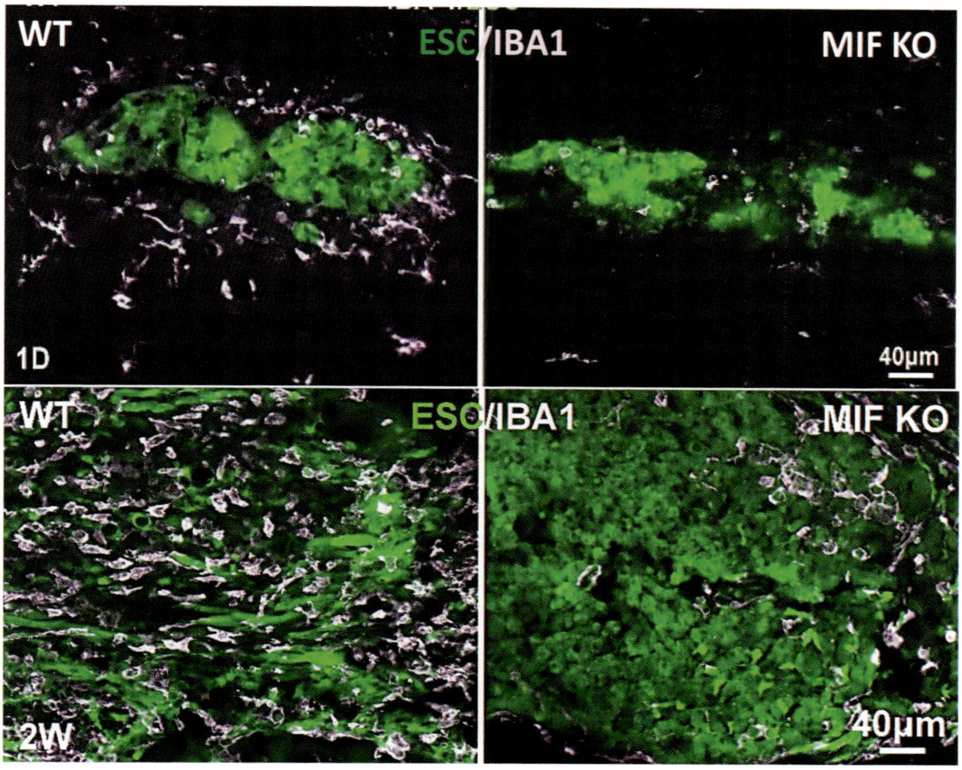

Fig. 3. Macrophage recruitment during teratoma progression. Macrophages in the longitudinal (coronal) sections of spinal cord at 1 day (upper panel) and 2 weeks (lower panel) after ESC transplantation in WT (left panel) and MIF KO mice (right panel) were detected by IBA-1 antibodies.

between inflammation and teratoma development. Although our findings suggest that MIF produced by bone marrow-derived cells contributes to an ESC microenvironment that leads to teratoma growth by promoting angiogenesis, our study also raises additional questions for future study. It would be of interest to examine whether other resident cells participate in teratoma growth since bone marrow-derived macrophages are not the only paracrine cell types present.

The mechanisms by which ESCs induce macrophage infiltration and activation are unclear. We showed that ESCs still have the ability to recruit bone marrow-derived cells, although they are not oncogenically transformed. ESCs are known to express monocyte chemoattractant protein-1 (MCP-1 or CCL2),[27] which acts through its receptor CCR2 to induce the migration and activation of macrophages and thus tumor progression.[24,25,28] Macrophage released MIF also may act as a noncanonical ligand for CXCR2 and CXCR4[29,30] and further attract CXCR2$^+$ and CXCR4$^+$ macrophages, which have been shown to mediate proangiogenic effects in various models of angiogenesis.[31,32] In addition, MIF can promote monocyte recruitment *in vivo* not only via the induction of endothelial release of MCP-1[33] but also by the promotion of expression of adhesion molecules.[34] More detailed studies will be necessary to determine whether additional molecules either alone or in combination contribute to macrophage recruitment to the site of ESC implantation.

3.2 *The role of MIF in ESC-induced angiogenesis*

Tumor angiogenesis, the formation of new blood vessels from host-derived, preexisting vasculature occurs in response to the increasing demand for nutrients and oxygen by proliferating tumor cells.[35] The tumor microenvironment is a complex system of many cell types, including endothelial cells, pericytes, smooth-muscle cells, and infiltrated inflammatory cells.[36] Tumor angiogenesis results not only from the interaction of tumor cells with endothelial cells and pericytes, but also from surrounding inflammatory cells that have a crucial role in directing and supporting the neoformation of blood cells.[37] Bone marrow-derived cells are known to contribute to tumor neovascularization,[38–40] and the recruitment of bone marrow-derived endothelial and pericyte progenitors participates in tumor vascular development.[41–43] One of the important activities of MIF in promoting tumorigenesis and metastasis is stimulating angiogenesis.[44,45] It would be reasonable to speculate that MIF could be a mediator in angiogenesis during teratoma development. We found that ESCs recruited bone marrow-derived macrophages that delivered MIF to stimulate host endothelial cell proliferation. Angiogenesis during ESC proliferation and teratoma progression did not require the contribution of bone marrow-derived endothelial, ESC-differentiated endothelial cells or bone marrow-derived pericyte progenitors. These results are consistent with the data of Purhonen *et al.*, who observed that bone marrow-derived cells do not contribute to vascular endothelium and are not needed for tumor growth.[46] We also demonstrated that MIF plays a role in teratoma angiogenesis by regulating the recruitment/differentiation of pericytes from implanted ESCs and host cells. Infiltrated macrophages produced appreciable amounts of MIF that not only regulated angiogenesis by directly interacting with endothelial cells but also enhanced ESC proliferation directly. In addition to producing MIF, macrophages promoted angiogenesis and tissue remodeling by the secretion of VEGF and MMP-9.[47–49] MMP-9 produced by bone marrow-derived cells initiates the angiogenic switch leading to tumor growth and progression.[47,50,51] It has been shown that MIF directly induced MMP-9 expression in vascular smooth muscle cells and macrophages.[52] Therefore, MIF may coordinate with MMP-9 and VEGF to drive angiogenesis and promote teratoma progression.

4. Strategies for Inhibiting Teratoma Development

4.1 *Targeting ESCs*

The tumorigenic properties of ESCs restrict their potential usefulness in clinical cell transplantation. Currently, the only way to ensure that teratomas do not form is to differentiate the ESCs in advance, enrich for the desired cell type, and screen for the presence of undifferentiated cells.[1] It has been reported that when mouse, primate, and human ESCs differentiate to a sufficient extent *in vitro* before transplantation, there is no teratoma formation following implantation into the brain.[53–56] A certain degree of differentiation *in vitro* appears essential prior to grafting in order to avoid teratomas. Therefore, removal of undifferentiated ESCs and proliferating cells from the culture before transplantation is potentially a powerful approach to reducing the risk of tumor growth. Although these methods sound very promising, certain obstacles remain with regard to their clinical use. This strategy alone, however, cannot completely reduce teratoma formations. At present no selection method exists that can yield a 100% pure population of differentiated ESCs, and the presence of even a few contaminating undifferentiated ESCs carries a risk of teratoma development. To effectively reduce the risk of teratomas following grafting of ESCs or ESC-derived cells, it is important to develop strategies to achieve the safe transplantation of these cells.

4.2 *Targeting MIF*

A better understanding of the regulation and function of different types of cells in the tumorigenicity of ESCs may yield useful therapies for the safe transplantation of ESCs. Given the important role of MIF in teratoma development, selective inhibition of MIF activity would inhibit angiogenesis and teratoma growth. Anti-MIF monoclonal antibody is known to neutralize MIF bioactivity and protect mice from lethal bacterial or virus infection.[57–59] MIF activity could be inhibited by small molecule inhibitors. One of these inhibitors, ISO-1 [(S,R)-3-(4-Hydroxyphenyl)-4,5-dihydro-5-isoxazole acetic acid methyl ester] was the result of a structure-based design approach to the MIF catalytic site pocket and inhibits both the intrinsic tautomerase as well as the cytokine actions of MIF.[60] The therapeutic implications of small molecule MIF antagonism has been demonstrated in models of sepsis, diabetes mellitus-like pancreatic islet inflammation, arthritis, etc.[61] MIF also inhibits mesenchymal stem cell (MSC) migration and homing, and the therapeutic effect of MSCs by systemic delivery is inefficient.[62,63] It is of interest that inhibition of MIF activity by ISO-1 restored MSC migration *in vitro.* We demonstrated that mice receiving a neutralizing anti-MIF antibody showed a significant inhibition of blood vessel formation and teratoma progression (Fig. 4). It is likely that small molecule MIF antagonists would be active in suppressing teratoma development after ESC transplantation by inhibiting the activity of MIF protein and its cytokine actions. Moreover, it is of further interest that human MIF is encoded by a functionally polymorphic locus (22q111)[64,65] and that *MIF* promoter polymorphisms associated with increased systemic MIF expression have been linked to increased inflammatory disease severity.[65,66] Thus, teratoma development may also have an important genetic basis that could affect the clinical selection of patients for therapy.

In addition to inhibiting MIF expression, the targeting of the host microenvironment of the transplantation site rather than ESCs directly could be an efficient approach for suppressing angiogenesis and teratoma progression without affecting the pluripotency of ESCs. Nonsteroidal anti-inflammatory

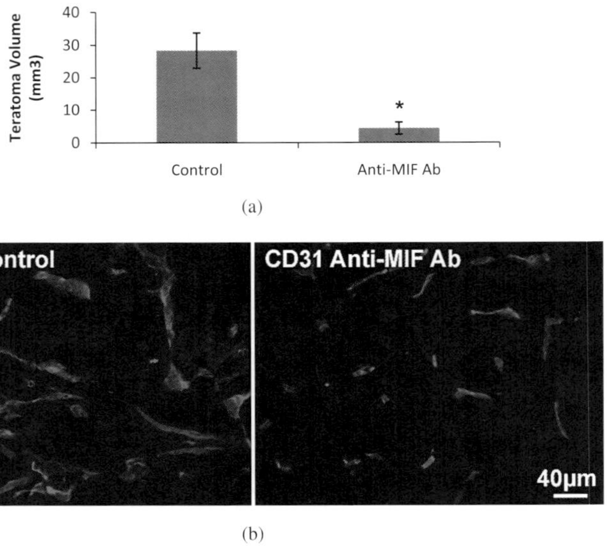

Fig. 4. Targeting MIF with a neutralizing anti-MIF antibody suppresses teratoma growth. (a) Average teratoma size at 3 weeks after ESC transplantation in control (IgG_1 isotype) and treated groups (anti-MIF mAb) (*$p < 0.05$). (b) Representative photographs of teratoma sections at 3 weeks post-injection revealing CD31 staining in control and anti-MIF mAb treated groups.

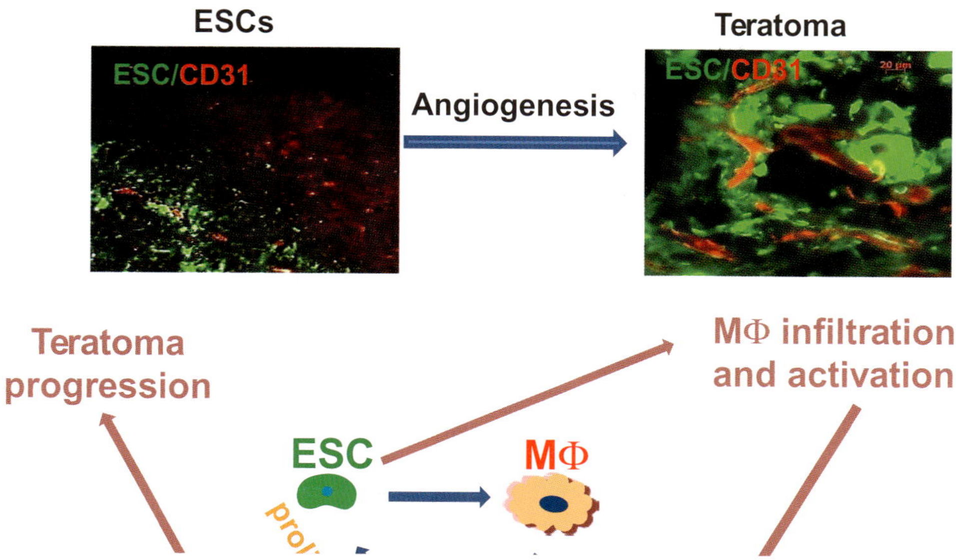

Fig. 5. A diagram illustrating the interaction between ESCs and macrophages that promotes teratoma development through stimulation of angiogenesis.

agents such as COX-2 inhibitors may be candidates for this purpose as they inhibit angiogenesis by direct effects on endothelial cells.[67] VEGF and MMP-9 may also be the valuable therapeutic targets in teratomas.

5. Conclusion

In summary, the microenvironmental niches of ESCs are crucial for teratoma initiation and progression. ESCs produce chemokines such as MCP-1 that induce the infiltration of macrophages and then induce M2 activation. These activated macrophages express multiple angiogenic growth factors and proteinases, such as MIF, VEGF, and MMP-9, which promote angiogenesis. Teratoma progression depends on the development of blood vessels (Fig. 5). Therapeutic targeting of MIF or other angiogenic factors in the ESC microenvironment may offer more effective approaches to inhibit teratoma development for treatment of many diseases by using ESCs and iPS cells.

References

1. Murry CE, Keller G. (2008) Differentiation of embryonic stem cells to clinically relevant populations: Lessons from embryonic development. *Cell* **132**: 661–680.

2. Takahashi K, Yamanaka S. (2006) Induction of pluripotent stem cells from mouse embryonic and adult fibroblast cultures by defined factors. *Cell* **126:** 663–676.
3. Takahashi K, Tanabe K, Ohnuki M *et al.* (2007) Induction of pluripotent stem cells from adult human fibroblasts by defined factors. *Cell* **131:** 861–872.
4. Bjorklund LM, Sanchez-Pernaute R, Chung S *et al.* (2002) Embryonic stem cells develop into functional dopaminergic neurons after transplantation in a Parkinson rat model. *Proc Natl Acad Sci USA* **99:** 2344–2349.
5. Roy NS, Cleren C, Singh SK *et al.* (2006) Functional engraftment of human ES cell-derived dopaminergic neurons enriched by coculture with telomerase-immortalized midbrain astrocytes. *Nat Med* **12:** 1259–1268.
6. Yang D, Zhang ZJ, Oldenburg M *et al.* (2008) Human embryonic stem cell-derived dopaminergic neurons reverse functional deficit in Parkinsonian rats. *Stem Cells* **26**: 55–63.
7. Dressel R, Schindehutte J, Kuhlmann T *et al.* (2008) The tumorigenicity of mouse embryonic stem cells and *in vitro* differentiated neuronal cells is controlled by the recipients' immune response. *PLoS One* **3:** e2622.
8. Fair JH, Cairns BA, Lapaglia MA *et al.* (2005) Correction of factor IX deficiency in mice by embryonic stem cells differentiated *in vitro. Proc Natl Acad Sci USA* **102:** 2958–2963.
9. Cao F, Lin S, Xie X *et al.* (2006) *In vivo* visualization of embryonic stem cell survival, proliferation, and migration after cardiac delivery. *Circulation* **113:** 1005–1014.
10. Nussbaum J, Minami E, Laflamme MA *et al.* (2007) Transplantation of undifferentiated murine embryonic stem cells in the heart: Teratoma formation and immune response. *FASEB J* **21:** 1345–1357.
11. Kolossov E, Bostani T, Roell W *et al.* (2006) Engraftment of engineered ES cell-derived cardiomyocytes but not BM cells restores contractile function to the infarcted myocardium. *J Exp Med* **203:** 2315–2327.
12. Yu J, Vodyanik MA, Smuga-Otto K *et al.* (2007) Induced pluripotent stem cell lines derived from human somatic cells. *Science* **318:** 1917–1920.
13. Nakagawa M, Koyanagi M, Tanabe K *et al.* (2008) Generation of induced pluripotent stem cells without Myc from mouse and human fibroblasts. *Nat Biotechnol* **26:** 101–106.
14. Blum B, Benvenisty N. (2008) The tumorigenicity of human embryonic stem cells. *Adv Cancer Res* **100:** 133–158.
15. Blum B, Benvenisty N. (2009) The tumorigenicity of diploid and aneuploid human pluripotent stem cells. *Cell Cycle* **8:** 3822–3830.
16. Calvanese V, Horrillo A, Hmadcha A *et al.* (2008) Cancer genes hypermethylated in human embryonic stem cells. *PLoS One* **3:** e3294.

17. Blum B, Bar-Nur O, Golan-Lev T *et al.* (2009) The anti-apoptotic gene survivin contributes to teratoma formation by human embryonic stem cells. *Nat Biotechnol* **27:** 281–287.
18. Goerdt S, Orfanos CE. (1999) Other functions, other genes: Alternative activation of antigen-presenting cells. *Immunity* **10:** 137–142.
19. Mantovani A, Sozzani S, Locati M *et al.* (2002) Macrophage polarization: Tumor-associated macrophages as a paradigm for polarized M2 mononuclear phagocytes. *Trends Immunol* **23:** 549–555.
20. Gordon S. (2003) Alternative activation of macrophages. *Nat Rev Immunol* **3:** 23–35.
21. Gordon S, Taylor PR. (2005) Monocyte and macrophage heterogeneity. *Nat Rev Immunol* **5:** 953–964.
22. Martinez FO, Helming L, Gordon S. (2009) Alternative activation of macrophages: An immunologic functional perspective. *Annu Rev Immunol* **27:** 451–483.
23. Balkwill F, Charles KA, Mantovani A. (2005) Smoldering and polarized inflammation in the initiation and promotion of malignant disease. *Cancer Cell* **7:** 211–217.
24. Pollard JW. (2004) Tumour-educated macrophages promote tumour progression and metastasis. *Nat Rev Cancer* **4:** 71–78.
25. Lewis CE, Pollard JW. (2006) Distinct role of macrophages in different tumor microenvironments. *Cancer Res* **66:** 605–612.
26. Mantovani A, Allavena P, Sica A. (2004) Tumour-associated macrophages as a prototypic type II polarised phagocyte population: Role in tumour progression. *Eur J Cancer* **40:** 1660–1667.
27. Guo Y, Graham-Evans B, Broxmeyer HE. (2006) Murine embryonic stem cells secrete cytokines/growth modulators that enhance cell survival/anti-apoptosis and stimulate colony formation of murine hematopoietic progenitor cells. *Stem Cells* **24:** 850–856.
28. Fujimoto H, Sangai T, Ishii G *et al.* (2009) Stromal MCP-1 in mammary tumors induces tumor-associated macrophage infiltration and contributes to tumor progression. *Int J Cancer* **125:** 1276–1284.
29. Bernhagen J, Krohn R, Lue H *et al.* (2007) MIF is a noncognate ligand of CXC chemokine receptors in inflammatory and atherogenic cell recruitment. *Nat Med* **13:** 587–596.
30. Weber C, Kraemer S, Drechsler M *et al.* (2008) Structural determinants of MIF functions in CXCR2-mediated inflammatory and atherogenic leukocyte recruitment. *Proc Natl Acad Sci USA* **105:** 16278–16283.
31. Raman D, Baugher PJ, Thu YM *et al.* (2007) Role of chemokines in tumor growth. *Cancer Lett* **256:** 137–165.

32. Vandercappellen J, Van Damme J, Struyf S. (2008) The role of CXC chemokines and their receptors in cancer. *Cancer Lett* **267:** 226–244.
33. Gregory JL, Morand EF, McKeown SJ *et al.* (2006) Macrophage migration inhibitory factor induces macrophage recruitment via CC chemokine ligand 2. *J Immunol* **177:** 8072–8079.
34. Cheng Q, McKeown SJ, Santos L *et al.* (2010) Macrophage migration inhibitory factor increases leukocyte-endothelial interactions in human endothelial cells via promotion of expression of adhesion molecules. *J Immunol* **185:** 1238–1247.
35. Lewis CE, De Palma M, Naldini L. (2007) Tie2-expressing monocytes and tumor angiogenesis: Regulation by hypoxia and angiopoietin-2. *Cancer Res* **67:** 8429–8432.
36. Albini A, Sporn MB. (2007) The tumour microenvironment as a target for chemoprevention. *Nat Rev Cancer* **7:** 139–147.
37. Ribatti D, Nico B, Crivellato E *et al.* (2007) Macrophages and tumor angiogenesis. *Leukemia* **21:** 2085–2089.
38. De Palma M, Venneri MA, Roca C *et al.* (2003) Targeting exogenous genes to tumor angiogenesis by transplantation of genetically modified hematopoietic stem cells. *Nat Med* **9:** 789–795.
39. De Palma M, Naldini L. (2006) Role of haematopoietic cells and endothelial progenitors in tumour angiogenesis. *Biochim Biophys Acta 1766:* 159–166.
40. Shaked Y, Ciarrocchi A, Franco M *et al.* (2006) Therapy-induced acute recruitment of circulating endothelial progenitor cells to tumors. *Science* **313:** 1785–1787.
41. Lyden D, Hattori K, Dias S *et al.* (2001) Impaired recruitment of bone-marrow-derived endothelial and hematopoietic precursor cells blocks tumor angiogenesis and growth. *Nat Med* **7:** 1194–1201.
42. Nolan DJ, Ciarrocchi A, Mellick AS *et al.* (2007) Bone marrow-derived endothelial progenitor cells are a major determinant of nascent tumor neovascularization. *Genes Dev* **21:** 1546–1558.
43. Reyes M, Dudek A, Jahagirdar B *et al.* (2002) Origin of endothelial progenitors in human postnatal bone marrow. *J Clin Invest* **109:** 337–346.
44. Bifulco C, McDaniel K, Leng L *et al.* (2008) Tumor growth-promoting properties of macrophage migration inhibitory factor. *Curr Pharm Des* **14:** 3790–3801.
45. Chesney J, Metz C, Bacher M *et al.* (1999) An essential role for macrophage migration inhibitory factor (MIF) in angiogenesis and the growth of a murine lymphoma. *Mol Med* **5:** 181–191.
46. Purhonen S, Palm J, Rossi D *et al.* (2008) Bone marrow-derived circulating endothelial precursors do not contribute to vascular endothelium and are not needed for tumor growth. *Proc Natl Acad Sci USA* **105:** 6620–6625.
47. Coussens LM, Tinkle CL, Hanahan D *et al.* (2000) MMP-9 supplied by bone marrow-derived cells contributes to skin carcinogenesis. *Cell* **103:** 481–490.

48. Lewis JS, Landers RJ, Underwood JC *et al.* (2000) Expression of vascular endothelial growth factor by macrophages is up-regulated in poorly vascularized areas of breast carcinomas. *J Pathol* **192:** 150–158.
49. Huang S, Van Arsdall M, Tedjarati S *et al.* (2002) Contributions of stromal metalloproteinase-9 to angiogenesis and growth of human ovarian carcinoma in mice. *J Natl Cancer Inst* **94:** 1134–1142.
50. Giraudo E, Inoue M, Hanahan D. (2004) An amino-bisphosphonate targets MMP-9-expressing macrophages and angiogenesis to impair cervical carcinogenesis. *J Clin Invest* **114:** 623–633.
51. Ahn GO, Brown JM. (2008) Matrix metalloproteinase-9 is required for tumor vasculogenesis but not for angiogenesis: Role of bone marrow-derived myelomonocytic cells. *Cancer Cell* **13:** 193–205.
52. Kong YZ, Yu X, Tang JJ *et al.* (2005) Macrophage migration inhibitory factor induces MMP-9 expression: Implications for destabilization of human atherosclerotic plaques. *Atherosclerosis* **178:** 207–215.
53. Kawasaki H, Suemori H, Mizuseki K *et al.* (2002) Generation of dopaminergic neurons and pigmented epithelia from primate ES cells by stromal cell-derived inducing activity. *Proc Natl Acad Sci USA* **99:** 1580–1585.
54. Takagi Y, Takahashi J, Saiki H *et al.* (2005) Dopaminergic neurons generated from monkey embryonic stem cells function in a Parkinson primate model. *J Clin Invest* **115:** 102–109.
55. Kawasaki H, Mizuseki K, Nishikawa S *et al.* (2000) Induction of midbrain dopaminergic neurons from ES cells by stromal cell-derived inducing activity. *Neuron* **28:** 31–40.
56. Tabar V, Panagiotakos G, Greenberg ED *et al.* (2005) Migration and differentiation of neural precursors derived from human embryonic stem cells in the rat brain. *Nat Biotechnol* **23:** 601–606.
57. Arjona A, Foellmer HG, Town T *et al.* (2007) Abrogation of macrophage migration inhibitory factor decreases West Nile virus lethality by limiting viral neuroinvasion. *J Clin Invest* **117:** 3059–3066.
58. Calandra T, Echtenacher B, Roy DL *et al.* (2000) Protection from septic shock by neutralization of macrophage migration inhibitory factor. *Nat Med* **6:** 164–170.
59. Donnelly SC, Haslett C, Reid PT *et al.* (1997) Regulatory role for macrophage migration inhibitory factor in acute respiratory distress syndrome. *Nat Med* **3:** 320–323.
60. Lubetsky JB, Dios A, Han J *et al.* (2002) The tautomerase active site of macrophage migration inhibitory factor is a potential target for discovery of novel anti-inflammatory agents. *J Biol Chem* **277:** 24976–24982.
61. Morand EF, Leech M, Bernhagen J. (2006) MIF: A new cytokine link between rheumatoid arthritis and atherosclerosis. *Nat Rev Drug Discov* **5**: 399–410.

62. Barrilleaux BL, Fischer-Valuck BW, Gilliam JK *et al.* (2010) Activation of CD74 inhibits migration of human mesenchymal stem cells. *In Vitro Cell Dev Biol Anim* **46:** 566–572.
63. Fischer-Valuck BW, Barrilleaux BL, Phinney DG *et al.* (2010) Migratory response of mesenchymal stem cells to macrophage migration inhibitory factor and its antagonist as a function of colony-forming efficiency. *Biotechnol Lett* **32:** 19–27.
64. Donn RP, Shelley E, Ollier WE *et al.* (2001) A novel 5′-flanking region polymorphism of macrophage migration inhibitory factor is associated with systemic-onset juvenile idiopathic arthritis. *Arthritis Rheum* **44:** 1782–1785.
65. Baugh JA, Chitnis S, Donnelly SC *et al.* (2002) A functional promoter polymorphism in the macrophage migration inhibitory factor (MIF) gene associated with disease severity in rheumatoid arthritis. *Genes Immun* **3:** 170–176.
66. Radstake TR, Sweep FC, Welsing P *et al.* (2005) Correlation of rheumatoid arthritis severity with the genetic functional variants and circulating levels of macrophage migration inhibitory factor. *Arthritis Rheum* **52:** 3020–3029.
67. Jones MK, Wang H, Peskar BM *et al.* (1999) Inhibition of angiogenesis by nonsteroidal anti-inflammatory drugs: Insight into mechanisms and implications for cancer growth and ulcer healing. *Nat Med* **5:** 1418–1423.

PART V

Atherogenesis and Cardiovascular Disease

V-1

MIF in Atherosclerosis

Heidi Noels*, Jürgen Bernhagen† and Christian Weber‡,§

1. Introduction

Macrophage migration inhibitory factor (MIF) has been recognized as a structurally and functionally unique cytokine with important proinflammatory functions in diverse inflammation-associated diseases, including atherosclerosis. Besides its involvement in proliferative and antiapoptotic responses, MIF-triggered cellular signaling has been linked to proinflammatory protein production, in addition to leukocyte recruitment and arrest through interaction with the chemokine receptors CXCR2 and CXCR4. In this chapter, MIF's structural and biological properties are discussed in the context of its atheroprogressive behavior. Furthermore, we will highlight and clarify the intensive search for new MIF inhibitors as promising therapeutics in the treatment of inflammatory diseases, such as atherosclerosis.

MIF was originally identified as an inhibitor of random macrophage migration.[1] Later, this inhibitory effect was reinterpreted as a desensitization result,[2,3] as a new study clearly showed MIF to specifically direct leukocyte recruitment and arrest.[2] In addition to these chemokine-like functions (see Sec. 4), the MIF exerts multiple other proinflammatory roles (see Sec. 5). Thus, it is not surprising that MIF is a key factor in various acute and chronic inflammatory diseases, such as rheumatoid arthritis,[4] sepsis,[5,6] asthma,[7] and acute respiratory distress syndrome.[8] Moreover, MIF has been identified as an atheroprogressive mediator,[2,9] which can also be linked to the inflammatory processes underlying both the development and progression of atherosclerotic plaques.[10,11] Disturbed lipid balances trigger the activation of the

*Institute for Molecular Cardiovascular Research (IMCAR), RWTH Aachen University, Aachen, Germany.

†Institute of Biochemistry and Molecular Cell Biology, RWTH Aachen University, Aachen, Germany.

‡Institute for Cardiovascular Prevention, Ludwig-Maximilians University, Munich, Germany.

§Corresponding Author. Email: christian.weber@med.uni-muenchen.de endothelial cells (ECs) of the vessel wall, resulting in the production of pro-inflammatory cytokines. Specific cytokines with chemotactic properties, so-called chemokines, then mediate the recruitment and transmigration of inflammatory cells from the blood stream into the inflamed vessel, where they sustain and extend the initial inflammation. While early atherosclerotic lesions are mainly composed of lipid-laden macrophages, excessive cholesterol uptake induces these foam cells to die, creating an extensive acellular necrotic core with deposited cholesterol crystals. A fibrous cap composed of smooth muscle cells (SMCs) and collagen fibers initially covers and stabilizes the plaques, but the activation of proteases by sustained inflammatory signaling gradually induces fibrous cap weakening. Such unstable plaques are prone to rupture, and the associated formation of thrombotic occlusions in the blood stream could ultimately lead to heart attack, stroke or peripheral organ dysfunction, by hindering the normal oxygen supply to the heart, brain or periphery.

The crucial role of chemokines in driving atherosclerotic inflammatory processes through leukocyte recruitment, arrest, and vessel infiltration has triggered intensive chemokine research. At first sight, the chemokine system appears rather redundant, given that around 50 human chemokines can interact with about 20 chemokine receptors. In addition, many chemokines have broad cellular sources and are able to bind multiple receptors. However, specificities in cell type-, site-, and disease stage-specific recruitment capacities have been identified, in addition to a preferential involvement of chemokines in specific steps of the recruitment cascade.[12] Thus, the apparent redundancy of the chemokine system has to be re-evaluated as a system with great robustness, ensuring a highly orchestrated recruitment of inflammatory cells into inflamed vessels. In this chapter, we will focus on the chemokine-like protein MIF as a recently emerged key factor in the vascular inflammatory reactions underlying atherosclerosis.

2. Structural and Functional Classification of MIF

MIF is an evolutionary conserved protein with an abundant expression in humans and nonprimate mammals. Furthermore, MIF homologues have been identified in avians, fish, plants (*Arabidopsis thaliana*), the nematode *Caenorhabditis elegans*, cyanobacteria, ticks, and parasites.[13,14]

X-ray crystallographic analysis identified that the MIF monomer, composed of two antiparallel α-helices positioned along a four-stranded β-sheet, is able to form a barrel-shaped homotrimer, which is stabilized by the interaction of an additional two β-strands of one subunit with the β-sheets of the neighboring subunits.[15] However, the true physiological oligomerization state of MIF remains unclear, and probably balances between monomers, dimers, and trimers, depending on the local MIF concentration. Biochemical studies did indeed identify MIF monomers and dimers as the main MIF species at physiological concentrations (ng/ml), while mostly MIF trimers were found at higher MIF levels (>10 μg/ml).[13,16,17]

Notwithstanding its important function as a cytokine, MIF cannot be structurally classified in any of the known cytokine families, due to broad differences in amino acid sequence and tertiary structure.[5,15,18,19] Likewise, structurally, MIF does not belong to one of the four classical chemokine subclasses characterized by the conserved N-terminal cysteine motif (C, CC, CXC or CX_3C),[20] despite MIF's ability to interact with the chemokine receptors CXCR2 and CXCR4 (see Sec. 4).[2] Nevertheless, MIF displays structural similarities with CXC chemokines known to bind CXCR2, such as CXCL1 and CXCL8. As such, MIF harbors a pseudo-(E)LR-motif composed of two nonadjacent amino acids in exposed neighboring loops, with an identical distance to that in the ELR-motif characterizing the canonical CXCR2 ligands.[2,21] Furthermore, an N-like loop in MIF, structurally similar to the N-loop of classical CXC chemokines, was recently discovered to interact with CXCR2, indicating that the two-site binding model underlying chemokine receptor activation[22] also applies to MIF.[23] In addition, the MIF monomer shares considerable structural similarities with CXCL8 dimers.[15] This too could underlie the ability of MIF to bind CXCR2, given the recognized chemokine receptor dimerization upon ligand interaction and the recent appreciation of chemokine ligand dimerization as a way to modulate and fine-tune interactions with their receptors.[24] In conclusion, MIF cannot be classified as a classical chemokine, but structurally displays critical chemokine elements underlying its chemokine-like functions.

Interestingly, also other inflammatory mediators have been identified to act as nonclassical chemokine receptor ligands. For example, an N-terminal cleavage fragment of tyrosyl-tRNA synthetase (TyrRS) has been shown to induce a CXCR1-dependent neutrophil chemotaxis through the exposure of a critical ELR-motif, which cannot be found back in the native full-length TyrRS due to a different spatial presentation.[25–27] Furthermore, the autoantigenic histidyl- and asparaginyl-RS, which are released under conditions of apoptosis, mediate the migration of T cells, monocytes, and immature dendritic cells through binding of the CCR5 and CCR3 receptors, respectively.[28] Similarly, immature dendritic cells and memory T cells are recruited by the human antimicrobial peptides β-defensin-1 and -2 through CCR6.[29,30] Although further studies are required to identify the true structural characteristics underlying the ability of these inflammatory mediators to interact with chemokine receptors, the existence of such proteins mediating leukocyte chemotaxis through molecular mimicry with classical chemokines is being increasingly appreciated. A formal extension of the current chemokine classification scheme by the addition of CXC chemokine-like (CXCLL) and CC chemokine-like (CCLL) ligands should therefore be taken into consideration.[31]

Also in a functional sense, MIF cannot be regarded as a classical cytokine, as it exerts important intracellular roles in addition to its extracellular cytokine/chemokine function (see Sec. 4). Moreover, MIF distinguishes itself from conventional cytokines in that it harbors enzymatic potential. As such, MIF shares a conserved CXXC motif with thiol-protein oxidoreductases and shows redox activity *in vitro*, suggesting that intracellular MIF is involved in cellular redox homeostasis.[32,33] Furthermore, the MIF trimer exposes a tautomerase pocket with high similarity to the catalytic site of a family of bacterial isomerases/tautomerases and the human D-dopachrome tautomerase.[13,34] The biological significance of MIF's tautomerase function remains unclear, however, as until now only synthetic substrates have been identified.[13,35] In addition, it remains disputed whether or not this tautomerase activity contributes to MIF's inflammatory properties. Small molecular weight (SMW) drugs targeting the tautomerase activity of MIF have been observed to confer protection against MIF's proinflammatory effects *in vitro* and *in vivo*.[36,37] However, these protective effects are expected to be caused by changes in the tertiary or quaternary structure of MIF, instead of being a direct consequence of a defective tautomerase function.[38,39]

3. MIF Expression and Secretion

MIF displays a broad tissue and cellular expression pattern. Although MIF was originally identified as a T cell-derived cytokine, it was later appreciated that monocytes and macrophages constitute the main cellular sources of MIF protein production in the immune system. In addition, neutrophils, dendritic cells, B cells, eosinophils, mast cells, and basophils synthesize MIF. MIF is also expressed in many other tissues and cell types, including ECs, SMCs, epithelial cells, fibroblasts, keratinocytes, cardiomyocytes, neurons, and various tumor cells.[13] Many of these cells have been shown to contain large intracellular stores of MIF protein, of which the secretion is tightly regulated by specific inflammatory stimuli and stress factors. Macrophages, for example, rapidly secrete MIF upon stimulation with lipopolysaccharide (LPS), bacterial exotoxins, and cytokines such as tumor necrosis factor (TNF)- α and interferon (IFN)- γ.[40,41] Ischemia/reperfusion challenge triggers cardiomyocytes to secrete MIF,[42,43] and ECs were also shown to rapidly release MIF from intracellular stores upon hypoxic conditions.[44] This first phase of MIF secretion is followed by an induction of MIF expression, which restores the intracellular MIF content and allows a second phase of MIF release.[44] Interestingly, many proatherogenic factors trigger MIF synthesis. While CD40 ligand (CD40L) and angiotensin II (AngII) drive MIF expression in monocytes/macrophages,[45] oxidized low-density lipoprotein (oxLDL)[9] and hypoxia[44,45] have been shown to induce MIF expression in both monocytes/macrophages and ECs.

Despite the observance of a tightly regulated MIF secretion, the underlying mechanisms remain to be clarified. So far, it is known that MIF release does not occur through the classical endoplasmic reticulum (ER)/Golgi pathway, as MIF lacks an N-terminal signal sequence mediating the translocation into the ER.[18] Instead, pharmacological inhibitors of the ATP binding cassette (ABC) transporter subfamily 1 (ABCA1) were found to severely inhibit MIF secretion from LPS-stimulated monocytes, indicating that MIF release requires the function of an ABCA1 transporter protein for at least one step of the secretion cascade.[46] Furthermore, a proteomics-based secretome analysis of keratinocytes revealed a role for caspase-1 (CASP1) in the nonconventional secretion of MIF and other leaderless proteins.[47] In addition, a specific function of the Golgi-associated protein p115 in monocytic MIF release was identified through silencer RNA (siRNA)-mediated knock-down studies. MIF and p115, which is involved in Golgi vesicle trafficking, interact in the cytosol and are co-secreted from stimulated monocytes, but the associated mechanism requires further clarification.[48] Finally, MIF secretion seems to be negatively regulated by the MIF-binding protein Jun-activation domain-binding protein 1 (JAB1), also known as COP9 signalosome subunit 5 (CSN5), as evidenced from siRNA interference experiments in fibroblasts and tumor cells.[49]

4. MIF Receptors in Relation with MIF's Chemokine-Like Functions

The initial association of MIF with an increased leukocyte accumulation in multiple inflammatory diseases[50–56] was later explained by the important chemokine-like functions of MIF. The first evidence for a role of MIF in inflammatory cell adhesion was the observation that MIF- or oxLDL-induced monocyte arrest to aortic ECs could be reduced when the ECs were preincubated with an MIF-blocking antibody.[56] Later, MIF was shown to be exposed on the EC surface, where it mediated the adhesion of monocytes, neutrophils, and T cells through stimulation of the chemokine receptors CXCR2 (on monocytes, neutrophils) and CXCR4 (on T cells).[2] Similar to the mechanism underlying leukocyte arrest by immobilized classical chemokines,[57,58] MIF binding to CXCR2/4 initiates a $G_{\alpha i}$-coupled signaling pathway that induces the activation of leukocytic β1/2-integrins, resulting in their interaction with integrin ligands on activated ECs. In addition to this leukocyte arrest function for membrane-bound MIF, *in vitro* transwell migration assays and an *in vivo* MIF-induced peritonitis model revealed soluble MIF acting as a direct chemoattractant for monocytes/neutrophils and T cells through the CXCR2 and CXCR4 receptors, respectively. Again, $G_{\alpha i}$ protein and phosphoinositide-3 kinase (PI3K)-involved signaling initiated by CXCR-receptor triggering were shown to be involved.[2]

Interestingly, further research revealed that MIF also interacts with the previously unrecognized CXCR2/CD74 receptor complex.[2] CD74 is the cell surface-expressed form of the MHC class II-associated invariant chain and was the first identified high-affinity binding receptor for MIF.[59] As CD74 lacks known intracellular signaling domains, MIF-induced CD74 signal propagation requires the recruitment of additional signaling molecules, such as the transmembrane protein CD44 and Src kinases.[60] Although the absence of CD74 on neutrophils underlines that MIF is able to bind to CXCR2 alone, ectopic CD74 expression in CD74-deficient promyelocytic cells and antibody-mediated CD74-blocking clearly showed CD74 to enforce CXCR2-mediated MIF-induced monocyte arrest, potentially through the amplification of CXCR2-induced downstream signals or through the induction of additional signaling pathways.[2]

More recently, MIF-responsive CXCR4/CD74 receptor complexes were also identified in monocytes, T cells, and fibroblasts.[61,62] In the future, it will be interesting to unravel the precise composition of these different MIF receptor complexes, in addition to the cellular and environmental conditions in which each complex is formed. As such, one could speculate as to the involvement of CD44 as a signal-transducing co-receptor in the CXCR2/CD74 and CXCR4/CD74 complexes, and it is even more intriguing whether or not heterotrimeric CXCR2/CXCR4/CD74 or heterotetrameric CXCR2/CXCR4/CD74/CD44 complexes exist. In addition, the specific stochiometry of such receptor complexes needs further evaluation. While the CXCR2 and CXCR4 receptor are accepted to form homodimers upon ligand binding, the precise composition of CD74/CD44 or of CD74 in complex with CXCR2 and CXCR4 remains unclear. Nonetheless, the existence and importance of chemokine dimerization is being increasingly appreciated, as it constitutes an interesting approach to fine-tuning leukocyte arrest in order to ensure a highly controlled cell-, site-, and disease stage-specific inflammatory cell adhesion.[12,63] Thus, the identification of a complex formation between a seven helix-spanning G-protein coupled receptor (CXCR) and a single helix-spanning type II membrane protein (CD74) could even further add to the robustness of the chemokine system by supplementary fine-tuning the initiated signaling pathways and cellular responses. In addition, unraveling the composition and biological significance of such receptor complexes could offer an interesting new therapeutic approach to treating associated diseases, as it could minimize unwanted side effects by specifically targeting disease-associated chemokine signaling, while leaving the physiological activities of chemokines more or less unaffected (see also Sec. 7).

Complementary to this direct involvement of MIF in leukocyte chemotaxis through its CXCR2/4 binding potential, MIF also indirectly contributes to leukocyte recruitment and arrest through MIF-induced expression of canonical chemokines and adhesion molecules. For example, MIF triggers the CCL2-CCR2-mediated monocyte adhesion and transmigration in the cremaster microvasculature by inducing CCL2 secretion from ECs.[64] In addition, intercellular adhesion molecule (ICAM)-1 and vascular cell adhesion molecule (VCAM)-1 are upregulated in ECs upon MIF stimulation.[50,52,54]

In conclusion, MIF can be regarded as a key factor regulating leukocyte recruitment and arrest, due to the combination of important direct and indirect chemotactic functions.

5. MIF-Induced Cellular Signaling

MIF is an unusual cytokine in that it exerts extracellular as well as intracellular functions, which both contribute to the proinflammatory behavior of MIF. Multiple signaling pathways have been shown to be modulated by extracellular MIF. As such, MIF induces both a transient and sustained activation of the mitogen-activated protein kinases (MAPK) extracellular signal-regulated 1 (ERK1) and ERK2.[60,65] Sustained MIF-triggered ERK1/ERK2 activation occurs through the MIF receptor complex CD74/CD44 and protein kinase A, and is linked with cell proliferation and enhanced proinflammatory phospholipase A2 (PLA2) activity.[60,66] Given that PLA2 and the associated arachidonic acid production constitute the key targets of the anti-inflammatory glucocorticoids used in the treatment of inflammatory diseases, the MIF-induced upregulation of PLA2 activity can at least partly explain the observed inhibitory effect of MIF on glucocorticoid therapeutics.[67–69]

In contrast to ERK1/2, both stimulatory and inhibitory effects of MIF have been observed on signaling through the MAPK Jun N-terminal kinase (JNK), depending on the cell type and cellular conditions studied. While MIF seems to hinder JNK signaling and JNK-mediated apoptosis in cells preconditioned to stress,[70,71] a fast and transient activation of the JNK/activator protein 1 (AP1) pathway was observed upon MIF triggering of resting T cells and fibroblasts. This stimulatory effect occurred through the CXCR4/CD74 receptor complex and induced a rapid Src/PI3K/JNK/AP1-mediated expression of the proinflammatory and proangiogenic chemokine CXCL8.[62]

In addition, the protein kinase AKT is activated upon MIF stimulation through CD74/Src/PI3K signaling, and the resulting AKT-mediated phosphorylation of the proapoptotic proteins BAD and FOXO3A has been linked to cell survival.[49] Comparably, the increased cell survival of chronic lymphocytic leukemia B cells could be related to their enhanced expression of CD74 compared to normal B cells, resulting in an increased MIF/CD74-triggered NF-κB activation and associated interleukin (IL)-8 production, promoting cell survival.[72]

Furthermore, MIF was shown to interfere with macrophage apoptosis through a negative regulation of the proapoptotic protein p53 at different levels: while extracellular MIF prevents p53 accumulation through the CD74/CD44-mediated upregulation of cyclooxygenase (COX)-2,[73] intracellular binding of MIF to p53 attenuates p53 activity by stabilizing p53-MDM2 protein complexes.[74] The cytosolic interaction of MIF with the proapoptotic BCL2 family member BIM could further add to MIF's antiapoptotic properties, as observed in fibroblasts.[75]

On the other hand, intracellular interaction of MIF with JAB1/CSN5 has been shown to block the JAB1-mediated activation of JNK/AP1. Also the JAB1-induced degradation of the cyclin-dependent kinase inhibitor p27Kip1 was shown to be negatively regulated by intra- and extracellular MIF, implying MIF in cell cycle arrest.[70]

In contrast, a recent genome-wide microarray analysis implied MIF in cell cycle progression via multiple pathways, including the activation of MAPK, PI3K/AKT, NF-κB, and c-Myc signaling, and the inhibition of TGF-β and p53-dependent pathways.[76] Similarly, MIF was shown to be involved in the activation of the transcription factor E2F1, important in cell cycling.[77]

In conclusion, many signaling pathways have been identified to be regulated by extra- or intracellular MIF, which in combination underlie MIF's proinflammatory and antiapoptotic properties. Nonetheless, MIF has also been connected to cell cycle arrest functions. It is expected that in the future, more MIF interaction partners and MIF-controlled signaling events will be identified, allowing an even more detailed description of MIF's pleiotropic inflammatory behavior.

6. MIF as an Important Player in Atherosclerosis

6.1 *MIF expression in atherosclerotic lesions*

Initial observations in hypercholesterolemic rabbits indicated a high *de novo* MIF expression in vascular ECs and SMCs of developing atherosclerotic plaques, indicating an important role for MIF in atherogenesis. Also, monocytes adhering to the endothelium and macrophages in early fatty streaks were associated with high MIF concentrations, which decreased again, however, in the foam cells of more advanced atherosclerotic lesions, possibly due to lower functional activities of these macrophages. Similarly, MIF expression in SMCs appeared to be transient, as almost no MIF could be detected in SMCs of foam cell-rich and advanced lesions.[52]

MIF cytoplasmic and nuclear expression was also detected in all cell types of mouse atherosclerotic plaques, with a predominant MIF expression in foam cells compared to ECs, SMCs, and lymphocytes.[54] Furthermore, MIF was found to be upregulated during accelerated lesion formation associated with carotid artery injury of atherosclerosis-prone mice: while MIF was exclusively expressed in the endothelium of uninjured vessels, EC denudation induced MIF levels in the medial SMCs surrounding the denuded area already after 24 hrs. Two weeks after injury, MIF was found to be predominantly expressed by macrophage-derived foam cells and ECs, in contrast to the negligible MIF concentrations in neointimal SMCs.[55,56]

Finally, MIF expression has also been associated with human atherosclerosis.[9,45] While only low levels of MIF could be detected in ECs and SMCs of healthy arteries, atherosclerotic lesions presented themselves with a significantly higher MIF expression in all cell types involved, including ECs, SMCs, macrophages, and T cells.[9] Interestingly, and in contrast to the findings in rabbits, MIF expression even increased during the progression of atherosclerotic lesions, suggesting an important role for MIF in plaque destabilization.[9,45]

6.2 *Mouse models confirm a role for MIF in atheroprogression and injury-induced neointima formation*

The proatherogenic role of MIF was further confirmed in different mouse models. Atherosclerosis-prone LDL receptor-deficient (*Ldlr*$^{-/-}$) mice with a genetic deletion of *Mif* developed smaller and less-progressed lesions upon high-fat diet, which was associated with a reduced cell proliferation and an attenuated protease expression.[53] In addition, apolipoprotein E-deficient (*ApoE*$^{-/-}$) mice that were treated with a MIF-blocking antibody during normal chow diet showed a clear decrease in intimal macrophages and in inflammatory mediator expression both in the circulation and on a local aortic level. Although this was not associated with a significant reduction in atherosclerotic lesion formation, these data do confirm a role for MIF in murine vascular inflammation.[54] Promising results were obtained in a study of *ApoE*$^{-/-}$ mice with established atherosclerotic lesions, in which the treatment with a MIF-neutralizing antibody induced a true regression of plaque size and a decrease in plaque inflammation.[2] These data clearly underline MIF's role in the progression of atherosclerosis and identify MIF blockade as a potential therapy to treat advanced atherosclerotic plaques.

In addition to its role in native or diet-induced atherosclerosis, MIF also contributes to accelerated vessel injury-induced neointima formation, which is frequently associated with therapeutic revascularization methods, such as balloon angioplasty or stent implantation, aimed at opening up narrowed atherosclerotic vessels.[78] An active role for MIF in injury-induced vascular inflammation and associated neointima formation was confirmed after experimental angioplasty inducing endothelial denudation in *Ldlr*$^{-/-}$ mice. One week after injury, the neointimal inflammatory cell influx and the medial cell proliferation were significantly decreased by the neutralization of circulating MIF with a MIF-blocking antibody, in contrast to a clearly increased cell apoptosis in both the neointima and the media. This resulted in a significant reduction in neointima formation already four weeks after injury, along with a marked decrease in neointimal cell proliferation.[55] In a comparable model of wire-induced carotid artery injury in *ApoE*$^{-/-}$ mice, MIF antibody blocking similarly reduced the neointimal macrophage content three weeks after injury. Although the lesion size was not significantly reduced, lesions were found to be more stable in the absence of circulating MIF, as characterized by their increased SMC and collagen content.[56] Finally, microarray profiling studies in a mouse model of low blood flow-induced carotid artery media-intima thickening identified enhanced MIF expression as a genetic susceptibility factor in vascular remodeling.[79]

Together, these animal studies reveal MIF as an important proinflammatory protein involved in native and injury-induced atherosclerosis, with a special emphasis on its role in lesion progression and destabilization. Although this could pave the way for new strategies in the treatment of atherosclerosis (see Sec. 7), it also warrants the frequent use of glucocorticoids in the treatment of inflammatory diseases. Not only does MIF override the anti-inflammatory effects of glucocorticoids,[68,69] glucocorticoid-induced MIF expression could underlie the accelerated atherosclerosis observed in relation with glucocorticoid therapy.[4]

6.3 *Epidemiological studies correlate MIF with atherosclerotic disease severity*

Single nucleotide polymorphisms (SNPs) in the human MIF gene have been correlated with the severity of cardiovascular disease. A G-to-C transition at position –173 in the MIF promoter region, which was reported to be associated with an increased susceptibility to diverse inflammatory diseases,[80] could also be identified as a risk factor for coronary heart disease (CHD),[81] although this did not hold for rheumatoid arthritis patients.[82] Furthermore, although Herder and colleagues could not detect a similar correlation of CHD with circulating MIF levels,[81] a more recent study did find a predictive association of augmented MIF plasma levels with increased coronary failure in patients with CHD and impaired glucose tolerance or type 2 diabetes mellitus.[83] In addition, increased myocardial infarction risk was observed in Czech female patients carrying the GG genotype of the MIF SNP rs1007888, which resides in the 3′ flanking region of the gene.[84]

Taken together, these epidemiologic studies further support a proatherogenic role of human MIF.

6.4 *Atherosclerosis-related stimuli driving MIF expression*

Diverse proinflammatory mediators have been described to induce cellular MIF expression and/or secretion. For example, the proinflammatory and proatherogenic cytokines TNF-α and IFN-γ rapidly induce MIF release from intracellular stores in monocytes and macrophages,[40] while CD40L, AngII, oxLDL, and hypoxia trigger monocytes/macrophages to synthesize MIF.[45] Also ECs and SMCs upregulate MIF expression in response to oxLDL,[9,85] whereas hypoxic conditioning of ECs, an important factor driving intraplaque angiogenesis, has been shown to induce both a rapid release of MIF and a second phase of MIF production.[44,45]

Together, these data can explain the observed upregulation of MIF expression in all cell types present in atherosclerotic lesions (see Sec. 6.1).

6.5 MIF-triggered biological responses in the context of atherogenesis (Fig. 1)

As a hallmark of atherosclerosis, monocytes adhere to the inflamed endothelium and transmigrate into the vessel, where they transform into lipid-laden macrophages and foam cells through the uptake of modified lipids. The high monocyte accumulation associated with MIF expression in atherosclerotic plaques can be explained by the important direct and indirect chemotactic functions of MIF, as discussed in Sec. 4. Furthermore, MIF enhances the uptake and degradation of oxLDL by macrophages,[86] which itself upregulates macrophage MIF expression.[9] Such a positive amplification loop could highly stimulate intracellular cholesterol accumulation and could provide an explanation for the decreased foam cell content in the neointimal carotid artery lesions of hypercholesterolemic *ApoE*$^{-/-}$ mice treated with a MIF-blocking antibody.[56]

In addition, MIF's stimulatory effect on macrophages to produce inflammatory cytokines (including TNF-α,[40,68,87] IL-1β,[50,68] IL-6,[68] and IL-8[68]) and other inflammatory mediators, such as nitric oxide (NO)[87] and inducible nitric oxide synthase (iNOS),[50] could further add to MIF's proinflammatory and proatherogenic behavior.

Also, MIF's capacity to interfere with the activation-induced, p53-mediated apoptosis of macrophages (as discussed in Sec. 5)[73] can be expected to strengthen MIF's proatherogenic role, as *p53*-deficient *ApoE*$^{-/-}$ mice display an increased aortic plaque formation, associated with an increased inflammatory cell proliferation and a reduced apoptosis of both inflammatory cells and SMCs.[88] In addition, MIF's antiapoptotic effect could contribute to the observed increase in neointimal cell apoptosis upon arterial injury in the presence of a MIF-blocking antibody.[55]

An additional mechanism underlying MIF's role in atherogenesis could relate to its interaction with JAB1, which was confirmed in human atherosclerotic lesions.[9] Given that MIF negatively regulates JAB1 functions[70] and that JAB1 has been identified as a negative regulator of the NF-κB signaling pathway,[89] one could speculate about a potential co-regulatory role of the JAB1/MIF complex in NF-κB signaling in the context of atherosclerosis. This could have important effects on the development and progression of atherosclerosis, as NF-κB is a key transcription factor with important inflammatory properties in all stages of this disease.[90,91]

Diverse other mechanisms have been proposed to underlie the intriguing effect of MIF on the progression and destabilization of atherosclerotic lesions. First of all, MIF-deficient aortic SMCs were reported to show a

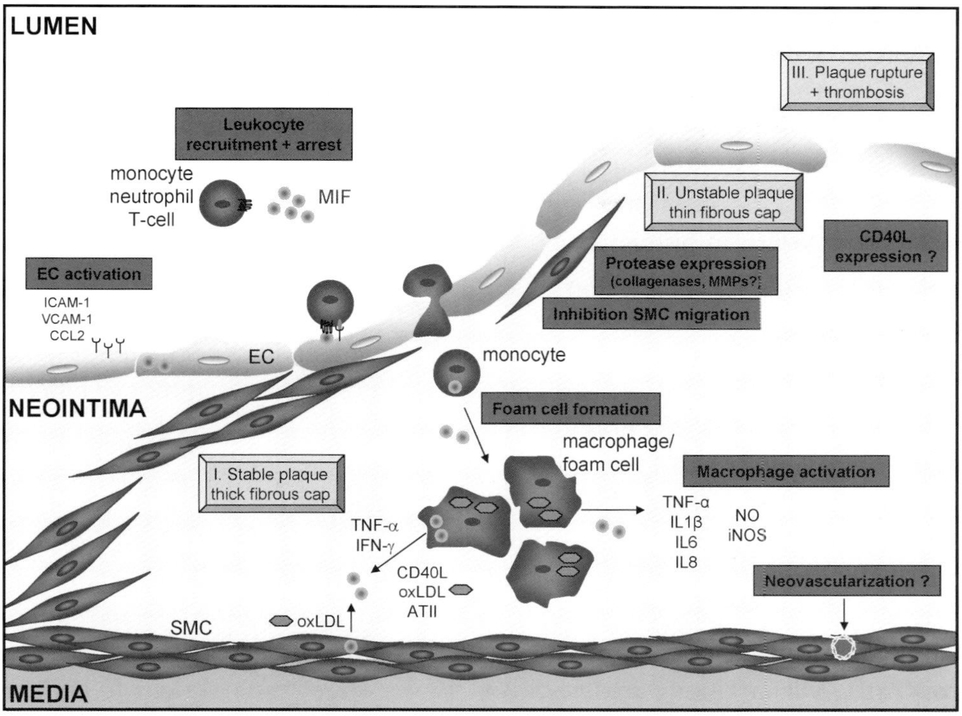

Fig. 1. MIF expression and proinflammatory properties in the context of atherosclerosis. The expression and secretion of MIF is induced by many stimuli associated with atherogenesis, explaining the high MIF levels in all cell types present in atherosclerotic plaques, including ECs, SMCs, monocytes/macrophages, and T cells. On the other hand, MIF exerts pleiotropic inflammatory functions, contributing to the development and progression of atherosclerosis. First of all, MIF triggers the recruitment and arrest of monocytes/neutrophils and T cells through its interaction with the chemokine receptors CXCR2 and CXCR4, respectively. MIF also induces ECs to upregulate the expression of adhesion molecules (ICAM-1, VCAM-1) and canonical chemokines (CCL2), which contributes to its chemotactic potential. Next, MIF stimulates foam cell formation and triggers macrophages to produce proinflammatory mediators, sustaining local vessel inflammation. By inhibiting PDGF-BB-induced SMC migration and promoting protease activation, MIF also undermines the stability of a protective fibrous cap. In addition, MIF has been associated with CD40L expression and could trigger intraplaque neovascularization through its angiogenic capacities. Together, these functions can explain MIF's recognized role in the progression and destabilization of atherosclerotic lesions. See the text for more details.

reduced expression and activity of elastin- and collagen-specific cysteine proteases, which was linked to the observed decrease in lesion progression in *Ldlr*$^{-/-}$ *Mif*$^{-/-}$ mice.[53] Although this study did not detect an effect of *Mif* deletion on the expression or activity of matrix metalloproteinases (MMPs) in SMCs, others have correlated MIF with an increased expression of MMPs, such as MMP1[92] and MMP9.[93] In addition to a strong co-localization of MIF with these MMPs in human vulnerable lesions, a direct role of MIF in MMP expression was confirmed *in vitro* upon MIF-mediated stimulation of macrophages (for MMP9) and SMCs (for MMP1/9), which was also associated with an enhanced collagenase activity.[92,93] However, the speculated role for MMP1 and MMP9 in plaque progression was contradicted in mouse atherosclerosis studies. Hypercholesterolemic mice with a macrophage-specific overexpression of MMP1 unexpectedly displayed smaller and less-progressed lesions upon high-fat diet,[94] while *ApoE*$^{-/-}$ mice with a genetic *Mmp9* deficiency revealed a similar atheroprotective and plaque-stabilizing role for MMP9.[95] Thus, the strong MMP1/9 expression observed in vulnerable human lesions[92,93] was suggested to be the result of a healing response rather than the driving force behind plaque destabilization.[95] Analogous with MMP1/9, MIF antibody treatment of *ApoE*$^{-/-}$ mice was shown to reduce aortic MMP2 expression.[54] However, the role of MMP2 in plaque progression has not been investigated. Until now, only an involvement of MMP12 in MIF-mediated plaque destabilization could be hypothesized, based on a simultaneous upregulation of MIF and MMP12 expression in abdominal aortic aneurysms,[96] in combination with an earlier and more stable plaque phenotype in hyperlipidemic *Mmp12*-deficient *ApoE*$^{-/-}$ mice.[95] Nonetheless, a direct correlation of MIF and MMP12 expression *in vitro* or in atherosclerotic lesions has so far not been reported.

Given that SMC migration is essential for plaque stabilization through the formation of a protective fibrous cap, the *in vitro* observation that MIF blocks the long-term platelet-derived growth factor (PDGF)-BB-induced migration of SMCs could represent a second mechanism underlying MIF's plaque destabilizing effect.[97] In contrast, the effect of MIF on SMC proliferation remains disputed. Although the initial observation of a MIF stimulatory effect on SMC proliferation *in vitro*[55] corresponds with a decreased medial cell proliferation upon MIF antibody administration in a mouse vascular injury model[55] and with a reduced SMC proliferation in atherosclerotic lesions of *Ldlr*$^{-/-}$ *Mif*$^{-/-}$ mice,[53] others could not detect a mitogenic effect of exogenous MIF on vascular SMCs.[97]

Moreover, the reduced aortic CD40L expression in hyperlipidemic *ApoE*$^{-/-}$ mice treated with an MIF-blocking antibody suggests a role for MIF in regulating CD40L expression.[54] Thus, enhanced CD40/CD40L signaling could represent a third mechanism strengthening MIF's involvement in plaque destabilization.[98] In the same study, MIF blockade was also associated with a reduced aortic expression of the proinflammatory cytokines TNF-α and IL12, and of the transcriptional regulators C/EBP-β and phospho-c-Jun. Given the involvement of C/EBPs in the expression of inflammatory cytokines and the presence of AP1 binding sites in the promoter region of adhesion molecules (e.g., E-selectin) and collagenase, these findings can further explain MIF's proatherogenic properties.[54]

Finally, the identification of MIF as a proangiogenic factor in matrigel angiogenesis assays and diverse tumor models[99–101] triggers speculation about a potential role of MIF in lesion destabilization through intraplaque neovascularization. However, additional studies of severely advanced plaques are required in order to investigate such a possible connection.

In conclusion, MIF's proatherogenic properties can be explained by the pleiotropic proinflammatory functions of MIF. It can be hoped that, in the future, interference with some of these functions will prove to be effective in the treatment of atherosclerosis.

7. Therapeutic Targeting of MIF in Inflammation and Atherosclerosis

The undisputed role of MIF in inflammatory diseases, including atherosclerosis, proposes MIF as an interesting target for the development of new therapeutics. In particular, chemical SMW drugs or peptide antagonists constitute an attractive method to block specialized MIF functions, as they can be orally administrated, are relatively cost-efficient, and are associated with a lower antigenicity in comparison with the expensive protein-based therapeutics such as neutralizing antibodies or soluble protein receptors. The key question is: Which MIF function should best be targeted, and in which way? The fact that MIF is an unusual cytokine in many aspects offers diverse options to be explored.[102]

As such, SMW drugs could be used to inhibit nonconventional MIF secretion (see Sec. 3). However, care should be taken to avoid a simultaneous blockade of cellular secretion of other so-called leaderless proteins that may follow the same secretion mechanism.[103–105] In addition, merely targeting ABCA1, p115 or CASP1 seems premature, as this could interfere with other cellular processes given the multifunctional aspects of these proteins. Thus, more research is required before MIF secretion can be regarded as an attractive therapeutic approach to blocking circulating MIF functions.

On the other hand, one could choose to target MIF/MIF receptor interactions to block MIF's involvement in leukocyte recruitment and arrest (see Secs. 2 and 4). As an alternative to neutralizing anti-MIF antibodies, MIF receptor binding could be blocked by the use of soluble MIF receptors. Although this would be feasible for the MIF receptor CD74 given its extended extracellular domain, the beneficence of such an approach was recently questioned by the observation that CD74 underlies important cardioprotective effects of MIF in myocardial ischemia/reperfusion injury.[43,71] As an attractive alternative, chemical or peptide SMW drugs could be explored to target the chemokine-like functions of MIF. Practical experience in the application of such SMW drugs targeting chemokine receptors has been obtained in the context of diverse inflammatory and immune diseases in addition to cancer, although not always with promising results.[106] An interesting approach would be the development of peptide-based therapeutics targeting the pseudo-(E)LR and the N-loop-like motifs in MIF that were identified to be critical for the MIF-CXCR2 interaction.[23,63,107] Although in theory MIF/CXCR4 complexes could also be targeted after identifying the critical MIF or CXCR4 motifs involved in this interaction, the broad expression profile of CXCR4 on all leukocyte subsets and many other cell types in the body and the important homeostatic functions of CXCL12/CXCR4 signaling would make it much more difficult to design drugs targeting specific MIF-driven cellular processes with few side effects.

Finally, the identification of a conserved catalytic tautomerase and redox motif in MIF, and the observation that SMW drugs targeting the tautomerase catalytic pocket interfere with MIF's biological functions (see Sec. 2),[36,37,39] provide an additional approach to inhibiting specific MIF functions through SMW drug-mediated, active-site directed targeting or allosteric regulation of these catalytic activities.[39]

In conclusion, MIF can be considered as a promising target protein in the development of new therapeutics for the treatment of atherosclerosis and other inflammation-associated diseases. Although much progress has been made in the search for new inhibitors targeting MIF through different mechanisms, considerable care should be taken in all these approaches to fully unravel all the effects of new inhibitors on the pleiotropic MIF functions, to avoid negative side effects by disturbing untargeted functions through, e.g., conformational changes or influences on MIF-protein interactions. Furthermore, the association of MIF inhibition with a possible immunosuppression in patients remains to be investigated, although it is already promising that *Mif*-deficient mice do not display a high risk of infection under normal conditions.

8. Concluding Remarks

Substantial evidence from human and mouse studies supports an important role for MIF in vascular inflammation and the associated development and progression of atherosclerotic lesions. Through a multitude of intra- and extracellular activities, MIF controls decisive checkpoints such as EC activation, leukocyte recruitment and arrest, macrophage activation, and foam cell formation, in addition to SMC migration and integrity of the extracellular matrix through protease expression regulation (Fig. 1). Over the years, much progress has been made to unravel the underlying mechanisms. In addition to the recognition of the chemokine receptors CXCR2 and CXCR4 as binding receptors for MIF, structural MIF motifs required for CXCR2 receptor binding were identified, binding partners for MIF were revealed, and intracellular pathways regulated by MIF were characterized. Additional research is now necessary to address the new questions arising: Which heteromultimeric receptor complexes are able to bind MIF, and what is their individual role in the diverse MIF functions? What are the precise interaction sites of MIF with these individual receptor complexes, and which MIF stochiometry is involved? What is the best strategy to target specific functions of MIF? Nevertheless, researchers are hopeful that MIF does indeed provide promising novel strategies in the treatment of atherosclerosis and diverse other inflammatory diseases. Regardless, the search for new MIF-targeting therapeutics has already made a good start.

9. Acknowledgments

This work was supported by the Alexander von Humboldt Foundation (to HN), the Deutsche Forschungsgemeinschaft (FOR 809, IRTG 1508) (to JB and CW), the European Research Council (to CW) and by Foundation Leducq (to CW).

References

1. David JR. (1966) Delayed hypersensitivity *in vitro*: Its mediation by cell-free substances formed by lymphoid cell-antigen interaction. *Proc Natl Acad Sci USA* **56**: 72–77.
2. Bernhagen J, Krohn R, Lue H *et al.* (2007) MIF is a noncognate ligand of CXC chemokine receptors in inflammatory and atherogenic cell recruitment. *Nat Med* **13**: 587–596.

3. Frascaroli G, Varani S, Blankenhorn N *et al.* (2009) Human cytomegalovirus paralyzes macrophage motility through down-regulation of chemokine receptors, reorganization of the cytoskeleton, and release of macrophage migration inhibitory factor. *J Immunol* **182**: 477–488.
4. Morand EF, Leech M, Bernhagen J. (2006) MIF: A new cytokine link between rheumatoid arthritis and atherosclerosis. *Nat Rev Drug Discov* **5**: 399–410.
5. Bernhagen J, Calandra T, Mitchell RA *et al.* (1993) MIF is a pituitary-derived cytokine that potentiates lethal endotoxaemia. *Nature* **365**: 756–759.
6. Calandra T, Echtenacher B, Le Roy D *et al.* (2000) Protection from septic shock by neutralization of macrophage migration inhibitory factor (MIF). *Nat Med* **6**: 164–169.
7. Mizue Y, Ghani S, Leng L *et al.* (2005) Role for macrophage migration inhibitory factor in asthma. *Proc Natl Acad Sci USA* **102**: 14410–14415.
8. Donnelly SC, Haslett C, Reid PT *et al.* (1997) Regulatory role for macrophage migration inhibitory factor in acute respiratory distress syndrome. *Nat Med* **3**: 320–323.
9. Burger-Kentischer A, Goebel H, Seiler R *et al.* (2002) Expression of macrophage migration inhibitory factor in different stages of human atherosclerosis. *Circulation* **105**: 1561–1566.
10. Hansson GK, Libby P. (2006) The immune response in atherosclerosis: A double-edged sword. *Nat Rev Immunol* **6**: 508–519.
11. Weber C, Zernecke A, Libby P. (2008) The multifaceted contributions of leukocyte subsets to atherosclerosis: Lessons from mouse models. *Nat Rev Immunol* **8**: 802–815.
12. Zernecke A, Weber C. (2010) Chemokines in the vascular inflammatory response of atherosclerosis. *Cardiovasc Res* **86**: 192–201.
13. Calandra T, Roger T. (2003) Macrophage migration inhibitory factor: A regulator of innate immunity. *Nat Rev Immunol* **3**: 791–800.
14. Kim S, Miska KB, Jenkins MC *et al.* (2010) Molecular cloning and functional characterization of the avian macrophage migration inhibitory factor (MIF). *Dev Comp Immunol* **34**: 1021–1032.
15. Sun HW, Bernhagen J, Bucala R, Lolis E. (1996) Crystal structure at 2.6-Å resolution of human macrophage migration inhibitory factor. *Proc Natl Acad Sci USA* **93**: 5191–5196.
16. Mischke R, Kleemann R, Brunner H, Bernhagen J. (1998) Cross-linking and mutational analysis of the oligomerization state of the cytokine macrophage migration inhibitory factor (MIF). *FEBS Lett* **427**: 85–90.
17. Philo JS, Yang TH, LaBarre M. (2004) Re-examining the oligomerization state of macrophage migration inhibitory factor (MIF) in solution. *Biophys Chem* **108**: 77–87.

18. Weiser WY, Temple DM, Witek-Gianotti JS *et al.* (1989) Molecular cloning of cDNA encoding a human macrophage migration inhibtion factor. *Proc Natl Acad Sci USA* **86**: 7522–7526.
19. Muhlhahn P, Bernhagen J, Czisch M *et al.* (1996) NMR characterization of structure, backbone dynamics, and glutathione binding of the human macrophage migration inhibitory factor (MIF). *Protein Sci* **5**: 2095–2103.
20. Murphy PM, Baggiolini M, Charo IF *et al.* (2000) International union of pharmacology. XXII. Nomenclature for chemokine receptors. *Pharmacol Rev* **52**: 145–176.
21. Hebert CA, Vitangcol RV, Baker JB. (1991) Scanning mutagenesis of interleukin-8 identifies a cluster of residues required for receptor binding. *J Biol Chem* **266**: 18989–18994.
22. Rajagopalan L, Rajarathnam K. (2006) Structural basis of chemokine receptor function — A model for binding affinity and ligand selectivity. *Biosci Rep* **26**: 325–339.
23. Kraemer S, Lue H, Zernecke A *et al.* (2011) MIF-chemokine receptor interactions in atherogenesis are dependent on an N-loop-based 2-site binding mechanism. *FASEB J* **25**: 894–906.
24. Weber C, Koenen RR. (2006) Fine-tuning leukocyte responses: Towards a chemokine 'interactome'. *Trends Immunol* **27**: 268–273.
25. Wakasugi K, Schimmel P. (1999) Two distinct cytokines released from a human aminoacyl-tRNA synthetase. *Science* **284**: 147–151.
26. Wakasugi K, Slike BM, Hood J *et al.* (2002) Induction of angiogenesis by a fragment of human tyrosyl-tRNA synthetase. *J Biol Chem* **277**: 20124–20126.
27. Yang XL, Skene RJ, McRee DE, Schimmel P. (2002) Crystal structure of a human aminoacyl-tRNA synthetase cytokine. *Proc Natl Acad Sci USA* **99**: 15369–15374.
28. Howard OM, Dong HF, Yang D *et al.* (2002) Histidyl-tRNA synthetase and asparaginyl-tRNA synthetase, autoantigens in myositis, activate chemokine receptors on T lymphocytes and immature dendritic cells. *J Exp Med* **196**: 781–791.
29. Yang D, Chertov O, Bykovskaia SN *et al.* (1999) Beta-defensins: Linking innate and adaptive immunity through dendritic and T cell CCR6. *Science* **286**: 525–528.
30. Hoover DM, Boulegue C, Yang D *et al.* (2002) The structure of human macrophage inflammatory protein-3alpha/CCL20. Linking antimicrobial and CC chemokine receptor-6-binding activities with human beta-defensins. *J Biol Chem* **277**: 37647–37654.
31. Noels H, Bernhagen J, Weber C. (2009) Macrophage migration inhibitory factor: A noncanonical chemokine important in atherosclerosis. *Trends Cardiovasc Med* **19**: 76–86.

32. Kleemann R, Kapurniotu A, Frank RW *et al.* (1998) Disulfide analysis reveals a role for macrophage migration inhibitory factor (MIF) as thiol-protein oxidoreductase. *J Mol Biol* **280**: 85–102.
33. Thiele M, Bernhagen J. (2005) Link between macrophage migration inhibitory factor and cellular redox regulation. *Antioxid Redox Signal* **7**: 1234–1248.
34. Esumi N, Budarf M, Ciccarelli L *et al.* (1998) Conserved gene structure and genomic linkage for D-dopachrome tautomerase (DDT) and MIF. *Mamm Genome* **9**: 753–757.
35. Rosengren E, Bucala R, Aman P *et al.* (1996) The immunoregulatory mediator macrophage migration inhibitory factor (MIF) catalyzes a tautomerization reaction. *Mol Med* **2**: 143–149.
36. Dios A, Mitchell RA, Aljabari B *et al.* (2002) Inhibition of MIF bioactivity by rational design of pharmacological inhibitors of MIF tautomerase activity. *J Med Chem* **45**: 2410–2416.
37. Senter PD, Al-Abed Y, Metz CN *et al.* (2002) Inhibition of macrophage migration inhibitory factor (MIF) tautomerase and biological activities by acetaminophen metabolites. *Proc Natl Acad Sci USA* **99**: 144–149.
38. Ouertatani-Sakouhi H, El-Turk F, Fauvet B *et al.* (2009) A new class of isothiocyanate-based irreversible inhibitors of macrophage migration inhibitory factor. *Biochemistry* **48**: 9858–9870.
39. Ouertatani-Sakouhi H, El-Turk F, Fauvet B *et al.* (2010) Identification and characterization of novel classes of macrophage migration inhibitory factor (MIF) inhibitors with distinct mechanisms of action. *J Biol Chem* **285**: 26581–26598.
40. Calandra T, Bernhagen J, Mitchell RA, Bucala R. (1994) The macrophage is an important and previously unrecognized source of macrophage migration inhibitory factor. *J Exp Med* **179**: 1895–1902.
41. Calandra T, Spiegel LA, Metz CN, Bucala R. (1998) Macrophage migration inhibitory factor is a critical mediator of the activation of immune cells by exotoxins of Gram-positive bacteria. *Proc Natl Acad Sci USA* **95**: 11383–11388.
42. Takahashi M, Nishihira J, Shimpo M *et al.* (2001) Macrophage migration inhibitory factor as a redox-sensitive cytokine in cardiac myocytes. *Cardiovasc Res* **52**: 438–445.
43. Miller EJ, Li J, Leng L *et al.* (2008) Macrophage migration inhibitory factor stimulates AMP-activated protein kinase in the ischaemic heart. *Nature* **451**: 578–582.
44. Simons D, Grieb G, Hristov M *et al.* (2011) Hypoxia-induced endothelial secretion of macrophage migration inhibitory factor and role in endothelial progenitor cell recruitment. *J Cell Mol Med* **15**: 668–678.

45. Schmeisser A, Marquetant R, Illmer T *et al.* (2005) The expression of macrophage migration inhibitory factor 1alpha (MIF 1alpha) in human atherosclerotic plaques is induced by different proatherogenic stimuli and associated with plaque instability. *Atherosclerosis* **178**: 83–94.
46. Flieger O, Engling A, Bucala R *et al.* (2003) Regulated secretion of macrophage migration inhibitory factor is mediated by a non-classical pathway involving an ABC transporter. *FEBS Lett* **551**: 78–86.
47. Keller M, Ruegg A, Werner S, Beer HD. (2008) Active caspase-1 is a regulator of unconventional protein secretion. *Cell* **132**: 818–831.
48. Merk M, Baugh J, Zierow S *et al.* (2009) The Golgi-associated protein p115 mediates the secretion of macrophage migration inhibitory factor. *J Immunol* **182**: 6896–6906.
49. Lue H, Thiele M, Franz J *et al.* (2007) Macrophage migration inhibitory factor (MIF) promotes cell survival by activation of the Akt pathway and role for CSN5/JAB1 in the control of autocrine MIF activity. *Oncogene* **26**: 5046–5059.
50. Lan HY, Bacher M, Yang N *et al.* (1997) The pathogenic role of macrophage migration inhibitory factor in immunologically induced kidney disease in the rat. *J Exp Med* **185**: 1455–1465.
51. Gregory JL, Leech MT, David JR *et al.* (2004) Reduced leukocyte-endothelial cell interactions in the inflamed microcirculation of macrophage migration inhibitory factor-deficient mice. *Arthritis Rheum* **50**: 3023–3034.
52. Lin SG, Yu XY, Chen YX *et al.* (2000) *De novo* expression of macrophage migration inhibitory factor in atherogenesis in rabbits. *Circ Res* **87**: 1202–1208.
53. Pan JH, Sukhova GK, Yang JT *et al.* (2004) Macrophage migration inhibitory factor deficiency impairs atherosclerosis in low-density lipoprotein receptor-deficient mice. *Circulation* **109**: 3149–3153.
54. Burger-Kentischer A, Gobel H, Kleemann R *et al.* (2006) Reduction of the aortic inflammatory response in spontaneous atherosclerosis by blockade of macrophage migration inhibitory factor (MIF). *Atherosclerosis* **184**: 28–38.
55. Chen Z, Sakuma M, Zago AC *et al.* (2004) Evidence for a role of macrophage migration inhibitory factor in vascular disease. *Arterioscler Thromb Vasc Biol* **24**: 709–714.
56. Schober A, Bernhagen J, Thiele M *et al.* (2004) Stabilization of atherosclerotic plaques by blockade of macrophage migration inhibitory factor after vascular injury in apolipoprotein E-deficient mice. *Circulation* **109**: 380–385.
57. Weber C. (2003) Novel mechanistic concepts for the control of leukocyte transmigration: Specialization of integrins, chemokines, and junctional molecules. *J Mol Med* **81**: 4–19.
58. Laudanna C, Alon R. (2006) Right on the spot. Chemokine triggering of integrin-mediated arrest of rolling leukocytes. *Thromb Haemost* **95**: 5–11.

59. Leng L, Metz CN, Fang Y *et al.* (2003) MIF signal transduction initiated by binding to CD74. *J Exp Med* **197**: 1467–1476.
60. Shi X, Leng L, Wang T *et al.* (2006) CD44 is the signaling component of the macrophage migration inhibitory factor-CD74 receptor complex. *Immunity* **25**: 595–606.
61. Schwartz V, Lue H, Kraemer S *et al.* (2009) A functional heteromeric MIF receptor formed by CD74 and CXCR4. *FEBS Lett* **583**: 2749–2757.
62. Lue H, Dewor M, Leng L *et al.* (2011) Activation of the JNK signalling pathway by macrophage migration inhibitory factor (MIF) and dependence on CXCR4 and CD74. *Cell Signal* **23**: 135–144.
63. Koenen RR, Weber C. (2010) Therapeutic targeting of chemokine interactions in atherosclerosis. *Nat Rev Drug Discov* **9**: 141–153.
64. Gregory JL, Morand EF, McKeown SJ *et al.* (2006) Macrophage migration inhibitory factor induces macrophage recruitment via CC chemokine ligand 2. *J Immunol* **177**: 8072–8079.
65. Lue H, Kapurniotu A, Fingerle-Rowson G *et al.* (2006) Rapid and transient activation of the ERK MAPK signalling pathway by macrophage migration inhibitory factor (MIF) and dependence on JAB1/CSN5 and Src kinase activity. *Cell Signal* **18**: 688–703.
66. Mitchell RA, Metz CN, Peng T, Bucala R. (1999) Sustained mitogen-activated protein kinase (MAPK) and cytoplasmic phospholipase A2 activation by macrophage migration inhibitory factor (MIF). Regulatory role in cell proliferation and glucocorticoid action. *J Biol Chem* **274**: 18100–18106.
67. Goppelt-Struebe M, Rehfeldt W. (1992) Glucocorticoids inhibit TNF alpha-induced cytosolic phospholipase A2 activity. *Biochim Biophys Acta* **1127**: 163–167.
68. Calandra T, Bernhagen J, Metz CN *et al.* (1995) MIF as a glucocorticoid-induced modulator of cytokine production. *Nature* **377**: 68–71.
69. Donnelly SC, Bucala R. (1997) Macrophage migration inhibitory factor: A regulator of glucocorticoid activity with a critical role in inflammatory disease. *Mol Med Today* **3**: 502–507.
70. Kleemann R, Hausser A, Geiger G *et al.* (2000) Intracellular action of the cytokine MIF to modulate AP-1 activity and the cell cycle through Jab1. *Nature* **408**: 211–216.
71. Qi D, Hu X, Wu X *et al.* (2009) Cardiac macrophage migration inhibitory factor inhibits JNK pathway activation and injury during ischemia/reperfusion. *J Clin Invest* **119**(12): 3807–3816.
72. Binsky I, Haran M, Starlets D *et al.* (2007) IL-8 secreted in a macrophage migration-inhibitory factor- and CD74-dependent manner regulates B cell chronic lymphocytic leukemia survival. *Proc Natl Acad Sci USA* **104**: 13408–13413.

73. Mitchell RA, Liao H, Chesney J *et al.* (2002) Macrophage migration inhibitory factor (MIF) sustains macrophage proinflammatory function by inhibiting p53: Regulatory role in the innate immune response. *Proc Natl Acad Sci USA* **99**: 345–350.
74. Jung H, Seong HA, Ha H. (2008) Critical role of cysteine residue 81 of macrophage migration inhibitory factor (MIF) in MIF-induced inhibition of p53 activity. *J Biol Chem* **283**: 20383–20396.
75. Liu L, Chen J, Ji C *et al.* (2008) Macrophage migration inhibitory factor (MIF) interacts with Bim and inhibits Bim-mediated apoptosis. *Mol Cells* **26**: 193–199.
76. Liu L, Ji C, Chen J *et al.* (2008) A global genomic view of MIF knockdown-mediated cell cycle arrest. *Cell Cycle* **7**: 1678–1692.
77. Liao H, Bucala R, Mitchell RA. (2003) Adhesion-dependent signaling by macrophage migration inhibitory factor (MIF). *J Biol Chem* **278**: 76–81.
78. Bittl JA. (1996) Advances in coronary angioplasty. *N Engl J Med* **335**: 1290–1302.
79. Korshunov VA, Nikonenko TA, Tkachuk VA *et al.* (2006) Interleukin-18 and macrophage migration inhibitory factor are associated with increased carotid intima-media thickening. *Arterioscler Thromb Vasc Biol* **26**: 295–300.
80. Donn RP, Shelley E, Ollier WE, Thomson W. (2001) A novel 5′-flanking region polymorphism of macrophage migration inhibitory factor is associated with systemic-onset juvenile idiopathic arthritis. *Arthritis Rheum* **44**: 1782–1785.
81. Herder C, Illig T, Baumert J *et al.* (2008) Macrophage migration inhibitory factor (MIF) and risk for coronary heart disease: Results from the MONICA/KORA Augsburg case-cohort study, 1984–2002. *Atherosclerosis* **200**: 380–388.
82. Palomino-Morales R, Gonzalez-Juanatey C, Vazquez-Rodriguez TR *et al.* (2010) Lack of association between macrophage migration inhibitory factor-173 gene polymorphism with disease susceptibility and cardiovascular risk in rheumatoid arthritis patients from northwestern Spain. *Clin Exp Rheumatol* **28**: 68–72.
83. Makino A, Nakamura T, Hirano M *et al.* (2010) High plasma levels of macrophage migration inhibitory factor are associated with adverse long-term outcome in patients with stable coronary artery disease and impaired glucose tolerance or type 2 diabetes mellitus. *Atherosclerosis* **213**: 573–578.
84. Tereshchenko IP, Petrkova J, Mrazek F *et al.* (2009) The macrophage migration inhibitory factor (MIF) gene polymorphism in Czech and Russian patients with myocardial infarction. *Clin Chim Acta* **402**: 199–202.
85. Chen L, Yang G, Zhang X *et al.* (2009) Induction of MIF expression by oxidized LDL via activation of NF-κB in vascular smooth muscle cells. *Atherosclerosis* **207**: 428–433.

86. Atsumi T, Nishihira J, Makita Z, Koike T. (2000) Enhancement of oxidised low-density lipoprotein uptake by macrophages in response to macrophage migration inhibitory factor. *Cytokine* **12**: 1553–1556.
87. Bernhagen J, Mitchell RA, Calandra T *et al.* (1994) Purification, bioactivity, and secondary structure analysis of mouse and human macrophage migration inhibitory factor (MIF). *Biochemistry* **33**: 14144–14155.
88. Mercer J, Figg N, Stoneman V *et al.* (2005) Endogenous p53 protects vascular smooth muscle cells from apoptosis and reduces atherosclerosis in ApoE knockout mice. *Circ Res* **96**: 667–674.
89. Schweitzer K, Bozko PM, Dubiel W, Naumann M. (2007) CSN controls NF-kappaB by deubiquitinylation of IkappaBalpha. *EMBO J* **26**: 1532–1541.
90. Collins T, Cybulsky MI. (2001) NF-kappaB: Pivotal mediator or innocent bystander in atherogenesis? *J Clin Invest* **107**: 255–264.
91. de Winther MP, Kanters E, Kraal G, Hofker MH. (2005) Nuclear factor kappaB signaling in atherogenesis. *Arterioscler Thromb Vasc Biol* **25**: 904–914.
92. Kong YZ, Huang XR, Ouyang X *et al.* (2005) Evidence for vascular macrophage migration inhibitory factor in destabilization of human atherosclerotic plaques. *Cardiovasc Res* **65**: 272–282.
93. Kong YZ, Yu X, Tang JJ *et al.* (2005) Macrophage migration inhibitory factor induces MMP-9 expression: Implications for destabilization of human atherosclerotic plaques. *Atherosclerosis* **178**: 207–215.
94. Lemaitre V, O'Byrne TK, Borczuk AC *et al.* (2001) ApoE knockout mice expressing human matrix metalloproteinase-1 in macrophages have less advanced atherosclerosis. *J Clin Invest* **107**: 1227–1234.
95. Johnson JL, George SJ, Newby AC, Jackson CL. (2005) Divergent effects of matrix metalloproteinases 3, 7, 9, and 12 on atherosclerotic plaque stability in mouse brachiocephalic arteries. *Proc Natl Acad Sci USA* **102**: 15575–15580.
96. Verschuren L, Lindeman JH, van Bockel JH *et al.* (2005) Up-regulation and coexpression of MIF and matrix metalloproteinases in human abdominal aortic aneurysms. *Antioxid Redox Signal* **7**: 1195–1202.
97. Schrans-Stassen BH, Lue H, Sonnemans DG *et al.* (2005) Stimulation of vascular smooth muscle cell migration by macrophage migration inhibitory factor. *Antioxid Redox Signal* **7**: 1211–1216.
98. Lievens D, Eijgelaar WJ, Biessen EA *et al.* (2009) The multi-functionality of CD40L and its receptor CD40 in atherosclerosis. *Thromb Haemost* **102**: 206–214.
99. Amin MA, Volpert OV, Woods JM *et al.* (2003) Migration inhibitory factor mediates angiogenesis via mitogen-activated protein kinase and phosphatidylinositol kinase. *Circ Res* **93**: 321–329.

100. Chesney J, Metz C, Bacher M *et al.* (1999) An essential role for macrophage migration inhibitory factor (MIF) in angiogenesis and the growth of a murine lymphoma. *Mol Med* **5**: 181–191.
101. Ogawa H, Nishihira J, Sato Y *et al.* (2000) An antibody for macrophage migration inhibitory factor suppresses tumour growth and inhibits tumour-associated angiogenesis. *Cytokine* **12**: 309–314.
102. Garai J, Lorand T. (2009) Macrophage migration inhibitory factor (MIF) tautomerase inhibitors as potential novel anti-inflammatory agents: Current developments. *Curr Med Chem* **16**: 1091–1114.
103. Nickel W. (2010) Pathways of unconventional protein secretion. *Curr Opin Biotechnol* **21**: 621–626.
104. Nickel W, Rabouille C. (2009) Mechanisms of regulated unconventional protein secretion. *Nat Rev Mol Cell Biol* **10**: 148–155.
105. Nickel W, Seedorf M. (2008) Unconventional mechanisms of protein transport to the cell surface of eukaryotic cells. *Annu Rev Cell Dev Biol* **24**: 287–308.
106. Horuk R, Proudfoot AE. (2009) Drug discovery targeting the chemokine system — Where are we? *Front Biosci (Elite Ed)* **1**: 209–219.
107. Weber C, Kraemer S, Drechsler M *et al.* (2008) Structural determinants of MIF functions in CXCR2-mediated inflammatory and atherogenic leukocyte recruitment. *Proc Natl Acad Sci USA* **105**: 16278–16283.

V-2

MIF in Cardiovascular Disease

Edward J. Miller*,§, Dake Qi†, Ji Li‡ and Lawrence H. Young†

1. Introduction

MIF is emerging as an important modulator of disease pathogenesis in the cardiovascular system, through both its proinflammatory effects and novel actions on cardiac signaling and metabolism. MIF plays a significant role in the pathogenesis of inflammatory diseases and vascular inflammation is a well-recognized mediator of atherosclerosis. MIF is an important regulator of systemic inflammation and is now thought to participate in atherosclerosis when chronically elevated. While the chronic inflammatory effects of MIF predominate at the systemic level, novel data from our laboratory and others indicates that endogenous MIF expressed in the heart has acute physiologic actions during stress that are not related to inflammation and appear to be protective against injury.

This chapter reviews the effects of MIF in the cardiovascular system and considers the following evolving paradigms. First, the balance of MIF's effects in cardiovascular disease depends upon the duration and nature of the stimulus causing its release. Second, that MIF, like certain other autocrine/paracrine factors, activates potent prosurvival pathways in solid organs that are independent of their systemic inflammatory effects. Understanding the role of MIF in the cardiovascular system is the result of the contributions of several investigators, including but not limited to those summarized in Table 1.

*Department of Internal Medicine, Section of Cardiovascular Medicine, Boston University School of Medicine, Boston, MA 02481, USA.
†Department of Internal Medicine, Section of Cardiovascular Medicine, Yale University School of Medicine, New Haven, CT, USA.
‡Department of Pharmacology and Toxicology, University at Buffalo-SUNY, Bufffalo, NY, USA.
§Corresponding author: Email: ejmiller@bu.edu

Table 1. Diverse effects of MIF in the cardiovascular system.

	Hypothesis	Proposed Intervention	References
Atherosclerosis	MIF promotes vascular inflammation and atherosclerosis	MIF inhibition	1, 4, 6
Sepsis	MIF excess contributes to cardiac dysfunction in sepsis	MIF inhibition	13, 16, 18
Acute myocardial ischemia	MIF activates cell survival and inhibits cell death pathways	MIF activation	15, 25, 32

2. MIF and Atherosclerosis

The pathogenesis of atherosclerosis reflects a complex interplay of inflammation, lipid deposition, and thrombosis. Atherosclerosis is initiated by endothelial injury leading to the recruitment of inflammatory cells and the incorporation of cholesterol into macrophages, which then become foam cells. Vascular inflammation can also transform stable atherosclerotic plaques into more vulnerable plaques, which are susceptible to ulceration, exposing their lipid-rich core, tissue factor, and other prothrombotic proteins, leading to arterial thrombosis. Coronary arterial thrombosis can further obstruct the vascular lumen of atherosclerotic vessels, leading to acute coronary syndromes, including acute myocardial infarction.

2.1 *MIF expression in atherosclerotic plaque*

Circulating monocytes and tissue macrophages promote atherosclerosis. Early evidence that MIF had a role in atherosclerosis came from Burger-Kentischer *et al.*, who showed that MIF is present in atheromas in human internal mammary arteries.[1] They also observed increased expression of MIF in human vascular endothelial cells exposed to oxidized LDL. MIF expression was also increased in endothelial cells, smooth muscle cells, and tissue macrophages within human atherosclerotic abdominal aortic aneurysms,[2] further supporting the hypothesis that MIF might have a pathogenic role in atherogenesis.

2.2 *MIF inhibition in experimental atherosclerosis*

These clinical observations spurred several investigators to examine whether MIF might play a causal role in the development of atherosclerotic plaques.

Experimental studies evaluated the role of MIF in the progression of atherosclerosis in animal models. MIF blockade with a neutralizing antibody had beneficial effects in both LDL receptor-deficient (*Ldlr*$^{-/-}$)[3] and apolipoprotein E-deficient (*ApoE*$^{-/-}$) mice[4] after vascular injury, and reduced spontaneous atherogenesis.[5] Blockade of MIF also reduced neointimal formation,[3] the number of activated macrophages,[4] and markers of intimal inflammation such as TNF, IL-12, and ICAM-1[5] in the vessel wall. Furthermore, MIF-deficient mice (*Mif*$^{-/-}$) crossbred with *Ldlr*$^{-/-}$ mice also displayed reduced atheroma formation and less smooth muscle cell proliferation, when fed a high cholesterol diet.[6]

MIF appears to promote atherogenesis through interaction with the chemokine receptors CXCR2 and CXCR4.[7] CXCR2 signaling mediates monocyte adhesion in vulnerable plaques.[8] MIF also might render atherosclerotic plaques more vulnerable to rupture by inducing the expression of matrix metalloproteinase-9, which causes breakdown of extracellular matrix.[9] Thus, these studies raise the possibility that MIF might promote both atherogenesis and plaque instability if chronically elevated as it is often in conditions such as diabetes.[10]

2.3 *MIF and clinical cardiovascular risk*

Chronic elevations of serum inflammatory markers, such as CRP, are associated with adverse events in patients at risk for cardiovascular disease. Similarly, serum MIF concentration has been proposed as a potential biomarker for cardiovascular risk. However, recent data suggest that serum MIF does not accurately predict risk for acute coronary syndromes or myocardial infarction.[11] Whether MIF serum concentrations might prove to be more reliable predictors of cardiac risk in specific populations or if measured in concert with other biomarkers remains to be determined. However, as discussed in more detail below, MIF expression is determined in part by *MIF* genotype, and women with higher expression *MIF* promoter alleles appear to be at increased risk for acute coronary syndromes.[11]

In summary, current evidence suggests that MIF has a pathogenic role in promoting atherogenesis. Whether therapies directed at blocking MIF action on the vasculature are antiatherogenic or prevent acute coronary syndromes in patients remains uncertain. Although MIF might well have a pathogenic role in atherosclerosis, a single measurement of serum MIF concentrations is not a reliable biomarker of cardiac risk. Initial studies suggest that MIF promoter genotype might be more predictive of risk, but more extensive research is needed to assess its clinical role in various populations.

3. MIF Expression and Effects in the Heart

With such strong emphasis on the role of MIF as a regulator of inflammation, it perhaps was quite unexpected to find that MIF was expressed in myocytes within the heart and even less anticipated that MIF would have physiologic actions in the heart unrelated to inflammation.

3.1 *Cardiac MIF expression and secretion*

The initial identification of MIF in the heart came from studies by Takahashi, who demonstrated that MIF is present in cardiomyocytes and is released in response to hypoxia or peroxide exposure.[12] They also found that inhibition of protein kinase C reduced MIF release during peroxide exposure, suggesting that this kinase might trigger cardiac MIF release during oxidative stress. MIF expression in the heart was also found to increase following lipopolysaccharide (LPS) exposure[13] and burn injury[14] in murine models. We showed that MIF is released from the mouse heart following ischemia/reperfusion.[15] These observations led us to postulate that endogenous MIF might have autocrine/paracrine effects in the heart, and we have now found that MIF indeed influences cardiac signaling, metabolism, and function independent of its proinflammatory effects.

3.2 *MIF effects on cardiac function in sepsis*

MIF appears to function as a negative modulator of cardiac function during severe inflammatory stress, such as sepsis or burn injury, in animal models. Anti-MIF neutralizing antibody preserves cardiac function in rats following burn injury.[14] Perfusion of isolated hearts with recombinant MIF also decreases cardiac contractile function. In addition, the administration of glucan phosphate reduced myocardial MIF expression and decreased myocardial dysfunction and apoptosis in a murine model of sepsis.[16] The source of MIF responsible for cardiac dysfunction during sepsis is uncertain, but some evidence suggests that lung-derived MIF contributes to the myocardial depression observed in sepsis in the mouse model.[17] LPS also increases the expression of a number of cytokines, including IL-1 and IL-6, which may act together with MIF to cause myocardial depression during sepsis.[18]

4. MIF in Myocardial Ischemia

Reductions in coronary blood flow lead to reversible myocardial ischemia when brief, but induce myocardial necrosis if the ischemia is persistent and severe. The effects of ischemia in the heart and in other solid organs depends not only on duration and severity, but also on the degree of compensatory cellular responses. Cardiomyocytes possess several mechanisms to defend against ischemic injury, including activation of prosurvival stress kinases, adaptive metabolic responses to maintain energetic charge, and alterations in gene transcription/translation. These adaptive mechanisms can enhance cell survival and limit the degree of irreversible cellular injury.

The role of MIF in the ischemic heart was initially thought to be as a marker of ischemia. Takahashi *et al.* observed increased serum MIF concentrations in patients during acute myocardial infarction.[12] Although they did not define the source of MIF in the clinical studies, their additional experimental results suggested that cardiomyocytes might be partially responsible. MIF release during myocardial ischemia was later confirmed in a mouse model of coronary ligation.[19] These results have triggered an interest in its potential autocrine/paracrine role in the setting of ischemia, where local extracellular concentrations might be expected to increase significantly in the absence of washout of MIF from the heart.

4.1 *MIF, glucose metabolism, and the AMPK signaling pathway*

Our group began investigating the effects of MIF on cardiomyocyte glucose metabolism during myocardial ischemia, based on interesting observations that MIF activated glycolysis in skeletal muscle cells.[20] During myocardial ischemia, an increase in glucose uptake and metabolism provide ATP for critical cellular processes that promote survival. We had previously found that glucose uptake during ischemia was mediated by the translocation of GLUT glucose transporters to the sarcolemma.[21] We had also discovered that GLUT4 translocation was triggered by the stress kinase, AMP-activated protein kinase (AMPK),[22,23] and hypothesized that MIF might activate AMPK.

Several lines of evidence now link clearly MIF to the AMPK pathway in the heart. We initially showed that MIF activates AMPK in a dose- and time-dependent manner, and that MIF neutralizing antibodies blunted AMPK activation during hypoxia in rat heart muscles.[15] In complementary studies, we demonstrated impaired activation of AMPK and cardiac glucose metabolism in *Mif*$^{-/-}$ mice during ischemia. Blunted AMPK activation was observed not only *in vivo*, where hearts are exposed to both circulating MIF and endogenous cardiac MIF, but also in isolated perfused *Mif*$^{-/-}$ hearts.[15] The latter results indicate the importance of endogenous cardiac MIF in activation of the AMPK pathway in the heart during ischemia (Fig. 1).

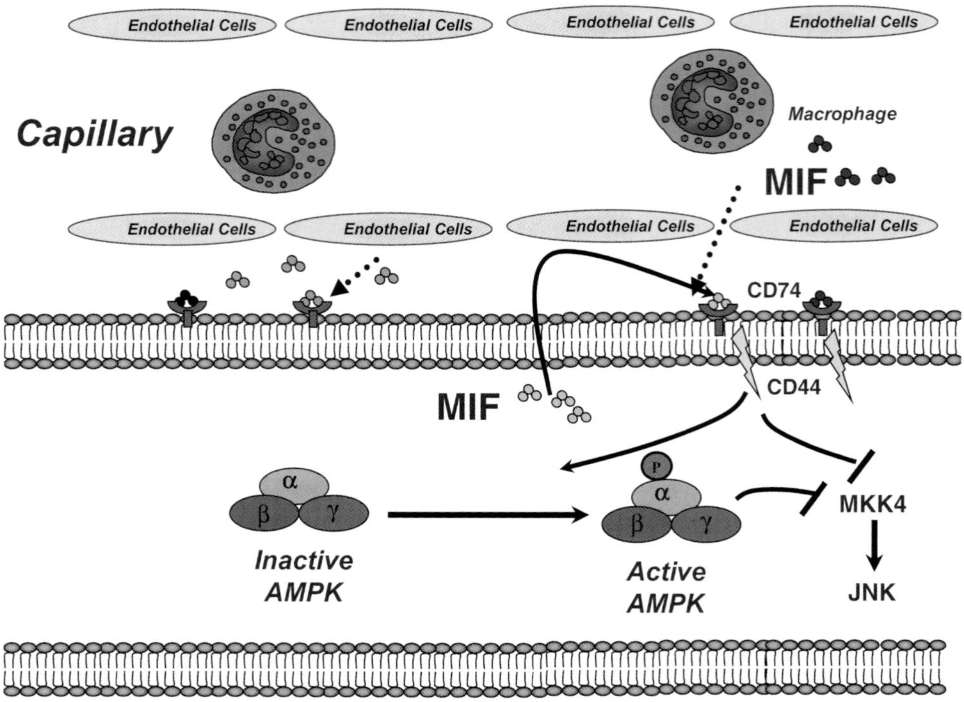

Fig. 1. MIF in the ischemic heart: Ischemia triggers MIF secretion and extracellular MIF activates the CD74/CD44 complex, amplifying the activation of the cardioprotective AMPK pathway and inhibiting the cell death JNK pathway in the heart.[15,25,28]

4.2 *MIF and cardiac injury during ischemia-reperfusion*

Because AMPK activation is central to the cardiomyocyte defense against ischemia, we examined whether *Mif*$^{-/-}$ mice might also have more injury after ischemia-reperfusion. Similar to transgenic mice with inactivated (kinase-dead) AMPK,[24] *Mif*$^{-/-}$ mice also demonstrated impaired recovery of contractile function and increased myocardial necrosis after ischemia-reperfusion. Greater injury was observed both during coronary occlusion-reperfusion *in vivo* and in isolated perfused *Mif*$^{-/-}$ hearts,[15] once again indicating an important role for endogenous cardiac MIF.

Subsequent reports have confirmed the cardioprotective effect of endogenous MIF during ischemia/reperfusion in the mouse model.[25] However, additional recent data raise the possibility that the duration of ischemia might be an important determinant of MIF action in this setting. Longer durations of ischemia that are associated with more necrosis promote more inflammation in the heart during subsequent reperfusion, and in this setting, a proinflammatory effect of circulating MIF might be detrimental.[26] Thus, the balance of MIF signaling and proinflammatory effects of MIF may vary depending on the duration of ischemia and the individual's predilection to inflammation. More clinical studies will be needed to determine the net effect of MIF in the setting of solid organ ischemia-reperfusion.

5. MIF and JNK Pathway Signaling in the Heart

Reperfusion following ischemia has an essential role in maintaining myocardial survival by delivering oxygen and nutrients that re-establish energy generation in the heart. However, reperfusion also leads to the formation of reactive oxygen species that activate signaling pathways that trigger reperfusion injury. One of the pathways that promotes apoptosis and cell death is the c-Jun N-terminal kinase (JNK), a mitogen-activated protein (MAP) kinase that is activated during ischemia-reperfusion.[27]

5.1 *Endogenous MIF and JNK pathway activation during ischemia-reperfusion*

JNK is activated in the heart by the upstream MAP kinase kinases (MKK4 and 7), and there is emerging evidence that endogenous cardiac MIF has an important role in modulating the JNK pathway during myocardial ischemia-reperfusion. We recently observed augmented activation of MKK4 and JNK, as well as downstream c-Jun, during reperfusion in hearts from *Mif*$^{-/-}$ mice (Fig. 1).[28]

In the absence of MIF, the excess activation of JNK promoted activation of BAD and increased both apoptosis and necrosis following ischemia-reperfusion.[28] The mechanism through which MIF inhibits the JNK pathway during reperfusion is uncertain. It is possible that MIF activation of AMPK could play a role, since activated AMPK inhibits JNK activation in other cell types during glucose deprivation[29] and hydrogen peroxide exposure.[30] However, it is interesting that we observed that the administration of MIF specifically during reperfusion inhibited JNK activation and ameliorated injury in the *Mif*$^{-/-}$ hearts.[28] These results would indicate that activation of AMPK during ischemia is not primarily responsible for the inhibition of JNK activation, but do not exclude a potential role for AMPK activation during reperfusion. In contrast to the findings in intact hearts submitted to ischemia-reperfusion, Ojamaa's laboratory demonstrated that recombinant MIF activates both JNK and p38 MAPK in cultured neonatal cardiomyocytes.[31] Thus, the effects of MIF once again depend on the physiologic setting as well as potentially on the concentration and duration of MIF exposure.

6. Oxidoreductase Activity of MIF in the Heart

In addition to extracellular MIF signaling through its cell surface receptor CD74, an additional intracellular oxidoreductase action of MIF has been postulated to be operative in the heart. *Mif*$^{-/-}$ hearts have increased oxidative stress after ischemia-reperfusion, which the Ojamaa group has attributed to the absence of intracellular MIF thiol-protein oxidoreductase activity during ischemia/reperfusion.[32] Cardiac *Mif*$^{-/-}$ fibroblasts accumulate increased reactive oxygen species when exposed to oxidative stress, and MIF repletion with adenoviral transfer ameliorated this problem. Although there is a possibility that MIF might have an intracellular oxidoreductase function, the exact molecular mechanism has yet to be defined, and it is uncertain whether cysteine residues are adequately exposed in the intact structure to serve this function. In addition, to the extent that MIF indeed has an important intracellular role, one might predict that *Cd74*$^{-/-}$ hearts should be protected by intracellular MIF and less sensitive to ischemic injury compared to *Mif*$^{-/-}$ hearts. However, recent evidence suggests that *Cd74*$^{-/-}$ and *Mif*$^{-/-}$ hearts have similar vulnerability to ischemia-reperfusion, indicating the need for further investigation to address the oxidoreductase hypothesis.

7. Cardiac MIF and Aging

Recent data suggest that MIF expression in the heart is reduced in aged mice and might contribute to their impaired tolerance during ischemia-reperfusion.[25] MIF expression is regulated by HIF-1α, which is also down-regulated in the hearts of aged mice. The reduced MIF expression was associated with impaired activation of AMPK during ischemia, which predisposes to injury in this model. When isolated hearts from aged mice were perfused with supplemental MIF, or MIF expression is increased in the intact heart by adenoviral-mediated expression, the impaired AMPK response and ischemic tolerance associated with aging was improved.[25] Thus, Li and colleagues hypothesize that aging is a state of relative MIF deficiency that is pathogenic in the cardiovascular system.

8. Genetic *MIF* Polymorphisms: Potential Role in the Heart

Transcription of the human *MIF* gene (GeneID 4282) is regulated by genetic variations in the *MIF* promoter, including the −173G/C single-nucleotide polymorphism and a sequence of tetranucleotide repeats at −794 (−794 $CATT_{5-8}$).[33] These polymorphisms might have arisen under evolutionarily pressure in response to endemic infectious diseases such as malaria,[34] where they influence the host response to infection. The *MIF* promoter polymorphisms might also have relevance to cardiovascular disease, and this area has become a growing area of investigation.

8.1 *Role of MIF promoter genotype in atherosclerosis*

Because of the pathologic data suggesting that MIF might have a role in vascular inflammation, a number of investigators have examined whether the human *MIF* promoter genotypes are associated with coronary atherosclerosis. In a case-control study of patients undergoing coronary angiography, the –173C *MIF* allele was associated with a 2.5-fold higher risk of coronary artery disease in a Han Chinese population.[35] A case-cohort study from Germany found no clear relationship between circulating MIF concentrations and coronary artery disease but did show that women with the rs755622C and rs2070766G alleles had a 2.4-fold higher incidence of coronary artery disease.[11] Thus, there is some evidence that *MIF* genotype is associated with coronary artery disease, although the degree of influence may well vary according to specific *MIF* alleles, populations, ethnicities, and other comorbidities, such as diabetes mellitus or familial hypercholesterolemia. To date, there are no data on the relationship between *MIF* alleles and long-term cardiovascular disease event rates, so the ability of *MIF* genotype to predict the risk (or severity) of ischemic events is uncertain.

8.2 *Role of MIF promoter genotype in the response to hypoxia*

MIF promoter polymorphisms (–794 $CATT_{5-8}$) are known to impact the expression of MIF in various cell types. Our group was interested in defining whether these polymorphisms might influence MIF secretion during hypoxia and potentially alter the "set-point" of hypoxic AMPK activation in human cells. We found that human fibroblasts with low-expression *MIF* promoter alleles had both impaired MIF secretion and lower levels of AMPK activation during hypoxia.[15] These findings have potentially important clinical implications. First, they raise the possibility that patients with the low-expression allele might be more susceptible to ischemic injury in the heart. Second, they suggest that patients with the low-expression allele might stand to particularly benefit from either MIF (or AMPK directed therapy) during ischemic insults. Clearly, additional work is needed to address these hypotheses, but they are of interest and potentially have applicability to other solid organs, in addition to the heart.

9. Summary and Conclusions

MIF is indeed emerging as a "most interesting factor" in the cardiovascular system. It appears to have "yin-yang" effects, promoting vascular inflammation and atherosclerosis, while also exerting beneficial effects on cell survival pathways in the ischemic heart. While research over the last 5–10 years has provided significant advancement in the field, much work is still required to fully understand the role of MIF in cardiovascular disease.

References

1. Burger-Kentischer A, Goebel H, Seiler R *et al.* (2002) Expression of macrophage migration inhibitory factor in different stages of human atherosclerosis. *Circulation* **105**: 1561–1566.
2. Pan JH, Lindholt JS, Sukhova GK *et al.* (2003) Macrophage migration inhibitory factor is associated with aneurysmal expansion. *J Vasc Surg* **37**: 628–635.
3. Chen Z, Sakuma M, Zago AC *et al.* (2004) Evidence for a role of macrophage migration inhibitory factor in vascular disease. *Arterioscler Thromb Vasc Biol* **24**: 709–714.
4. Schober A, Bernhagen J, Thiele M *et al.* (2004) Stabilization of atherosclerotic plaques by blockade of macrophage migration inhibitory factor after vascular injury in apolipoprotein E-deficient mice. *Circulation* **109**: 380–385.
5. Burger-Kentischer A, Gobel H, Kleemann R *et al.* (2006) Reduction of the aortic inflammatory response in spontaneous atherosclerosis by blockade of macrophage migration inhibitory factor (MIF). *Atherosclerosis* **184**: 28–38.
6. Pan JH, Sukhova GK, Yang JT *et al.* (2004) Macrophage migration inhibitory factor deficiency impairs atherosclerosis in low-density lipoprotein receptor-deficient mice. *Circulation* **109**: 3149–3153.
7. Gregory JL, Morand EF, McKeown SJ *et al.* (2006) Macrophage migration inhibitory factor induces macrophage recruitment via CC chemokine ligand 2. *J Immunol* **177**: 8072–8079.
8. Schober A, Bernhagen J, Weber C. (2008) Chemokine-like functions of MIF in atherosclerosis. *J Mol Med* **86**: 761–770.
9. Kong YZ, Yu X, Tang JJ *et al.* (2005) Macrophage migration inhibitory factor induces MMP-9 expression: Implications for destabilization of human atherosclerotic plaques. *Atherosclerosis* **178**: 207–215.
10. Herder C, Klopp N, Baumert J *et al.* (2008) Effect of macrophage migration inhibitory factor (MIF) gene variants and MIF serum concentrations on the risk of type 2 diabetes: Results from the MONICA/KORA Augsburg case-cohort study, 1984–2002. *Diabetologia* **51**: 276–284.

11. Herder C, Illig T, Baumert J *et al.* (2008) Macrophage migration inhibitory factor (MIF) and risk for coronary heart disease: Results from the MONICA/KORA Augsburg case-cohort study, 1984–2002. *Atherosclerosis* **200**: 380–388.
12. Takahashi M, Nishihira J, Shimpo M *et al.* (2001) Macrophage migration inhibitory factor as a redox-sensitive cytokine in cardiac myocytes. *Cardiovasc Res* **52**: 438–445.
13. Garner LB, Willis MS, Carlson DL *et al.* (2003) Macrophage migration inhibitory factor is a cardiac-derived myocardial depressant factor. *Am J Physiol Heart Circ Physiol* **285**: H2500–H2509.
14. Willis MS, Carlson DL, Dimaio JM *et al.* (2005) Macrophage migration inhibitory factor mediates late cardiac dysfunction after burn injury. *Am J Physiol Heart Circ Physiol* **288**: H795—H804.
15. Miller EJ, Li J, Leng L *et al.* (2008) Macrophage migration inhibitory factor stimulates AMP-activated protein kinase in the ischaemic heart. *Nature* **451**: 578–582.
16. Ha T, Hua F, Grant D *et al.* (2006) Glucan phosphate attenuates cardiac dysfunction and inhibits cardiac MIF expression and apoptosis in septic mice. *Am J Physiol Heart Circ Physiol* **291**: H1910–H1918.
17. Lin X, Sakuragi T, Metz CN *et al.* (2005) Macrophage migration inhibitory factor within the alveolar spaces induces changes in the heart during late experimental sepsis. *Shock* **24**: 556–563.
18. Chagnon F, Metz CN, Bucala R, Lesur O. (2005) Endotoxin-induced myocardial dysfunction: Effects of macrophage migration inhibitory factor neutralization. *Circ Res* **96**: 1095–1102.
19. Yu CM, Lai KW, Chen YX *et al.* (2003) Expression of macrophage migration inhibitory factor in acute ischemic myocardial injury. *J Histochem Cytochem* **51**: 625–631.
20. Benigni F, Atsumi T, Calandra T *et al.* (2000) The proinflammatory mediator macrophage migration inhibitory factor induces glucose catabolism in muscle. *J Clin Invest* **106**: 1291–1300.
21. Young LH, Renfu Y, Russell RR *et al.* (1997) Low-flow ischemia leads to translocation of canine heart GLUT-4 and GLUT-1 glucose transporters to the sarcolemma *in vivo*. *Circulation* **95**: 415–422.
22. Russell RR, Bergeron R, Shulman GI, Young LH. (1999) Translocation of myocardial GLUT4 and increased glucose uptake through activation of AMPK by AICAR. *Am J Physiol* **277**: H643–H649.
23. Russell RR 3rd, Li J, Coven DL *et al.* (2004). AMP-activated protein kinase mediates ischemic glucose uptake and prevents postischemic cardiac dysfunction, apoptosis, and injury. *J Clin Invest* **114**: 495–503.
24. Russell R, Young L. (1998) Myocardial glucose uptake is stimulated by AICAR, an activator of AMP-activated protein kinase. *Diabetes* **47**: A271.

25. Ma H, Wang J, Thomas DP *et al.* (2010) Impaired macrophage migration inhibitory factor-AMP-activated protein kinase activation and ischemic recovery in the senescent heart. *Circulation* **122**: 282–292.
26. Gao XM, Liu Y, White D, Su Y *et al.* (2011) Deletion of macrophage migration inhibitory factor protects the heart from severe ischemia-reperfusion injury: A predominant role of anti-inflammation. *J Mol Cell Cardiol* **50**(6): 991–999.
27. Kaiser RA, Liang Q, Bueno O *et al.* (2005) Genetic inhibition or activation of JNK1/2 protects the myocardium from ischemia-reperfusion-induced cell death *in vivo*. *J Biol Chem* **280**: 32602–32608.
28. Qi D, Hu X, Wu X *et al.* (2009) Cardiac macrophage migration inhibitory factor inhibits JNK pathway activation and injury during ischemia/reperfusion. *J Clin Invest* **119**: 3807–3816.
29. Yun H, Kim HS, Lee S *et al.* (2009) AMP kinase signaling determines whether c-Jun N-terminal kinase promotes survival or apoptosis during glucose deprivation. *Carcinogenesis* **30**: 529–537.
30. Schulz E, Dopheide J, Schuhmacher S *et al.* (2008) Suppression of the JNK pathway by induction of a metabolic stress response prevents vascular injury and dysfunction. *Circulation* **118**: 1347–1357.
31. Dhanantwari P, Nadaraj S, Kenessey A *et al.* (2008) Macrophage migration inhibitory factor induces cardiomyocyte apoptosis. *Biochem Biophys Res Commun* **371**: 298–303.
32. Koga K, Kenessey A, Powell SR *et al.* (2011) Macrophage migration inhibitory factor provides cardioprotection during ischemia/reperfusion by reducing oxidative stress. *Antioxid Redox Signal* **14**(7): 1191–1202.
33. De Benedetti F, Meazza C, Vivarelli M *et al.* (2003) Functional and prognostic relevance of the −173 polymorphism of the macrophage migration inhibitory factor gene in systemic-onset juvenile idiopathic arthritis. *Arthritis Rheum* **48**: 1398–1407.
34. Zhong XB, Leng L, Beitin A *et al.* (2005) Simultaneous detection of microsatellite repeats and SNPs in the macrophage migration inhibitory factor (MIF) gene by thin-film biosensor chips and application to rural field studies. *Nucleic Acids Res* **33**: e121.
35. Shan ZX, Fu YH, Yu XY *et al.* (2006) Association of the polymorphism of macrophage migration inhibitory factor gene with coronary heart disease in Chinese population. *Zhonghua Yi Xue Yi Chuan Xue Za Zhi* **23**: 548–550.

PART VI

Neurophysiology and Neuropathology

VI-1

A Detrimental Role of MIF in Ischemic Brain Damage

Kate L. Lambertsen[†] and Tomas Deierborg[*]

1. Introduction

Stroke is the number one cause of death in China, the second most common cause of death in India, and in the Western world it is the third most common cause of death and the most common cause of long-term physical disability.[1,2] A stroke arises as a consequence of a blocked cerebral blood vessel (embolic or thrombotic stroke) or from a rupture of a cerebral blood vessel (hemorrhagic stroke). Cessation of blood flow and hence cessation of the supply of oxygen and nutrients to the occluded territory in the brain results in the development of an infarct. A cerebral infarct characteristically consists of a core of pan-necrosis with irreversible tissue damage and a border zone surrounding the infarct, the potentially salvageable peri-infarct or penumbra, in which the neurons degenerate within hours to days after the ischemic attack.

Inflammation appears to play an important role in the pathogenesis of ischemic stroke, and it is well known that the susceptibility of patients to stroke and the subsequent prognosis are influenced by both systemic and local inflammatory processes.[3,4] In experimental stroke there is a time-dependent activation of resident immune cells, especially microglial cells, and recruitment of activated inflammatory cells from the periphery, including neutrophils and monocytes/macrophages. In addition, it has become clear that growth factors, cytokines, chemokines, and other neuromodulatory factors synthesized by activated micro- and astroglial cells and infiltrating

*Corresponding author: Department of Experimental Medical Science, University of Lund, Lund, Sweden. Email: Tomas.deierborg@med.lu.se
†Department of Neurobiology Research, Institute of Molecular Medicine, University of Southern Denmark, Odense, Denmark.

blood-borne leukocytes are involved in controlling the sensitivity of the surviving neurons to glutamate and other potential neurotoxic stimuli.

In this chapter, we provide an overview of brain inflammation following stroke and in particular the importance of the proinflammatory cytokine macrophage migration inhibitory factor (MIF) and the contribution of MIF to ischemic injury, and we highlight recent findings regarding this cytokine in inflammation and stroke.

2. Rodent Models of Ischemic Stroke

Experimental stroke models intend to mimic pathological conditions similar to what is observed in human stroke, with the aim of studying basic cellular processes aimed at developing new therapies for stroke treatment. Rodents, and especially mice due to the increased availability of transgenic and knock-out (KO) strains, have become widely used in experimental stroke research and appear to be suitable approximations of ischemic stroke in humans for several reasons: their anatomy and physiology is well-known and similar to that of humans, their cerebral vasculature resembles that of humans, and further they also are easy to handle and available at a reasonable price.[5,6]

Ischemic stroke results from a reduction in regional cerebral blood flow, with the middle cerebral artery (MCA) being the artery most often affected in humans.[7] The most relevant and commonly utilized experimental stroke models involve occlusion of the MCA, transient occlusion (tMCAO) or permanent occlusion (pMCAO). tMCAO is normally induced by either (i) introduction of an intraluminal thread or wire filament into the MCA for a defined period of time (the model most commonly used), (ii) abluminal application of the potent vasoconstrictor endothelin-1 to the MCA, or (iii) introduction of an embolus into the cerebral circulation or photochemically inducing a MCA thrombosis.[8] The intraluminal model normally results in large ischemic infarcts, which besides the cortex also affects the striatum.[9] pMCAO often is induced by permanently occluding the distal part of the MCA using an electrocoagulator; this is a highly reproducible model associated with very low mortality.[10,11] For comparison, the permanent distal model only results in a cortical infarct within lamina I–VI of the frontal and parietal cortices.[10,11]

In addition, the permanent model of MCA occlusion generally results in infarcts with a relatively small penumbra. This makes the transient model, which results in a relatively large penumbra, more suitable to mimic the human situation, where cerebral vessel occlusion is seldom permanent and most human ischemic cases have spontaneous or thrombolytic-induced reperfusion.[12,13]

3. Inflammatory Responses in the Ischemic Stroke Brain

In addition to the early events caused by a compromised cerebral blood flow, there is also a delayed inflammatory response that takes place hours to weeks after ischemia. This includes both activation of resident microglia and astrocytes[14–17] and recruitment of blood-borne leukocytes, including macrophages[18,19] and polymorphonuclear granulocytes[20] into the brain.

In the acute phase (minutes to hours) after the cerebral event, proinflammatory mediators (including chemokines and cytokines) are released from the injured, necrotic tissue.[21–24] This induces the expression of adhesion molecules (selectins, cellular adhesion molecules and integrins) on cerebral endothelial cells and circulating leukocytes, thereby promoting the diapedesis of leukocytes from the circulation into the infarcted brain.[25] In the subacute phase (hours to days) that follows, infiltrating leukocytes contribute to the release of chemokines and cytokines by amplifying the brain-inflammatory response and causing a more extensive activation of resident cells and a further recruitment of infiltrating leukocytes, which eventually all lead to a disruption of the blood brain barrier (BBB), edema, and neuronal cell death.[26,27]

Microglia are strategically located in close vicinity to neurons in the gray matter and between fiber tracts in the white matter. In the resting state, microglia exhibit radially branched processes and constitutively express surface complement type 3 and Fc receptors. The fine processes of microglia are highly motile and appear to survey the microenvironment, whereas the soma itself is static.[28] Following cerebral ischemia, microglia are activated within minutes to hours.[17,29] This activation is associated with increased expression of complement type 3 receptors, MHC class I and II antigens,[15,30,31] galectin-3,[32] and transformation to a phagocytic phenotype.[15,31,33] Activated microglia in the brain become indistinguishable from blood-borne macrophages because of similar immunophenotypes and other histological characteristics for these cells.[34–36] It is known that microglial can enter the cell cycle and proliferate after acute brain injury.[37,38] Post-ischemia microglial proliferation peaks around 48–72 hrs after cerebral ischemia and may last for several weeks after the initial injury.[39,40] In contrast to the rapid resident microglial response, blood-derived leukocytes are recruited to the brain tissue from approximately 6–12 hrs and further increase in number up to 10–14 days after the insult.[14,19,22,41–43]

4. Peripheral Immunosuppression Following Ischemic Stroke

Stroke patients are commonly affected by pneumonia and infections to the urinary tract.[44] Pneumonia is known to be one of the most common causes of death after the subacute stroke phase, when the brain injury itself has stabilized. It was long believed that swallowing difficulty was an indirect reason for the high infection risk in stroke patients. A few years ago, researchers identified the underlying cause and demonstrated that a stroke can induce a downregulation of the peripheral immune system.[45,46] In one of these studies, a reduction in the proinflammatory cytokine profile from T helper cells, Th1 to Th2, together with lower cell counts of B, T, and NK cells in blood were reported after experimental stroke in mice.[45] Human stroke data confirms a post-stroke reduction in leukocyte numbers and changes in cytokine levels, with the number of T-lymphocytes observed to dramatically decrease in number 12 hrs after the stroke insult and to correlate with a subsequent risk of infections.[47–49] From a biological point of view, the rapid depression of the immune system could be beneficial for the evolution of brain injury. The brain communicates with the peripheral immune system via the autonomic nervous system and the HPA (hypothalamo-pituitary-adrenal) axis. The sympathetic nervous system may play an especially pivotal role in immunosuppression following stroke.[45]

MIF, through its well-known immunosuppressive action on glucocorticoids and its regulation by the HPA axis,[50] could play an important role in post-stroke immunomodulation. The proinflammatory role of MIF in infections, e.g., by increasing the level of tumor necrosis factor-alpha (TNF-α), interferon-gamma (IFN-γ), and interleukin (IL)-1β,[51] could potentiate the immune response against bacterial pathogens such as *Streptococcus pneumoniae*. MIF's role in modifiying the cytokine response to a more proinflammatory response indicates further that the role of MIF in post-stroke immunodepression could be beneficial.[52] The proinflammatory function of MIF in sepsis,[53] in rheumatoid arthritis,[54] and recently in myocardial ischemia and reperfusion,[55] further implicates MIF as a proinflammatory mediator in the peripheral immune system, where it could conceivably assist in promoting a proper immunological response to combat infections after stroke. Therefore, in the subacute phase of stroke, the peripheral role of MIF could have a function in orchestrating the immunological response to fight infections. As discussed below, this possible subacute role of MIF in the periphery is probably distinct from the role of MIF in the acute cell death phase of experimental stroke, where we have observed a clear detrimental role of MIF in terms of neurological behavior and infarct size,[23] as discussed below.

5. Spatial and Temporal Localization of MIF in the Injured Stroke Brain

We have recently reported that MIF is localized in the neuronal cell population, Beta-3$^+$ and NeuN$^+$ cells, in the naïve mouse brain.[23] In particular, parvalbumin$^+$ cortical neurons, a subtype of inhibitory cortical interneurons, display a high immunoreactivity for MIF. To a lower extent, MIF was also found in astrocytes [glial fibrillary acidic protein (GFAP)$^+$ cells]. To clarify the role of MIF in experimental stroke and its spatial and temporal expression, we studied MIF-KO mice and their wildtype littermates and subjected the mice to 45 min of tMCAO, resulting in injury to both the striatum and cortex.[23] We found that MIF expression in the brain during the first week following stroke was highly dynamic and that the cellular localization of MIF depended on the time after ischemia.

Already at 3 hrs after stroke, MIF immunoreactivty was lost in neurons located within the infarct core, and this was sustained for at least 7 days. Still, a small population of parvalbumin$^+$ cells could be found in the outer part of the ischemic territory up to 18 hrs after stroke and which at later time points was included in the infarcted area.

From 3 hrs to 3 days we observed a gradual increase in MIF immunoreactivity in the brain area surrounding the infarct; in particular, an upregulation of MIF was found in neurons. One week after experimental stroke, MIF was detected primarily in the glial population. MIF was present in glial-like cells surrounding the endothelial cells (CD31$^+$), in microglial cells located in the infarct core (galectin-3$^+$), and in reactive astrocytes (GFAP$^+$) surrounding the infarct core. MIF was not co-localized to oligodendrocytes (GST-π^+) or CD74$^+$ leukocytes (CD74 being the cellular receptor for MIF).[23]

In another model of parenchymal brain injury, transient global brain ischemia in rats that mimics the brain damage after cardiac arrest and succesful cardiopulmanary resuscitation, MIF mRNA and protein was upregulated in the injured rat brain, especially in neuronal processes located in the cortex.[56] Recently, a clinical study demonstrated a correlation between MIF expression, mRNA and protein in blood monocytes from stroke patients, and stroke severity, where MIF expression peaked after 24 hrs.[57] The same study further reported on an upregulation of MIF in the rat brain after focal cerebral ischemia.[57]

6. The Role of MIF in Brain Inflammation

Information about the role of MIF in regulating the inflammation in the brain is sparse. An examination of cytokine gene expression in the brain upon West

Nile virus infection in mice deficient in MIF revealed lower expression of several proinflammatory cytokines (e.g., TNF-α and IFN-γ).[58] Using another model of virus infection, Borna disease virus, Bacher and co-authors[59] reported high expression of MIF in astrocytic end-feets, which are an important part of the BBB controlling influx of substances and cells into the brain. In this context, the high MIF concentration in the astrocytic end-feets appeared to repress monocyte infiltration. Neverthless, the role of MIF in regulating the infiltration of immunological cells into tissue is not clear.[58,60]

7. Deleterious Function of MIF in Experimental Stroke and Its Inflammatory Role

To address the functional role of MIF following stroke, we used the same model, tMCAO, as described above. We found that MIF-KO mice when subjected to experimental stroke display a significant reduction in infarct volume (by 23%) compared to their wildtype littermates. We also found behavioral improvement as assessed by the grip strength test and the rotating pole test at 2 days after the ischemic insult, and importantly, the positive behavioral effect on the rotating pole was also evident after 7 days.[23] In view of the post-stroke immunosuppression described above, MIF abrogation may influence the mortality in the MIF KO mice due to infections.[45] However, we did not observe any difference in the mortality rate between MIF KO and wildtype littermate mice. Peripheral changes in blood were studied in MIF KO mice 2 and 7 days after the stroke injury by examining nine different Th1/Th2 cytokines (IL-1β, IFN-γ, IL-2, IL-4, IL-5, IL-10, IL-12, KC/CXCL-1, and TNF-α) (Inacio *et al.*, 2011c). Although stroke induced an increased expression in IL-1β protein levels 2 days post-stroke compared to sham-mice, no significant changes were detected between MIF KO and wildtype mice. Spleen weight, which is known to be reduced after stroke,[61] was also unchanged between MIF KO and wildtype mice. Stroke injury significantly increased the protein levels of IL-12 and CXCL-1 in the ipsilateral infarct, but again MIF KO mice did not show significant changes in cytokine expression when compared to wildtype mice. These data suggests that MIF may not be important for changing the polarity of Th1/Th2 mediated cytokine responses; this also is consistent with the lack of an MIF effect on cytokine levels of MIF in an experimental animal model of multiple sclerosis, experimental autoimmune encephalomyelitis (EAE).[60]

To further elucidate changes in neuroinflammation in MIF KO mice, we studied microgliosis 7 days after stroke by examining the microglial population using the markers CD68 (transmembrane glycoprotein/scavenger receptor) and MAC-2/galectin-3 (carbohydrate binding protein). Microglia become activated in the injured stroke brain within hours after the ischemic insult, and by one week post-ischemia there was an intense activation of micoglia.[17] The origin of the $CD68^+$ and galectin-3^+ cells are likely both from blood-borne leukocytes and from an expansion of the parenchymal microglial population.[62] It is believed that the infiltrating blood-borne cells have a more detrimental function in the injured brain compared to the parenchymal population.[22,63,64] Examination of the CD68 immunoreactivtiy in the stroke-injured brain of MIF KO and wildtype littermate mice revealed no difference in the CD68 microglial population. Interestingly, investigation of the galectin-3 immunoreactivity revealed increased levels in MIF KO mice. The density of galectin-3^+ cells was also higher in MIF KO mice. In view of the smaller infarcts present in the MIF KO mice, the magnitude of increased galectin-3 could be underestimated. Lalancette-Hebert and collaborators[40] found increased infarcts after selective ablation of the proliferative microglial population following experimental stroke in mice and suggested that the proliferative population of microglial cells were of a trophic and protective nature. Interestingly, they identified this population of microglial cells to be galectin-3^+ and that these cells also expressed the trophic factor, insulin-like growth factor 1 (IGF-1). The increased number of galectin-3^+ cells in MIF KO mice could indicate that MIF, by a direct or indirect mechanism, alters the microglial population to have a more protective/supportive function in the inflamed stroke-injured brain. Yan and collaborators additionally demonstated that galectin-3 can be important in the remodeling of the brain after stroke.[65] To further examine the inflammatory role of MIF in the stroke brain, we examined reactive astrocytes (GFAP+) that are positive for MIF 3–7 days post-ischemia in the peri-infarct area of the cortex and hippocampus. No effect of MIF on the $GFAP^+$ astrocyte population was discovered. Finally, the cell surface receptor of MIF, CD74, which is found especially on lymphocytes and antigen presenting cells, was studied without detection of any difference in $CD74^+$ cells between the two mice strains in the injured hemisphere.

8. A Direct Neuronal Role of MIF in Brain Ischemia

The majority of cells subside within the first hours after stroke during a time period when inflammation is limited. In view of the presence of MIF in neurons, MIF could exert a direct effect on neurons that may be more important in the acute ischemic phase than its expected role as a modulator of inflammation. Nishio and collaborators[66] previously examined the role of MIF in spinal cord injury and excitotoxicity, and reported on MIF's detrimental role following spinal cord injury, where fewer apoptotic neurons were found in MIF KO mice; this finding also correlated with better hind limb function. Using cerebellar granule neurons from MIF KO and wildtype mice, they found that MIF KO mice were more resistant to excitotoxicity, suggesting a direct neuronal effect of MIF. We have also examined the direct neuronal effect of MIF by using primary neuronal cultures from rat hippocampus, embryonic day 17, and subjected the cultures to *in vitro* ischemia, lethal oxygen glucose deprivation (OGD), with and without the MIF inhibitor ISO-1. When evaluating the neuronal death with lactate dehydrogenase (LDH) activity, we found a neuroprotective effect of MIF inhibition with ISO-1 24 hrs after OGD, suggesting that MIF does indeed have a direct detrimental role on neurons following stroke.[23] Using a shorter OGD time period, sublethal OGD, we observed an upregulation of MIF in the neurons that was in line with our *in vivo* data. This direct intraneuronal function of MIF could be important for neuronal cell death in the acute ischemic phase, where a reduction or an inhibition of MIF is beneficial. The mechanistic role of MIF in this context can only be speculated upon. A direct extracellular effect of MIF binding to its receptor CD74 is unlikely due to the lack of CD74 receptors on neurons (Inacio *et al.*, 2011c). Perhaps an intracellular role of MIF is governing this detrimental function. For example, MIF interaction with the p53-dependent pathway,[66] activation of the p38 MAPK and c-Jun N-terminal protein kinase (JNK), regulation of the mitochondria-associated apoptotic[67] or the extracellular signal-regulated (ERK) mitogen-activated protein (MAP) kinase pathways known to be involved in cell death regulation,[68] or of p27-dependent processes could be implicated in MIF-induced neuronal cell death.[69,70]

The cellular regulation of MIF in the acute ischemic brain could be related to its known regulation by hypoxia-inducible factor-1, which is a major regulatory pathway in hypoxia-ischemia. At later time-points, where a multitude of inflammatory factors converge, several other regulatory mechanisms might be involved,[71] and possibly a direct neuronal effect could be less important in the later inflammatory phases of brain ischemia.

9. Downregulation of MIF in the Ischemic Brain after Enriched Environment

An enriched environment is very different from the normal housing of rodents in standard cages.[72] Housing cages used to provide an enriched environment are multileveled with tubes, ladders and various items that provide rodents with multimodal stimulation. More animals are also housed together, which gives important social stimuli.[73] An enriched environment is known to improve sensorimotor function and cognitive performance through alterations in brain connectivity. A direct neuronal function of MIF has been suggested whereby MIF stimulates delayed rectifier K^+ currents[74] and inhibits the firing of hypothalamic neurons.[75] Our findings of a decreased MIF protein level in the cortex of rats housed in an enriched environment compared to standard housed rats indicate that MIF could have a role in neuronal plasticity.[72]

When stroke-injured rats (cortical injury in spontaneous hypertensive rats) were housed in an enriched environment from day 2–5 after stroke, a reduced level of MIF protein was observed in the peri-infarct region close to the injury that is known to be involved in neuronal plasticity after stroke.[76] Interestingly, in this peri-infarct region close to the cortical infarct, a three-fold increase in the number of parvalbumin$^+$ cells were found in injured rats housed in an enriched environment when compared to injured rats housed in standard cages, indicating a role of parvalbumin in neuronal reorganization/activity. The possible role of GABAergic parvalbumin neurons in enriched environment[77] and the expression of MIF in parvalbumin$^+$ neurons suggests that MIF could be implicated in the reorganization of spared neural network following stroke.

10. Conclusions

Inflammation after ischemic stroke is a central feature of neuropathology that is evident from the first hours after the insult to the more chronic phase of scar formation and infarct debri resolution. The inflammation following stroke is not merely a neuroninflammatory response specific for the injured brain, but it also influences systemic immunity by exerting a peripheral immunosuppressive effect. MIF is known to be critically involved in several inflammatory conditions and to act primarily as a proinflammtory factor. We have discussed herein our recent data regarding the role of MIF in experimental stroke and showed that MIF renders neurons and brain tissue more vulnerable to ischemic stroke. We conclude that MIF has a lesser role in regulating cytokine levels in the injured brain or in the blood following experimental stroke, but a more direct detrimental effect on neurons. Despite the lack of a cytokine regulating effect of MIF following stroke, MIF appears to be involved in regulating the microglia response, and MIF KO mice showed an alteration in microglial activation (galectin-3). The downregulated protein levels of MIF in the area around the infarct following exposure to an enriched environment further implies that MIF has a role beyond its cytotoxic effect on neurons and is disadvantageous for post-stroke plasticity. The collective MIF data reviewed in this chapter emphasize its relevance in stroke and confirm a detrimental role of MIF in the stroke-injured brain.

References

1. Murray J, Young J, Forster A. (2009) Measuring outcomes in the longer term after a stroke. *Clin Rehabil* **23**: 918–921.
2. Roger VL, Go AS, Lloyd-Jones DM *et al.* (2011) Heart disease and stroke statistics — 2011 update: A report from the American Heart Association. *Circulation* **123**: e18–e209.
3. Emsley HC, Hopkins SJ. (2008) Acute ischaemic stroke and infection: Recent and emerging concepts. *Lancet Neurol* **7**: 341–353.
4. McColl BW, Allan SM, Rothwell NJ. (2009) Systemic infection, inflammation and acute ischemic stroke. *Neuroscience* **158**: 1049–1061.
5. Yamori Y, Horie R, Handa H *et al.* (1976) Pathogenetic similarity of strokes in stroke-prone spontaneously hypertensive rats and humans. *Stroke* **7**: 46–53.
6. Durukan A, Strbian D, Tatlisumak T. (2008) Rodent models of ischemic stroke: A useful tool for stroke drug development. *Curr Pharm Des* **14**: 359–370.
7. Karpiak SE, Tagliavia A, Wakade CG. (1989) Animal models for the study of drugs in ischemic stroke. *Annu Rev Pharmacol Toxicol* **29**: 403–414.
8. Jin R, Yang G, Li G. (2010) Inflammatory mechanisms in ischemic stroke: Role of inflammatory cells. *J Leukoc Biol* **87**: 779–789.
9. Carmichael ST. (2005) Rodent models of focal stroke: Size, mechanism, and purpose. *NeuroRx* **2**: 396–409.
10. Xi GM, Wang HQ, He GH *et al.* (2004) Evaluation of murine models of permanent focal cerebral ischemia. *Chin Med J (Engl)* **117**: 389–394.
11. Kuraoka M, Furuta T, Matsuwaki T *et al.* (2009) Direct experimental occlusion of the distal middle cerebral artery induces high reproducibility of brain ischemia in mice. *Exp Anim* **58**: 19–29.
12. Mhairi Macrae I. (1992) New models of focal cerebral ischaemia. *Br J Clin Pharmacol* **34**: 302–308.
13. Lo EH. (2008) Experimental models, neurovascular mechanisms and translational issues in stroke research. *Br J Pharmacol* **153**(Suppl 1): S396–S405.
14. Clark RK, Lee EV, Fish CJ *et al.* (1993) Development of tissue damage, inflammation and resolution following stroke: An immunohistochemical and quantitative planimetric study. *Brain Res Bull* **31**: 565–572.
15. Morioka T, Kalehua AN, Streit WJ. (1993) Characterization of microglial reaction after middle cerebral artery occlusion in rat brain. *J Comp Neurol* **327**: 123–132.

16. Davies CA, Loddick SA, Stroemer RP *et al.* (1998) An integrated analysis of the progression of cell responses induced by permanent focal middle cerebral artery occlusion in the rat. *Exp Neurol* **154**: 199–212.
17. Lambertsen KL, Meldgaard M, Ladeby R, Finsen B. (2005) A quantitative study of microglial-macrophage synthesis of tumor necrosis factor during acute and late focal cerebral ischemia in mice. *J Cereb Blood Flow Metab* **25**: 119–135.
18. Yamagami S, Tamura M, Hayashi M *et al.* (1999) Differential production of MCP-1 and cytokine-induced neutrophil chemoattractant in the ischemic brain after transient focal ischemia in rats. *J Leukoc Biol* **65**: 744–749.
19. Clausen BH, Lambertsen KL, Babcock AA *et al.* (2008) Interleukin-1beta and tumor necrosis factor-alpha are expressed by different subsets of microglia and macrophages after ischemic stroke in mice. *J Neuroinflammation* **5**: 46.
20. Barone FC, Hillegass LM, Price WJ *et al.* (1991) Polymorphonuclear leukocyte infiltration into cerebral focal ischemic tissue: Myeloperoxidase activity assay and histologic verification. *J Neurosci Res* **29**: 336–345.
21. Clausen BH, Lambertsen KL, Meldgaard M, Finsen B. (2005) A quantitative *in situ* hybridization and polymerase chain reaction study of microglial-macrophage expression of interleukin-1beta mRNA following permanent middle cerebral artery occlusion in mice. *Neuroscience* **132**: 879–892.
22. Lambertsen KL, Clausen BH, Babcock AA *et al.* (2009) Microglia protect neurons against ischemia by synthesis of tumor necrosis factor. *J Neurosci* **29**: 1319–1330.
23. Inacio AR, Ruscher K, Leng L *et al.* (2011a) Macrophage migration inhibitory factor promotes cell death and aggravates neurologic deficits after experimental stroke. *J Cereb Blood Flow Metab* **31**: 1093–1106.
24. Mirabelli-Badenier M, Braunersreuther V, Viviani GL *et al.* (2011) CC and CXC chemokines are pivotal mediators of cerebral injury in ischaemic stroke. *Thromb Haemost* **105**(3): 409–420.
25. Yilmaz G, Granger DN. (2008) Cell adhesion molecules and ischemic stroke. *Neurol Res* **30**: 783–793.
26. Kriz J. (2006) Inflammation in ischemic brain injury: Timing is important. *Crit Rev Neurobiol* **18**: 145–157.
27. Amantea D, Nappi G, Bernardi G *et al.* (2009) Post-ischemic brain damage: Pathophysiology and role of inflammatory mediators. *FEBS J* **276**: 13–26.
28. Nimmerjahn A, Kirchhoff F, Helmchen F. (2005) Resting microglial cells are highly dynamic surveillants of brain parenchyma *in vivo*. *Science* **308**: 1314–1318.
29. Rupalla K, Allegrini PR, Sauer D, Wiessner C. (1998) Time course of microglia activation and apoptosis in various brain regions after permanent focal cerebral ischemia in mice. *Acta Neuropathol* **96**: 172–178.

30. Finsen BR, Jorgensen MB, Diemer NH, Zimmer J. (1993) Microglial MHC antigen expression after ischemic and kainic acid lesions of the adult rat hippocampus. *Glia* **7**: 41–49.
31. Lehrmann E, Christensen T, Zimmer J *et al.* (1997) Microglial and macrophage reactions mark progressive changes and define the penumbra in the rat neocortex and striatum after transient middle cerebral artery occlusion. *J Comp Neurol* **386**: 461–476.
32. Walther M, Kuklinski S, Pesheva P *et al.* (2000) Galectin-3 is upregulated in microglial cells in response to ischemic brain lesions, but not to facial nerve axotomy. *J Neurosci Res* **61**: 430–435.
33. Kato H, Kogure K, Liu XH *et al.* (1996) Progressive expression of immunomolecules on activated microglia and invading leukocytes following focal cerebral ischemia in the rat. *Brain Res* **734**: 203–212.
34. Davis EJ, Foster TD, Thomas WE. (1994) Cellular forms and functions of brain microglia. *Brain Res Bull* **34**: 73–78.
35. Campanella M, Sciorati C, Tarozzo G, Beltramo M. (2002) Flow cytometric analysis of inflammatory cells in ischemic rat brain. *Stroke* **33**: 586–592.
36. Zhang C, Lam TT, Tso MO. (2005) Heterogeneous populations of microglia/macrophages in the retina and their activation after retinal ischemia and reperfusion injury. *Exp Eye Res* **81**: 700–709.
37. Ladeby R, Wirenfeldt M, Dalmau I *et al.* (2005) Proliferating resident microglia express the stem cell antigen CD34 in response to acute neural injury. *Glia* **50**: 121–131.
38. Gowing G, Vallieres L, Julien JP. (2006) Mouse model for ablation of proliferating microglia in acute CNS injuries. *Glia* **53**: 331–337.
39. Denes A, Vidyasagar R, Feng J *et al.* (2007) Proliferating resident microglia after focal cerebral ischaemia in mice. *J Cereb Blood Flow Metab* **27**: 1941–1953.
40. Lalancette-Hebert M, Gowing G, Simard A *et al.* (2007) Selective ablation of proliferating microglial cells exacerbates ischemic injury in the brain. *J Neurosci* **27**: 2596–2605.
41. Schroeter M, Jander S, Witte OW, Stoll G. (1994) Local immune responses in the rat cerebral cortex after middle cerebral artery occlusion. *J Neuroimmunol* **55**: 195–203.
42. Schilling M, Besselmann M, Leonhard C *et al.* (2003) Microglial activation precedes and predominates over macrophage infiltration in transient focal cerebral ischemia: A study in green fluorescent protein transgenic bone marrow chimeric mice. *Exp Neurol* **183**: 25–33.
43. Tanaka R, Komine-Kobayashi M, Mochizuki H *et al.* (2003) Migration of enhanced green fluorescent protein expressing bone marrow-derived microglia/macrophage into the mouse brain following permanent focal ischemia. *Neuroscience* **117**: 531–539.

44. Davenport RJ, Dennis MS, Wellwood I, Warlow CP. (1996) Complications after acute stroke. *Stroke* **27**: 415–420.
45. Prass K, Meisel C, Hoflich C *et al.* (2003) Stroke-induced immunodeficiency promotes spontaneous bacterial infections and is mediated by sympathetic activation reversal by poststroke T helper cell type 1-like immunostimulation. *J Exp Med* **198**: 725–736.
46. Meisel C, Schwab JM, Prass K *et al.* (2005) Central nervous system injury-induced immune deficiency syndrome. *Nat Rev Neurosci* **6**: 775–786.
47. Haeusler KG, Schmidt WU, Fohring F *et al.* (2008) Cellular immunodepression preceding infectious complications after acute ischemic stroke in humans. *Cerebrovasc Dis* **25**: 50–58.
48. Vogelgesang A, Grunwald U, Langner S *et al.* (2008) Analysis of lymphocyte subsets in patients with stroke and their influence on infection after stroke. *Stroke* **39**: 237–241.
49. Klehmet J, Harms H, Richter M *et al.* (2009) Stroke-induced immunodepression and post-stroke infections: Lessons from the preventive antibacterial therapy in stroke trial. *Neuroscience* **158**: 1184–1193.
50. Calandra T, Bernhagen J, Metz CN *et al.* (1995) MIF as a glucocorticoid-induced modulator of cytokine production. *Nature* **377**: 68–71.
51. Roger T, Froidevaux C, Martin C, Calandra T. (2003) Macrophage migration inhibitory factor (MIF) regulates host responses to endotoxin through modulation of Toll-like receptor 4 (TLR4). *J Endotoxin Res* **9**: 119–123.
52. Koebernick H, Grode L, David JR *et al.* (2002) Macrophage migration inhibitory factor (MIF) plays a pivotal role in immunity against *Salmonella typhimurium*. *Proc Natl Acad Sci USA* **99**: 13681–13686.
53. Bernhagen J, Calandra T, Mitchell RA *et al.* (1993) MIF is a pituitary-derived cytokine that potentiates lethal endotoxaemia. *Nature* **365**: 756–759.
54. Morand EF, Leech M, Bernhagen J. (2006) MIF: A new cytokine link between rheumatoid arthritis and atherosclerosis. *Nat Rev Drug Discov* **5**: 399–410.
55. Gao XM, Liu Y, White D *et al.* (2011) Deletion of macrophage migration inhibitory factor protects the heart from severe ischemia-reperfusion injury: A predominant role of anti-inflammation. *J Mol Cell Cardiol* **50**: 991–999.
56. Yoshimoto T, Nishihira J, Tada M *et al.* (1997) Induction of macrophage migration inhibitory factor messenger ribonucleic acid in rat forebrain by reperfusion. *Neurosurgery* **41**: 648–653.
57. Wang L, Zis O, Ma G *et al.* (2009) Upregulation of macrophage migration inhibitory factor gene expression in stroke. *Stroke* **40**: 973–976.
58. Arjona A, Foellmer HG, Town T *et al.* (2007) Abrogation of macrophage migration inhibitory factor decreases West Nile virus lethality by limiting viral neuroinvasion. *J Clin Invest* **117**: 3059–3066.

59. Bacher M, Weihe E, Dietzschold B *et al.* (2002) Borna disease virus-induced accumulation of macrophage migration inhibitory factor in rat brain astrocytes is associated with inhibition of macrophage infiltration. *Glia* **37**: 291–306.
60. Kithcart AP, Cox GM, Sielecki T *et al.* (2010) A small-molecule inhibitor of macrophage migration inhibitory factor for the treatment of inflammatory disease. *FASEB J* **24**: 4459–4466.
61. Offner H, Subramanian S, Parker SM *et al.* (2006) Splenic atrophy in experimental stroke is accompanied by increased regulatory T cells and circulating macrophages. *J Immunol* **176**: 6523–6531.
62. Lambertsen KL, Deierborg T, Gregersen R *et al.* (2011) Differences in origin of reactive microglia in bone marrow chimeric mouse and rat after rransient global ischemia. *J Neuropathol Exp Neurol* **70**: 481–494.
63. Popovich PG, Guan Z, Wei P *et al.* (1999) Depletion of hematogenous macrophages promotes partial hindlimb recovery and neuroanatomical repair after experimental spinal cord injury. *Exp Neurol* **158**: 351–365.
64. Glezer I, Simard AR, Rivest S. (2007) Neuroprotective role of the innate immune system by microglia. *Neuroscience* **147**: 867–883.
65. Yan YP, Lang BT, Vemuganti R, Dempsey RJ. (2009) Galectin-3 mediates post-ischemic tissue remodeling. *Brain Res* **1288**: 116–124.
66. Nishio Y, Koda M, Hashimoto M *et al.* (2009) Deletion of macrophage migration inhibitory factor attenuates neuronal death and promotes functional recovery after compression-induced spinal cord injury in mice. *Acta Neuropathol* **117**: 321–328.
67. Dhanantwari P, Nadaraj S, Kenessey A *et al.* (2008) Macrophage migration inhibitory factor induces cardiomyocyte apoptosis. *Biochem Biophys Res Commun* **371**: 298–303.
68. Mitchell RA, Metz CN, Peng T, Bucala R. (1999) Sustained mitogen-activated protein kinase (MAPK) and cytoplasmic phospholipase A2 activation by macrophage migration inhibitory factor (MIF). Regulatory role in cell proliferation and glucocorticoid action. *J Biol Chem* **274**: 18100–18106.
69. Kleemann R, Grell M, Mischke R *et al.* (2002) Receptor binding and cellular uptake studies of macrophage migration inhibitory factor (MIF): Use of biologically active labeled MIF derivatives. *J Interferon Cytokine Res* **22**: 351–363.
70. Lue H, Kleemann R, Calandra T *et al.* (2002) Macrophage migration inhibitory factor (MIF): Mechanisms of action and role in disease. *Microbes Infect* **4**: 449–460.
71. Baugh JA, Gantier M, Li L *et al.* (2006) Dual regulation of macrophage migration inhibitory factor (MIF) expression in hypoxia by CREB and HIF-1. *Biochem Biophys Res Commun* **347**: 895–903.

72. Inacio AR, Ruscher K, Wieloch T. (2011b) Enriched environment downregulates macrophage migration inhibitory factor and increases parvalbumin in the brain following experimental stroke. *Neurobiol Dis* **41**: 270–278.
73. Johansson BB. (1996) Functional outcome in rats transferred to an enriched environment 15 days after focal brain ischemia. *Stroke* **27**: 324–326.
74. Matsuura T, Sun C, Leng L *et al.* (2006) Macrophage migration inhibitory factor increases neuronal delayed rectifier K+ current. *J Neurophysiol* **95**: 1042–1048.
75. Sun C, Li H, Leng L *et al.* (2004) Macrophage migration inhibitory factor: An intracellular inhibitor of angiotensin II-induced increases in neuronal activity. *J Neurosci* **24**: 9944–9952.
76. Murphy TH, Corbett D. (2009) Plasticity during stroke recovery: From synapse to behaviour. *Nat Rev Neurosci* **10**: 861–872.
77. Bertini G, Peng ZC, Fabene PF *et al.* (2002) Fos induction in cortical interneurons during spontaneous wakefulness of rats in a familiar or enriched environment. *Brain Res Bull* **57**: 631–638.
78. Inacio AR, Bucala R, Deierborg T (2011c) Lack of macrophage migration inhibitory factor in mice does not affect hallmarks of the inflammatory/immune response during the first week after strok. *J Neuroinflammation* **8**: 75

VI-2

Association of MIF with Autism Spectrum Disorders

Ivana Kawikova*, James F. Leckman*,
Astrid Morer† and Elena L. Grigorenko*,‡

1. Introduction

Autism spectrum disorders (ASD) have unknown etiology. The interactions of heterogeneous genetic and environmental factors have been implicated, and the role of the immune system has been suggested at the interface of inner and outer environments. Here we discuss the current evidence for a role of the immune system in ASD in relation to macrophage migration inhibitory factor (MIF), which is a critical molecule for the activation of innate immunity.

2. Autism Spectrum Disorders

Autism spectrum disorders (ASD) are complex neurodevelopmental disorders characterized by impairments in cognitive skills, social interactions, verbal and nonverbal communication, repetitive behaviors, and restricted interests.[1] Next to the core syndrome, many patients have additional disturbances, including aggression, tantrums, and self-injurious behavior. The onset of ASD is in early childhood, possibly even *in utero,* and it is diagnosed when deviation from the normal development of communication skills is noted, most commonly between 2–3 years of age.[2–5] ASD affects as many as 62 children per 10,000 in the general population, and the incidence has recently increased.[6,7]

The site of pathology in ASD is not entirely clear. In postmortem studies, the histologic changes noted below are quite variable, but most consistent

*Child Study Center, Yale School of Medicine, New Haven, CT, USA.
†Department of Child Adolescent Psychiatry, Hospital Clinic-Cibersam, Barcelona, Spain.
‡Corresponding author: Email: elenalgrigorenko@gmail.com

alterations were observed in the limbic system, cerebellum, related inferior olive nuclei, and in the vertical limb of the diagonal band of Broca.[8] The etiology remains undefined and appears to involve interactions of unknown environmental factors with genetic endowment.[9] Currently, no biological markers are available for the diagnosis of ASD. The diagnosis is established using psychological assessment tools and the current treatment relies mainly on behavioral modification. Risperidone, an atypical antipsychotic drug, has also been shown to reduce both core and behavioral symptoms of individuals with ASD.[10,11]

2.1 *Genetics of ASD*

Heritability has been estimated to occur in approximately 90% of ASD cases.[12] The nature of genetic transmission has not been elucidated and the existing evidence indicates that ASD is a highly complex, heterogeneous disorder. A growing number of new genes have recently been implicated. They include neuroligin 4-X linked (*NLGN4X*), methylcytosin binding protein 2 (*MECP2*), contactin-associated protein-like 2 (*CNTNAP2*), fragile X mental retardation protein 1 (*FMR1*), neuroligin ankyrin repeat 3 (*SHANK3*), neurexin contactin-associated protein like 4, attractin-like (*ATRNL1*) gene, CD38 (cyclic ADP ribose hydrolase), oxytocin receptor, glutamate transporter gene (*SLC1A1*), and fatty acid binding protein (FABP).[13]

Many of the proteins encoded by these ASD-related genes are involved in neuronal development and functioning, but some of the genes are also likely to affect immune functions. For example, *MECP2* is necessary for histone modification, which critically affects the functioning of *FOXP3*-expressing regulatory T cells.[14] These cells are essential for maintaining the balance between antimicrobial immune response and self-tolerance. Numbers of regulatory T cells are reduced in individuals with ASD, leaving the individuals more susceptible to the development of autoimmune disorders.[15,16] Another example may be *SHANK3*, which is expressed not only in brain areas that were implicated in autism, but in lymphocytes and the thymus as well.[17,18] Future studies will need to determine whether the parallel expression in the brain and the immune system might be related to the pathogenesis of autism.

2.2 *Role of the immune system in ASD*

The immune system was implicated in ASD by elevated levels of various immune parameters that will be described below. There are at least three possible scenarios of the relationship between ASD and enhanced immune activity. First, it is possible that the immune activity is unrelated to ASD symptoms and may be a result of the same genetic defects that affect neuronal cells and lead to neuronal dysfunction in ASD. Second, the immune system activation may reflect enhanced responsiveness to environmental stimuli, which could lead to autoimmunity due to mimicry between neuronal antigens in the brains of ASD individuals and epitopes of microbial pathogens that the child or its mother encountered previously. Third, it is possible that the enhanced immune activity is caused by ASD. The mechanism could involve chronic stress induced by the presence of chronic neuropsychiatric disorder. Though the first and third possibilities may be clinically relevant, the most important is the second option. There are currently several immune therapies available or under development for patients with autoimmune disorders, and a better understanding of the role of the immune system may open novel treatment avenues for ASD patients. The autoimmune pathogenesis is implicated by the above normal frequency of autoimmune disorders among families with ASD patients[19–21] and by the biological evidence of enhanced immune activity in both the central nervous system and peripheral blood of individuals with autism. However, no specific antigens stimulating autoimmune pathological processes have yet been identified.

2.2.1 *Signs of inflammation in brains of individuals with autism*

Studies of postmortem brains of individuals with autism revealed localized inflammation in the cortex and cerebellum. The inflammation was characterized by the elevated expression of multiple inflammatory mediators, activated microglia and macroglia, and perivascular localization of macrophages.[22] Decreased numbers of Purkinje cells in autism have been previously reported,[8] and Vargas and colleagues demonstrated intimate co-localization of Purkinje cells and complementary lytic complexes, suggesting that the immune system might be involved in their destruction. Remarkably, no lymphocytes were identified in the 11 specimens studied,[22] but in the same study, a 232-fold increase in interferon-gamma (IFN-γ) was found in the cerebrospinal fluid (CSF), which strongly suggests the presence of activated T lymphocytes in the central nervous system (CNS) of individuals with autism.[22] Li *et al.* confirmed the elevated levels of IFN-γ and additionally demonstrated an elevation in tumor necrosis factor-alpha (TNF-α), IL-6, IL-8, and GM-CSF.[23] Other studies have implicated the activity of the immune system in the CNS of individuals with autism by the elevation of soluble TNF receptor II and

TNF-α, as well as neopterin.[24,25] With regard to MIF, its expression or protein levels have yet to be evaluated in the CNS of individuals with autism.

The presence of these inflammatory markers in the brains of ASD individuals suggests enhanced activity of innate and T helper-1 immunity. A critical question is whether the inflammatory changes represent the clearance of degenerating neurons or an autoimmune process directed against originally healthy neuronal cells. The answer to this question may come from a better understanding of the overall functioning of the immune system in patients with ASD.

2.2.2 *Signs of enhanced activity of the immune system in peripheral blood systems?*

Analyses of more readily available blood specimens revealed elevations of IL-2,[26] IL-12,[27] and IFN-γ[27,28] in patients with ASD, implying a T helper-1-driven immune process. The increase in IFN-γ was not found in other studies.[29] Further cytokines associated with the activation of immune cells, including TNF-α, IL-1β, IL-4, IL-5, and IL-6, were not detected in the serum of patients with ASD by Zimmerman and colleagues,[30] but IL-1β, IL-6, IL-8, and IL-12p40 were recently reported to be elevated in a well-controlled study involving a significant number of individuals with ASD.[31] Consistent with elevated inflammatory markers, functional studies of *in vitro* stimulated peripheral blood mononuclear cells (PBMC) revealed increased expression or release of IFN-γ, TNF-α, and IL-12 in patients with ASD.[25,32,33] Recently, enhanced activity of natural killer cells also was reported.[34] These studies together suggest that similar to the findings in the CSF, there is enhanced activity of innate immunity and T helper-1 lymphocytes also in the peripheral blood.

B cell immunity of children with ASD is also affected, since antibody production is altered and the sera of children with ASD and their mothers contain brain-reactive antibodies.[35] Interestingly, several studies also suggested increased levels of IgG_2 and IgG_4,[36–38] as well as decreased levels of IgG_3 and IgM.[39] These discrepancies are not well understood but represent dysfunctions in the form of immune deficiencies and allergies. These possibilities need to be further investigated.

The markers of enhanced activity of the immune system appear to be accompanied by signs of decreased suppressive capacity of the immune system in individuals with ASD, as suggested by decreased levels of TGF-β (a cytokine typically released by regulatory T cells)[40,41] and decreased numbers of regulatory T cells (a subset of T lymphocytes that are required for the prevention of autoimmune disorders).[15,16]

Together, these studies suggest an imbalance between the effector branches of the immune system that are required for defense against infectious microorganisms and suppressor branches of the immune system that are necessary to maintain tolerance and prevent autoimmune disorders. Why only some children develop autoimmune responses as a consequence of common environmental pathogens is not clear, but genetic predisposition is likely involved in this vulnerability, as is the case in most autoimmune disorders. This notion is currently supported by several studies addressing the role of loci for histocompatibility lymphocyte antigens,[42–46] but genetic alterations in other immune parameters are likely to significantly affect functions of the immune system. We now consider the possible role of MIF in autism.

3. Role of MIF in Immunity

MIF is a critical regulator of both innate and adaptive immunity. In humans, MIF is encoded by a highly conserved gene on chromosome 22 (22q11.2).[47,48] A single RNA species of about 0.8 kB encodes the 114 amino-acid protein of 12.5 kDa.[49] MIF expression has been observed in all immune cell types, including monocytes, macrophages, dendritic cells, mast cells, granulocytes, and T and B lymphocytes.[50] MIF is also expressed in endocrine tissues, particularly in the hypothalamus, pituitary gland, and adrenal glands.[51–54] These organs constitute an axis that orchestrates the stress response.

MIF is constitutively expressed and stored in intracellular pools, making it possible to be rapidly released without *de novo* synthesis.[49] MIF binds to CD74, which is the cell surface form of the invariant chain of class II major histocompatibility complex, leading to activation of the ERK1/ERK2 signaling pathways. CD74 also forms complexes with the chemokine receptors CXCR2 and CXCR4.[55]

From the moment of its description as a soluble mediator, it became apparent that MIF regulates T cell-mediated responses,[56,57] and later it was shown that MIF is also important in innate and B cell immunity.[49] MIF has a dual mode of action. It is critical for the host's defense against infection, but at high concentrations, MIF causes cytokine dysregulation that can become highly detrimental in sepsis.[49]

Specific MIF gene (*MIF*) polymorphisms have been linked with susceptibility to inflammatory diseases. In humans, four polymorphisms have been reported: a 5–8-CATT tetranucleotide repeat at position −794 (−794 $CATT_{5-8}$) and three single-nucleotide polymorphisms at positions −173 (−173*G/C), +254 (+254*T/C), and +656 (+656*C/G).[58–60] The last two single nucleotide polymorphisms are located at introns and do not affect the coding sequence of the MIF gene. Thus, mainly –794 $CATT_{5–8}$ and –173*G/C have been examined in populations. The –794 $CATT_{5–8}$ mainly influences constitutive and stimulus-induced transcriptional activity such that transcription increases in an almost proportional fashion with repeat number (the $CATT_5$ is typically referred to as a "low-expression" allele and the $CATT_{6–8}$ as "higher expression" alleles.[58] The –794 $CATT_{5–8}$ was reported to be associated with the severity of rheumatic arthritis[58] and increased incidence of prostate cancer.[61] The –173*G/C SNP is located within the same haplotype block as the CATT site and may exert a regulatory function by means of linkage disequilibrium or functional interaction with the repeat. An increase in –173*G/C SNP was found among children with juvenile idiopathic arthritis compared to unrelated healthy individuals,[59] in patients with erythema nodosum associated with sarcoidosis,[62] in individuals with ulcerative colitis,[63] and in atopic, but not asthmatic, patients.[64]

3.1 *Evidence for the role of MIF in ASD*

Since MIF is a critical immune factor that regulates innate T cell and B cell immunity, and it is also related to the stress response axis, we were interested in whether *MIF* is linked to ASD. In a recent study, we addressed the possibility that –794 $CATT_{5–8}$ and –173*G/C SNPs of the MIF gene are associated with ASD. We analyzed two sets of families.[65] One set was recruited through the Yale Child Study Center (527 individuals from 152 families ascertained through probands with ASD); the other group was recruited through the Center of Child and Adolescent Psychiatry of the University of Groningen in the Netherlands (532 individuals from 183 families with a proband with ASD). Diagnosis of the probands was established using the Autism Diagnostic Interview (ADI) and the Autism Diagnostic Observation Schedule (ADOS). We tested the association of the alleles $CATT_5$, $CATT_6$, and $CATT_7$, with the ADI and ADOS scales for social impairment, communication impairment, and stereotypical behavior ($CATT_8$ is rare, and the number of individuals with that allele was too low to provide informative data). In the Yale sample, the strongest association was observed between –794 $CATT_6$ *MIF* alleles and the stereotypical components of ASD. The analysis of the Dutch sample confirmed these findings and also revealed the association of –173 SNP with social and communication impairment and stereotypical behavior scores. In both samples, there was also high linkage disequilibrium between the –794 CATT and –173 SNP. This genetic analysis linked SNPs of the *MIF* gene to ASD. We therefore compared levels of MIF in the plasma of probands with ASD and healthy siblings and found elevated levels in the probands (Fig. 1).

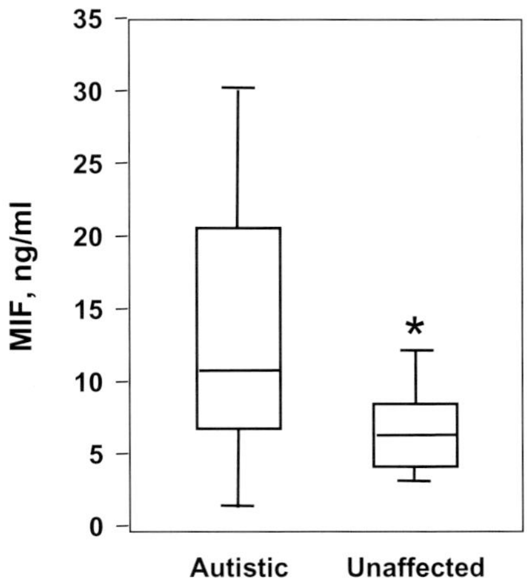

Fig. 1. Box plots of the plasma concentrations of MIF in probands and their unaffected siblings ($n = 10$ per group). The bottom, middle, and top lines of the box demarcate the 25th, 50th, and 75th percentiles, respectively. The vertical lines show the minimum and maximum values. $^{a}P = 0.0323$ for the autistic group versus unaffected siblings (t test, unpaired). From Ref. 29.

This study demonstrated that the SNP of the *MIF* gene is associated with ASD and that the levels of MIF protein are altered in individuals with ASD. The receptor of MIF is CD74, and it belongs to one of 48 molecules whose RNA expression levels in peripheral blood cells and protein levels in serum significantly correlated in individuals with ASF. These studies together suggest that MIF may play a significant role in the pathogenesis of ASD, but specific mechanisms await elucidation.

4. Conclusions

At present, no well-validated biological markers exist for ASD. MIF may play a role in the pathogenesis of the disorder and deserves further study. In particular, through prospective longitudinal studies, it will be important to determine if the levels of MIF in peripheral blood correlate to the clinical severity of ASD. In addition, advances in the development of animal models of ASD should allow investigators to address the role of MIF experimentally. Once inhibitors of MIF become available for clinical use, it will be important to determine if these agents show promise as novel interventions for these disorders.

Acknowledgments

The preparation of this article was supported by funds from the US National Institutes of Health, [NIH; award MH81756 (PI: Klin)], and from the Autism Speaks Foundation (PI: Vaccarino). Grantees undertaking such projects are encouraged to freely express their professional judgment. This article, therefore, does not necessarily represent the position or policies of the NIH and no official endorsement should be inferred. We express our gratitude to Ms. Mei Tan for her editorial assistance.

References

1. Volkmar F, Chawarska K, Klin A. (2005) Autism in infancy and early childhood. *Annu Rev Psychol* **56**: 315–336.
2. Baranek GT. (1999) Autism during infancy: A retrospective video analysis of sensory-motor and social behaviors at 9–12 months of age. *J Autism Develop Dis* **29**: 213–224.
3. Klin A, Chawarska K, Paul R *et al.* (2004) Autism in a 15-month-old child. *Am J Psychiat* **161**: 1981–1988.
4. Osterling JA, Dawson G, Munson JA. (2002) Early recognition of 1-year-old infants with autism spectrum disorder versus mental retardation. *Dev Psychopathol* **14**: 239–251.
5. Werner E, Dawson G, Osterling J *et al.* (2000) Brief report: Recognition of autism spectrum disorder before one year of age: A retrospective study based on home videotapes. *J Autism Develop Dis* **30**: 157–162.
6. Fombonne E. (2005) Epidemiology of autistic disorder and other pervasive developmental disorders. *J Clin Psychiat* **66**(Suppl 10): 3–8.
7. Yeargin-Allsopp M, Rice C, Karapurkar T *et al.* (2003) Prevalence of autism in a US metropolitan area. *JAMA* **289**: 49–55.
8. Bauman ML, Kemper TL. (2005) Neuroanatomic observations of the brain in autism: A review and future directions. *Int J Dev Neurosci* **23**: 183–187.
9. Geschwind DH. (2008) Autism: Many genes, common pathways? *Cell* **135**: 391–395.
10. McCracken JT, McGough J, Shah B *et al.* (2002) Risperidone in children with autism and serious behavioral problems. *N Engl J Med* **347**: 314–321.
11. McDougle CJ, Scahill L, Aman MG *et al.* (2005) Risperidone for the core symptom domains of autism: Results from the study by the autism network of the research units on pediatric psychopharmacology. *Am J Psychiat* **162**: 1142–1148.
12. Levitt P, Campbell DB. (2009) The genetic and neurobiologic compass points toward common signaling dysfunctions in autism spectrum disorders. *J Clin Invest* **119**: 747–754.

13. El-Fishawy P, State MW. (2010) The genetics of autism: Key issues, recent findings, and clinical implications. *Psychiat Clin N Am* **33**: 83–105.
14. Lal G, Bromberg JS. (2009) Epigenetic mechanisms of regulation of Foxp3 expression. *Blood* **114**: 3727–3735.
15. Ashwood P, Krakowiak P, Hertz-Picciotto I *et al.* (2010) Altered T cell responses in children with autism. *Brain Behav Immun* **25**(5): 840–849.
16. Mostafa GA, Al Shehab A, Fouad NR. (2009) Frequency of CD4+CD25high regulatory T cells in the peripheral blood of Egyptian children with autism. *J Child Neurol* **25**: 328–335.
17. Beri S, Tonna N, Menozzi G *et al.* (2007) DNA methylation regulates tissue-specific expression of Shank3. *J Neurochem* **101**: 1380–1391.
18. Redecker P, Bockmann J, Bockers TM. (2006) Expression of postsynaptic density proteins of the ProSAP/Shank family in the thymus. *Histochem Cell Biol* **126**: 679–685.
19. Comi AM, Zimmerman AW, Frye VH *et al.* (1999) Familial clustering of autoimmune disorders and evaluation of medical risk factors in autism. *J Child Neurol* **14**: 388–394.
20. Money J, Bobrow NA, Clarke FC. (1971) Autism and autoimmune disease: A family study. *J Autism Child Schizophr* **1**: 146–160.
21. Sweeten TL, Bowyer SL, Posey DJ *et al.* (2003) Increased prevalence of familial autoimmunity in probands with pervasive developmental disorders. *Pediatrics* **112**: e420.
22. Vargas DL, Nascimbene C, Krishnan C *et al.* (2005) Neuroglial activation and neuroinflammation in the brain of patients with autism. *Ann Neurol* **57**: 67–81.
23. Li X, Chauhan A, Sheikh AM *et al.* (2009) Elevated immune response in the brain of autistic patients. *J Neuroimmunol* **207**(1): 111–116.
24. Chez MG, Dowling T, Patel PB *et al.* (2007) Elevation of tumor necrosis factor-alpha in cerebrospinal fluid of autistic children. *Pediatr Neurol* **36**: 361–365.
25. Jyonouchi H, Geng L, Ruby A *et al.* (2005) Dysregulated innate immune responses in young children with autism spectrum disorders: Their relationship to gastrointestinal symptoms and dietary intervention. *Neuropsychobiology* **51**: 77–85.
26. Singh VK, Warren RP, Odell JD *et al.* (1991) Changes of soluble interleukin-2, interleukin-2 receptor, T8 antigen, and interleukin-1 in the serum of autistic children. *Clin Immunol Immunopathol* **61**: 448–455.
27. Singh VK. (1996) Plasma increase of interleukin-12 and interferon-g. Pathological significance in autism. *J Neuroimmunol* **66**: 143–145.
28. Croonenberghs J, Bosmans E, Deboutte D *et al.* (2002) Activation of the inflammatory response system in autism. *Neuropsychobiology* **45**: 1–6.
29. Sweeten TL, Posey DJ, Shankar S *et al.* (2004) High nitric oxide production in autistic disorder: A possible role for interferon-g. *Biol Psychiat* **55**: 434–437.

30. Zimmerman AW, Jyonouchi H, Comi AM *et al.* (2005) Cerebrospinal fluid and serum markers of inflammation in autism. *Pediat Neurol* **33**: 195–201.
31. Ashwood P, Krakowiak P, Hertz-Picciotto I *et al.* (2011) Elevated plasma cytokines in autism spectrum disorders provide evidence of immune dysfunction and are associated with impaired behavioral outcome. *Brain Behav Immun* **25**(1): 40–45.
32. Jyonouchi H, Sun S, Itokazu N. (2002) Innate immunity associated with inflammatory responses and cytokine production against common dietary proteins in patients with autism spectrum disorder. *Neuropsychobiology* **46**: 76–84.
33. Jyonouchi H, Sun S, Le H. (2001) Proinflammatory and regulatory cytokine production associated with innate and adaptive immune responses in children with autism spectrum disorders and developmental regression. *J Neuroimmunol* **120**: 170–179.
34. Enstrom AM, Lit L, Onore CE *et al.* (2009) Altered gene expression and function of peripheral blood natural killer cells in children with autism. *Brain Behav Immun* **23**: 124–133.
35. Wills S, Cabanlit M, Bennett J *et al.* (2007) Autoantibodies in autism spectrum disorders (ASD). *Ann N Y Acad Sci* **1107**: 79–91.
36. Croonenberghs J, Wauters A, Devreese K *et al.* (2002) Increased serum albumin, g globulin, immunoglobulin IgG, and IgG2 and IgG4 in autism. *Psychol Med* **32**: 1457–1463.
37. Trajkovski V, Ajdinski L, Spiroski M. (2004) Plasma concentration of immunoglobulin classes and subclasses in children with autism in the Republic of Macedonia: Retrospective study. *Croat Med J* **45**: 746–749.
38. Enstrom A, Krakowiak P, Onore C *et al.* (2009) Increased IgG4 levels in children with autism disorder. *Brain Behav Immun* **23**: 389–395.
39. Heuer L, Ashwood P, Schauer J *et al.* (2008) Reduced levels of immunoglobulin in children with autism correlates with behavioral symptoms. *Autism Res* **1**: 275–283.
40. Ashwood P, Enstrom A, Krakowiak P *et al.* (2008) Decreased transforming growth factor beta1 in autism: A potential link between immune dysregulation and impairment in clinical behavioral outcomes. *J Neuroimmunol* **204**: 149–153.
41. Okada K, Hashimoto K, Iwata Y *et al.* (2007) Decreased serum levels of transforming growth factor-beta1 in patients with autism. *Prog Neuro-psychoph* **31**: 187–190.
42. Guerini FR, Bolognesi E, Manca S *et al.* (2009) Family-based transmission analysis of HLA genetic markers in Sardinian children with autistic spectrum disorders. *Hum Immunol* **70**: 184–190.
43. Johnson WG, Buyske S, Mars AE *et al.* (2009) HLA-DR4 as a risk allele for autism acting in mothers of probands possibly during pregnancy. *Arch Pediat Adol Med* **163**: 542–546.

44. Lee LC, Zachary AA, Leffell MS *et al.* (2006) HLA-DR4 in families with autism. *Pediat Neurol* **35**: 303–307.
45. Torres AR, Maciulis A, Stubbs EG *et al.* (2002) The transmission disequilibrium test suggests that HLA-DR4 and DR13 are linked to autism spectrum disorder. *Hum Immunol* **63**: 311–316.
46. Warren RP, Odell JD, Warren WL *et al.* (1996) Strong association of the third hypervariable region of HLA-DR beta 1 with autism. *J Neuroimmunol* **67**: 97–102.
47. Bozza M, Kolakowski LF Jr, Jenkins NA *et al.* (1995) Structural characterization and chromosomal location of the mouse macrophage migration inhibitory factor gene and pseudogenes. *Genomics* **27**: 412–419.
48. Weiser WY, Temple PA, Witek-Giannotti JS *et al.* (1989) Molecular cloning of a cDNA encoding a human macrophage migration inhibitory factor. *Proc Natl Acad Sci USA* **86**: 7522–7526.
49. Calandra T, Roger T. (2003) Macrophage migration inhibitory factor: A regulator of innate immunity. *Nat Rev* **3**: 791–800.
50. Baugh JA, Bucala R. (2002) Macrophage migration inhibitory factor. *Crit Care Med* **30**: S27–S35.
51. Bacher M, Meinhardt A, Lan HY *et al.* (1997) Migration inhibitory factor expression in experimentally induced endotoxemia. *Am J Pathol* **150**: 235–246.
52. Bernhagen J, Calandra T, Mitchell RA *et al.* (1993) MIF is a pituitary-derived cytokine that potentiates lethal endotoxaemia. *Nature* **365**: 756–759.
53. Calandra T, Bernhagen J, Mitchell RA *et al.* (1994) The macrophage is an important and previously unrecognized source of macrophage migration inhibitory factor. *J Exp Med* **179**: 1895–1902.
54. Fingerle-Rowson G, Koch P, Bikoff R *et al.* (2003) Regulation of macrophage migration inhibitory factor expression by glucocorticoids *in vivo*. *Am J Pathol* **162**: 47–56.
55. Schwartz V, Lue H, Kraemer S *et al.* (2009) A functional heteromeric MIF receptor formed by CD74 and CXCR4. *FEBS Lett* **583**: 2749–2757.
56. Bloom BR, Bennett B. (1966) Mechanism of a reaction *in vitro* associated with delayed-type hypersensitivity. *Science* **153**: 80–82.
57. David JR. (1966) Delayed hypersensitivity *in vitro*: Its mediation by cell-free substances formed by lymphoid cell-antigen interaction. *Proc Natl Acad Sci USA* **56**: 72–77.
58. Baugh JA, Chitnis S, Donnelly SC *et al.* (2002) A functional promoter polymorphism in the macrophage migration inhibitory factor (MIF) gene associated with disease severity in rheumatoid arthritis. Genes Immun **3**: 170–176.
59. Donn R, Alourfi Z, De Benedetti F *et al.* (2002) Mutation screening of the macrophage migration inhibitory factor gene: Positive association of a functional

polymorphism of macrophage migration inhibitory factor with juvenile idiopathic arthritis. *Arthritis Rheum* **46**: 2402–2409.
60. Donn RP, Shelley E, Ollier WE *et al.* (2001) A novel 5′-flanking region polymorphism of macrophage migration inhibitory factor is associated with systemic-onset juvenile idiopathic arthritis. *Arthritis Rheum* **44**: 1782–1785.
61. Meyer-Siegler KL, Vera PL, Iczkowski KA *et al.* (2007) Macrophage migration inhibitory factor (MIF) gene polymorphisms are associated with increased prostate cancer incidence. *Genes Immun* **8**: 646–652.
62. Amoli MM, Donn RP, Thomson W *et al.* (2002) Macrophage migration inhibitory factor gene polymorphism is associated with sarcoidosis in biopsy proven erythema nodosum. *J Rheum* **29**: 1671–1673.
63. Nohara H, Okayama N, Inoue N *et al.* (2004) Association of the –173 G/C polymorphism of the macrophage migration inhibitory factor gene with ulcerative colitis. *J Gastroenterol* **39**: 242–246.
64. Shimizu T, Abe R, Ohkawara A *et al.* (1997) Macrophage migration inhibitory factor is an essential immunoregulatory cytokine in atopic dermatitis. *Biochem Biophys Res Commun* **240**: 173–178.
65. Grigorenko EL, Han SS, Yrigollen CM *et al.* (2008) Macrophage migration inhibitory factor and autism spectrum disorders. *Pediatrics* **122**: e438–e445.

Index